Etiology and Morphogenesis of Congenital Heart Disease

Toshio Nakanishi • Roger R. Markwald
H. Scott Baldwin • Bradley B. Keller
Deepak Srivastava • Hiroyuki Yamagishi
Editors

Etiology and Morphogenesis of Congenital Heart Disease

From Gene Function and Cellular
Interaction to Morphology

Editors
Toshio Nakanishi
Department of Pediatric Cardiology
Tokyo Women's Medical University
Tokyo, Japan

Roger R. Markwald
Cardiovascular Developmental Biology
  Center
Medical University of South Carolina
Clemson University
Charleston, SC, USA

H. Scott Baldwin
Department of Pediatrics (Cardiology)
Vanderbilt University
Nashville, TN, USA

Department of Cell and Developmental Biology
Vanderbilt University
Nashville, TN, USA

Bradley B. Keller
Pediatrics, Pharmacology and
  Toxicology, and Bioengineering
University of Louisville
Louisville, KY, USA

Deepak Srivastava
Gladstone Institute of Cardiovascular Disease
San Francisco, CA, USA

Department of Pediatrics and Department of
  Biochemistry and Biophysics
University of California
San Francisco, San Francisco, CA, USA

Hiroyuki Yamagishi
Pediatrics, Division of Pediatric
  Cardiology
Keio University School of Medicine
Tokyo, Japan

ISBN 978-4-431-54627-6   ISBN 978-4-431-54628-3   (eBook)
DOI 10.1007/978-4-431-54628-3

Library of Congress Control Number: 2015960929

Preface

This book is based on the 7th Takao International Symposium on the Etiology and Morphogenesis of Congenital Heart Disease held in August 2013 in Tokyo, Japan. The Takao symposium was first held in 1978 by Dr. Atsuyoshi Takao. At that time, Dr. Takao wanted to organize an international conference to stimulate and integrate research on the morphogenesis of the heart because not many conferences were focused on this aspect in those days. Dr. Takao did not want to use his name in the title of the symposium and the conference was simply called "Symposium on the Etiology and Morphogenesis of Congenital Heart Disease." Since the first symposium, the Takao symposium has been held every 5 years, and proceedings were published after each symposium, in 1980, 1984, 1990, 1995, 2000, and 2005. This book, published in 2016, is the first one to be officially named the "Takao" book. I would like to note that the Takao symposium and its publications have been supported by the Akemi fund from The Sankei Newspaper since its inception. The Akemi fund is based on donations from many persons from all over Japan to save children with congenital heart disease.

For those who are not familiar with Dr. Takao, he was a professor and the chief of pediatric cardiology at The Heart Institute of Japan at Tokyo Women's Medical University from 1972 to 1990. He passed away on August 8, 2006, at the age of 81.

Dr. Takao was very interested in the research of cardiac morphogenesis. In the 1970s, research in his laboratory was aimed at inducing heart disease by centrifugation, radiation, and various drugs such as retinoic acid. Some of those studies found that maternal administration of retinoic acid caused heterotaxy syndrome and transposition of the great arteries in mice. Thymic abnormalities and interruption of the aortic arch (type B) were also noted. However, Dr. Takao began to see the limitations of the methodologies in experimental teratology in clarifying the mechanisms underlying heart malformation.

In 1965, DiGeorge reported on an infant with hypoparathyroidism and recurrent infections along with three necropsy cases with absent thymus and parathyroid glands. From the early 1980s, deletion of chromosome 22q11 was recognized as a cause of DiGeorge syndrome and microdeletion of this region was confirmed in 1991 and 1992.

Conotruncal anomaly face syndrome was first reported by Kinouchi and Takao in 1976. In the early 1970s, Dr. Takao began to recognize a characteristic facial appearance with a flat nasal bridge, small mouth, nasal voice, high arched palate, and ear lobe abnormalities, along with mental retardation among patients with conotruncal anomalies, mainly tetralogy of Fallot. He named this combination "conotruncal anomaly face syndrome." Soon, he recognized the importance of molecular biology in clarifying the etiology and mechanisms of this syndrome. In the early 1980s, the molecular biology laboratory was established in his department. Microdeletion of chromosome 22q11.2 was confirmed also in conotruncal anomaly face syndrome.

Although Dr. Takao realized the importance of molecular biology in the late 1970s, it was not until after he retired in 1990, that he began to study molecular biology techniques. In 1990, he set up the International Molecular Cellular Immunological Research Center. He worked as the director of the Institute until 2006. In 2003, investigators of the Institute, including Atsuyoshi Takao and Rumiko Matsuoka, published a paper suggesting that the *TBX1* mutation was responsible for conotruncal anomaly face syndrome.

It is amazing that so much progress has been accomplished since the first Takao symposium, with path-breaking changes occurring in fields ranging from experimental teratology to regenerative medicine. As we understand more about the detailed molecular mechanisms underlying cardiac malformation, we begin to see some hope of manipulating these mechanisms to treat congenital heart disease, as noted by Dr. Markwald in his chapter in this book. Recent progress in cardiac regenerative medicine by using progenitor cells is largely based on the knowledge obtained from research on the etiology and morphogenesis of congenital heart disease. Basic researchers and cardiologists who have been working in this field should be proud that they have contributed so significantly, in the past nearly 40 years, to this valuable progress.

I hope this book sheds light on the direction in which we should proceed in the research field in the next 5 years, to ultimately save the lives of those who suffer from congenital and acquired heart diseases.

Shinjuku-ku, Tokyo, Japan   Toshio Nakanishi

Contents

The original version of this book was revised. An erratum to this book can be found at
https://doi.org/10.1007/978-4-431-54628-3_55

From Molecular Mechanism to Intervention for Congenital Heart Diseases, Now and the Future

Perspective

Toshio Nakanishi

Srivastava and his team have long since been consistently contributing to research on the molecular mechanisms underlying early heart development, especially on the signaling and transcriptional cascades that regulate cell fate in the heart tissue of the embryo. Early on, Srivastava recognized the importance of microRNA in regulating gene transcripts in the developing heart. Through their research, they were successful in directly converting fibroblasts into cardiomyocytes, a process in which microRNA may play an important role. Srivastava emphasizes the importance of the molecular biology of cardiac morphogenesis for the development of regenerative medicine. He also emphasizes the importance of cardiac fibroblasts as a source of cells for restoring damaged cardiac cells. There seems to be great difficulty in the conversion of fibroblasts to cardiac cells noninvasively in humans. It will be a while before the clinical use of these technologies becomes a reality in saving the lives of pediatric and adult patients with heart failure due to congenital heart disease.

Gittenberger-de Groot and her team emphasize the potential importance of epicardium in repairing damaged myocardium. They report that epicardial-derived cells can differentiate into various cell types such as fibroblasts, arterial smooth muscle cells, endothelial cells, and cardiomyocytes. Epicardial-derived cells contribute to the formation of the coronary arteries, semilunar valves, atrioventricular valves, and myocardium. It is thought that since epicardial-derived cells have the

T. Nakanishi (✉)
Department of Pediatric Cardiology, Heart Institute, Tokyo Women's Medical University, Tokyo, Japan

International Molecular Cellular Immunological Research Center, Tokyo, Japan
e-mail: pnakanis@hij.twmu.ac.jp

potential to develop into various types of cells, it may be possible to use them for the repair of ischemic myocardium. The methodology of using the epicardium or epicardial-derived cells to rescue the damaged heart tissue remains to be investigated.

Sekine, Shimizu, and Okano report a technology to generate a cell sheet efficiently by using a temperature-responsive culture dish. The cell sheet can then be layered to form the cardiac tissue. Clinical trials using this cardiac tissue to repair the damage caused by myocardial infarction or cardiomyopathy are ongoing. Long-term results of this technology will need to be studied in order to apply it to pediatric patients with dilated cardiomyopathy and severely reduced contractile function.

Tohyama and Fukuda report a technology to efficiently purify cardiomyocytes differentiated from induced pluripotent stem (iPS) cells derived from human T lymphocytes. They were able to transplant these cardiomyocytes into the heart. They have also made disease models, including that of long QT syndrome, using these cardiomyocytes. iPS technology has great potential because it can synthesize the cardiac tissue from a patient's own blood cells. However, even for diagnostic purposes, the establishment of stable techniques to generate cardiomyocytes reproducibly from iPS cells is difficult, at least at this moment. For clinical use, safety issues of iPS cells remain to be clarified.

Markwald and his team emphasize that central signaling pathways (or hubs) in which mutations disrupt fundamental cell functions play a key role in the development of congenital heart disease. More specifically, they note two signaling pathways as "hubs": the nodal kinase activated by extracellular ligands (such as periostin) and the cytoskeletal regulatory protein, filamin A. They report that disruption of nodal kinase pathways causes septal defects and malformation of atrioventricular valves. However, if we can manipulate these pathways after birth, we may be able to rescue these defects. Fibroblasts, which may be derived from the bone marrow, play an important role in forming the ventricular septum and atrioventricular valves, and their activity may be regulated by periostin and filamin A. Manipulation of periostin or filamin A may modify fibroblast activity and may be able to rescue septal and valve malformation.

Reprogramming Approaches to Cardiovascular Disease: From Developmental Biology to Regenerative Medicine

Deepak Srivastava

Abstract

Heart disease is a leading cause of death in adults and children. We, and others, have described complex signaling, transcriptional, and translational networks that guide early differentiation of cardiac progenitors and later morphogenetic events during cardiogenesis. We found that networks of transcription factors and miRNAs function through intersecting positive and negative feedback loops to reinforce differentiation and proliferation decisions. We have utilized a combination of major cardiac regulatory factors to induce direct reprogramming of cardiac fibroblasts into cardiomyocyte-like cells with global gene expression and electrical activity similar to cardiomyocytes. The in vivo efficiency of reprogramming into cells that are more fully reprogrammed was greater than in vitro and resulted in improved cardiac function after injury. We have also identified a unique cocktail of transcription factors and small molecules that reprogram human fibroblasts into cardiomyocyte-like cells and are testing these in large animals. Knowledge regarding the early steps of cardiac differentiation in vivo has led to effective strategies to generate necessary cardiac cell types for regenerative approaches and may lead to new strategies for human heart disease.

Keywords
Cardiac • Reprogramming • Regeneration • Transcription factors

D. Srivastava (✉)
Gladstone Institute of Cardiovascular Disease, San Francisco, CA 94158, USA

Department of Pediatrics and Department of Biochemistry and Biophysics, University of California, San Francisco, San Francisco, CA 94158, USA
e-mail: dsrivastava@gladstone.ucsf.edu

T. Nakanishi et al. (eds.), *Etiology and Morphogenesis of Congenital Heart Disease*,
DOI 10.1007/978-4-431-54628-3_1

1.1   Introduction

Heart disease remains the leading cause of death worldwide, despite improved treatments that have decreased death rates from cardiovascular diseases. Congenital heart malformations, the most common of all human birth defects, occur in nearly 1 % of the population worldwide, regardless of race. With more than one million survivors of congenital heart disease (CHD) in the United States, it is becoming apparent that genetic disruptions that predispose to developmental defects can have ongoing consequences in maintenance of specific cell types and cellular processes over decades. A more precise understanding of the causes of CHD is imperative for the recognition and potential intervention of progressive degenerative conditions among survivors of CHD.

Because cardiomyocytes (CMs) rarely regenerate in postnatal hearts, survivors of congenital and acquired heart disease often develop chronic heart failure. Unfortunately, end-stage heart failure can only be addressed by heart transplanta-tion, which is limited by the number of donor organs available, particularly in children. Alternative solutions include cellular therapy that replaces lost CMs either by transplanting CMs or inducing new CMs in situ at areas affected by the infarction.

In recent years it has become apparent that adult somatic cells can be converted into other types of cells through epigenetic reprogramming. With this technology, for example, somatic cells, such as fibroblasts, can be dedifferentiated into pluripo-tent stem cells by nuclear transfer [1] or with defined transcription factors [2]. Direct reprogramming of fibroblasts into the chief functional cells of different organs, including CMs, neurons, hepatocytes, hematopoietic cells, and endothelial cells, has been accomplished and holds great promise for regenerative medicine [3]. In particular, cardiac fibroblasts represent a large pool of cells in the adult heart [4] and thus may provide a reservoir of cells from which to generate new CMs through epigenetic reprogramming.

In 2010, we reported that mouse cardiac and dermal fibroblasts could be directly reprogrammed into induced CM-like cells (iCMs) in vitro by a combination of three developmental cardiac transcription factors, Gata4, Mef2c, and Tbx5 (GMT) [5]. Since then, other labs around the world have reported success in reprogramming mouse fibroblasts into iCMs with similar cocktails of reprogramming factors [6–10]. We and others have also directly converted human cardiac and dermal fibroblasts into cardiac cells [11–13]. Several review papers published from different labs illustrate the potential and challenges of this new avenue for cardiac regenerative medicine [14–19].

Here, we discuss the cardiac developmental biology discoveries that underpin direct reprogramming approaches and consider the opportunities and challenges of this technology for addressing cardiac regeneration.

1.2 Molecular Networks Regulate Cardiac Cell Fate

In vertebrate embryos, the heart is the first functional organ to form. The morphological development of the heart has been summarized in several reviews. During development, numerous signaling and transcriptional cascades regulate cell fate decisions in distinct heart fields. In addition, the Gata, Mef2, Hand, Nkx, and T-box family of transcription factors control expression of cardiac genes and direct the specification and differentiation of cardiac myocytes. These transcription factors have been frequently included in studies that directly convert fibroblasts into cardiac cells.

MicroRNAs (miRs)—small single-stranded noncoding RNAs that negatively regulate the stability of gene transcripts—also regulate cardiac gene expression [20, 21]. Transcription factors regulate miR expression, and in turn, miRs can modulate the activities of transcription factors through positive and negative feedback loops. One of the major regulators of cardiac lineage determination during heart development is miR-1. Expression of miR-1 in either mouse or human embryonic stem cells (ESCs) caused them to favor the muscle cell fate. In contrast, miR-133 promoted muscle progenitor expansion and prevented terminal differentiation, while another miRNA, miR-499, promoted the ventricular cell fate in human ESC-differentiated CMs and caused cardiac hypertrophy and enlarged hearts in miR-499 transgenic mice. These studies demonstrate that miRs cooperate with transcription factors to form an intertwined network that reinforces specific cell fate decisions and differentiation during cardiac development.

1.3 Cardiac Fibroblasts in the Normal and Remodeling Heart

Fibroblasts are mesenchymal cells that produce many extracellular matrix components in organs. Fibroblasts show heterogeneity based on morphology, glycogen pools, collagen production, cell surface markers, and global gene-expression profiles. Although the percentage of fibroblasts among the total cells in the heart varies between species, the large population of fibroblasts is quiescent and abundantly distributed in the interstitial and perivascular matrix in the normal heart.

Cardiac fibroblasts synthesize extracellular matrix to provide a 3D network for myocytes and other cells of the heart; they also regulate the biological and electrophysiological response of CMs during physiological and pathological development. In embryonic mouse hearts, cardiac fibroblasts induced proliferation of CMs via paracrine signals of fibronectin, collagen, and heparin-binding EGF-like growth factor; however, in adult mouse hearts, cardiac fibroblasts promoted hypertrophic maturation of CMs via beta1-integrin signaling [22]. Cardiac fibroblasts were also found to form intracellular electrical coupling and communicate with myocytes through gap junctions, suggesting that cardiac fibroblasts could conduct electric signaling between different regions of myocytes that are electrically isolated by connective tissue in the normal heart.

1.4   Direct Cardiac Reprogramming *In Vitro*

To identify the combination of reprogramming factors that could convert a cardiac fibroblast toward a cardiomyocyte-like state, we generated a transgenic mouse in which enhanced green fluorescent protein (EGFP) was driven by the alpha myosin heavy chain (αMHC) promoter [5, 22], providing a tool to screen through many potential regulators. We ultimately found that a combination of three transcription factors—Gata4, Mef2c, and Tbx5 (GMT)—could convert ~15 % of cardiac fibroblasts into αMHC-EGFP-positive cells, which we called induced CM-like cells (iCMs) [5]. iCMs formed sarcomere structures and displayed whole-transcriptome expression profiles that were shifted significantly toward the profile of CMs, even at the single-cell level. Although most in vitro iCMs were only partially reprogrammed, many of them could generate Ca^{2+} transients, and some started beating spontaneously 4–6 weeks after reprogramming. Using a lineage-tracing strategy in mice (e.g., Isl1-Cre-YFP and Mesp1-Cre-YFP), we did not observe activation of cardiac progenitor markers during GMT cardiac reprogramming [5], suggesting that GMT directly converted fibroblasts toward the cardiac cell fate without dedifferentiation back into a progenitor status. Song et al. [8] found that a basic helix-loop-helix transcription factor, Hand2, could help GMT to convert TTFs into functional beating iCMs.

1.5   Direct Cardiac Reprogramming *In Vivo*

The ultimate goal in generating new iCMs is to improve systolic function of damaged hearts and restore its normal structure and function. Therefore, we tested direct cardiac reprogramming in vivo with the hypothesis that the heart's native microenvironment would promote direct reprogramming of fibroblasts to CMs. We expected that direct cardiac reprogramming would be enhanced in the native heart and could improve the function of the damaged heart.

We delivered GMT into mouse hearts after acute MI via retroviruses. These viruses can only infect actively dividing cells and, thus, could deliver the reprogramming factors into non-myocytes (mostly fibroblasts), but not into CMs that exit the cell cycle after differentiation. Four weeks after introducing GMT, we found numerous genetically labeled reprogrammed cells within the scar area of mouse hearts, indicating newly born iCMs reprogrammed from cardiac fibroblasts (Fig. 1.1). We found that more than 50 % of in vivo-derived iCMs closely resembled endogenous ventricular CMs, had a rod shape, were binucleate, assembled sarcomeres, generated Ca^{2+} transients, and elicited ventricular-like action potentials and beating activity [23]. By microarray analysis, in vivo mouse GMT-iCMs showed similar global gene-expression profiles with mouse adult CMs, such that they were clustered as one type of cells [11]. The iCMs reprogrammed in vivo were electrically coupled with endogenous CMs, and no arrhythmias were observed in mice that received GMT reprogramming factors. Most importantly, introducing GMT in vivo reduced scar size and cardiac

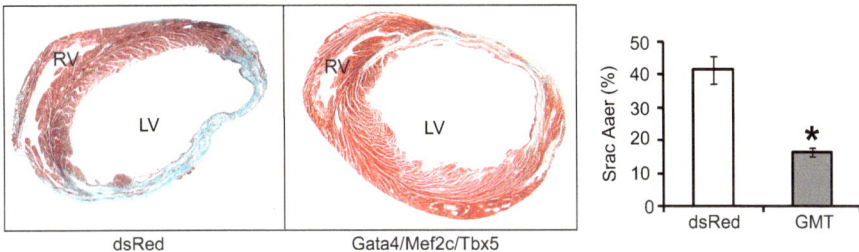

Fig. 1.1 In situ reprogramming of fibroblasts to cardiomyocytes. Representative histologic sections from mouse hearts treated with dsRed or Gata4/Mef2c/Tbx5 containing retroviral vectors injected into the myocardium after coronary ligation. Scar area quantification is indicated

dysfunction up to 12 weeks after coronary ligation [23]. Similarly, Inagawa et al. [24] successfully reprogrammed cardiac fibroblasts into iCMs in vivo by introducing GMT into the heart of immunosuppressed mice with single-polycistronic retrovirus, which contains GMT with self-cleaving 2A peptides. Furthermore, Song et al. [8] found that introducing GMT and Hand2 in vivo could directly convert cardiac fibroblasts into iCMs and also improved function and decreased scar.

1.6 Direct Cardiac Reprogramming in Human Fibroblasts

Establishing the technology of cardiac reprogramming in human cells was a necessary step toward considering clinical application. However, neither GMT nor GHMT, which reprogrammed iCMs from mouse fibroblasts, was able to reprogram human fibroblasts into iCMs in vitro [11–13]; however, inclusion of additional reprogramming factors resulted in successful reprogramming. Nam et al. [12] found that GHT, without MEF2C, but with another transcription factor, myocardin, and two muscle-specific miRNAs, miR-1 and miR-133, could reprogram human fibroblasts into iCMs. These reprogrammed iCMs expressed multiple cardiac genes, developed sarcomere-like structures, and generated Ca^{2+} transients with a small subset of the cells exhibiting spontaneous contractility after 11 weeks in culture. Wada R. et al. [13] reported that GMT with MESP1 and myocardin could activate cardiac gene expression in human neonatal and adult cardiac fibroblasts. In our study, pairing GMT with ESRRG and MESP1 induced global expression of cardiac genes and shifted the phenotype of human fibroblasts toward the CM-like state. Reprogrammed human iCMs were epigenetically stable and formed sarcomere structures, and some could generate Ca^{2+} transients and action potentials [11]. By comparing whole-transcriptome expression of 4-week and 12-week iCMs, we found that reprogramming human cells takes longer than mouse cells because of their progressive repression of fibroblast genes. Nonetheless, our analysis of orthologous gene expression indicated that at the global gene-expression level, human iCMs were reprogrammed at a level similar to mouse iCMs reprogrammed by GMT in vitro [11].

1.7   Challenges and Future Directions

The studies summarized here demonstrate that forced expression of three or four developmentally critical transcription factors can directly convert mouse cardiac fibroblasts into CM-like cells in vitro, although most of these iCMs were partially reprogrammed. These same factors generated more mature CM-like cells in the native heart, improved heart function, and reduced scar size in the mouse heart post-MI. Similarly, expression of five to seven reprogramming factors converted human fibroblasts into non-perfect iCMs in vitro. By comparing in vitro versus in vivo iCMs in mice [8, 23] and considering the similarities between human and mouse iCMs in vitro [11], we speculate that the cocktails recently identified in human fibroblasts may be sufficient to reprogram adult CM-like cells that are fully functional in the in vivo environment, such as in the pig or nonhuman primate heart, as is the case in mice. These studies have been driven by the deep knowledge of cardiac development gained over the last two decades and represent a valuable application of this knowledge for potential clinical development postnatally.

Successfully achieving cardiac reprogramming requires high expression and proper stoichiometry of reprogramming factors, healthy and non-senescent fibroblasts, and optimal conditions for cell culture. In addition, culture conditions, such as electric stimulation, might help facilitate and maintain the functional maturation of iCMs at late stages (i.e., after cardiac cell fate conversion), and the process may require some small-molecule compounds and growth factors to overcome epigenetic barriers at the early stages of in vitro reprogramming. The epigenetic barriers that prevent cardiac reprogramming in vitro remain unknown; however, the in vivo environment of the heart appears to overcome these epigenetic blocks. We speculate that secreted factors and direct cell-cell interactions, including mechanical and electrical, from myocytes and non-myocytes may work together to improve direct cardiac reprogramming.

Another elusive concept is the molecular mechanism that underlies direct cardiac reprogramming. What are the DNA targets of those reprogramming factors? How are those transcriptional changes epigenetically stabilized during reprogramming? By combining mechanism assays at whole-population and single-cell levels, we can gain a more integral and comprehensive understanding of how core transcription factors establish a self-reinforcing molecular network that controls cardiac cell fate.

While many challenges and hurdles remain in this blossoming research field, the high demand for regenerative medicine strategies for the heart emphasizes the significance of these efforts in discovering new therapeutic strategies. Observing the functional benefits of in vivo reprogramming in mouse heart and the promising and similar degree of reprogramming in mouse and human in vitro iCMs, we are endeavoring to translate direct cardiac reprogramming for future clinical applications.

References

1. Gurdon JB. The developmental capacity of nuclei taken from intestinal epithelium cells of feeding tadpoles. J Embryol Exp Morphol. 1962;10:622–40.
2. Takahashi K, Yamanaka S. Induction of pluripotent stem cells from mouse embryonic and adult fibroblast cultures by defined factors. Cell. 2006;126:663–76.
3. Sancho-Martinez I, Baek SH, Izpisua Belmonte JC. Lineage conversion methodologies meet the reprogramming toolbox. Nat Cell Biol. 2012;14:892–9.
4. Snider P, Standley KN, Wang J, Azhar M, Doetschman T, Conway SJ. Origin of cardiac fibroblasts and the role of periostin. Circ Res. 2009;105:934–47.
5. Ieda M, Fu JD, Delgado-Olguin P, Vedantham V, Hayashi Y, Bruneau BG, et al. Direct reprogramming of fibroblasts into functional cardiomyocytes by defined factors. Cell. 2010;142:375–86.
6. Jayawardena TM, Egemnazarov B, Finch EA, Zhang L, Payne JA, Pandya K, et al. MicroRNA-mediated in vitro and in vivo direct reprogramming of cardiac fibroblasts to cardiomyocytes. Circ Res. 2012;110:1465–73.
7. Protze S, Khattak S, Poulet C, Lindemann D, Tanaka EM, Ravens U. A new approach to transcription factor screening for reprogramming of fibroblasts to cardiomyocyte-like cells. J Mol Cell Cardiol. 2012;53:323–32.
8. Song K, Nam YJ, Luo X, Qi X, Tan W, Huang GN, et al. Heart repair by reprogramming non-myocytes with cardiac transcription factors. Nature. 2012;485:599–604.
9. Addis RC, Ifkovits JL, Pinto F, Kellam LD, Esteso P, Rentschler S, et al. Optimization of direct fibroblast reprogramming to cardiomyocytes using calcium activity as a functional measure of success. J Mol Cell Cardiol. 2013;60:97–106.
10. Christoforou N, Chellappan M, Adler AF, Kirkton RD, Wu T, Addis RC, et al. Transcription factors MYOCD, SRF, Mesp1 and SMARCD3 enhance the cardio-inducing effect of GATA4, TBX5, and MEF2C during direct cellular reprogramming. PLoS One. 2013;8:e63577.
11. Fu JD, Stone NR, Liu L, Spencer CI, Qian L, Hayashi Y, et al. Direct reprogramming of human fibroblasts toward a cardiomyocyte-like state. Stem Cell Rep. 2013;1:235–47.
12. Nam YJ, Song K, Luo X, Daniel E, Lambeth K, West K, et al. Reprogramming of human fibroblasts toward a cardiac fate. Proc Natl Acad Sci U S A. 2013;110:5588–93.
13. Wada R, Muraoka N, Inagawa K, Yamakawa H, Miyamoto K, Sadahiro T, et al. Induction of human cardiomyocyte-like cells from fibroblasts by defined factors. Proc Natl Acad Sci U S A. 2013;110:12667–72.
14. Yi BA, Mummery CL, Chien KR. Direct cardiomyocyte reprogramming: a new direction for cardiovascular regenerative medicine. Cold Spring Harb Perspect Med. 2013;3:a014050.
15. Srivastava D, Berry EC. Cardiac reprogramming: from mouse toward man. Curr Opin Genet Dev. 2013;23:574–8.

16. Qian L, Srivastava D. Direct cardiac reprogramming: from developmental biology to cardiac regeneration. Circ Res. 2013;113:915–21.
17. Nam YJ, Song K, Olson EN. Heart repair by cardiac reprogramming. Nat Med. 2013;19:413–5.
18. Addis RC, Epstein JA. Induced regeneration – the progress and promise of direct reprogramming for heart repair. Nat Med. 2013;19:829–36.
19. Muraoka N, Ieda M. Direct reprogramming of fibroblasts into myocytes to reverse fibrosis. Annu Rev Physiol. 2014;76:21–37.
20. Ivey KN, Srivastava D. MicroRNAs as regulators of differentiation and cell fate decisions. Cell Stem Cell. 2010;7:36–41.
21. Cordes KR, Srivastava D. MicroRNA regulation of cardiovascular development. Circ Res. 2009;104:724–32.
22. Ieda M, Tsuchihashi T, Ivey KN, Ross RS, Hong TT, Shaw RM, et al. Cardiac fibroblasts regulate myocardial proliferation through beta1 integrin signaling. Dev Cell. 2009;16:233–44.
23. Qian L, Huang Y, Spencer CI, Foley A, Vedantham V, Liu L, et al. In vivo reprogramming of murine cardiac fibroblasts into induced cardiomyocytes. Nature. 2012;485:593–8.
24. Inagawa K, Miyamoto K, Yamakawa H, Muraoka N, Sadahiro T, Umei T, et al. Induction of cardiomyocyte-like cells in infarct hearts by gene transfer of gata4, mef2c, and tbx5. Circ Res. 2012;111:1147–56.

The Arterial Epicardium: A Developmental Approach to Cardiac Disease and Repair

Adriana C. Gittenberger-de Groot, E.M. Winter, M.J. Goumans, M.M. Bartelings, and R.E. Poelmann

Abstract

The significance of the epicardium that covers the heart and the roots of the great arteries should not be underestimated as it is a major component with impact on development, disease, and repair. The epicardium differentiates from the proepicardial organ located at the venous pole (vPEO). The differentiation capacities of the vPEO into epicardium-derived cells (EPDCs) have been extensively described. A hitherto escaped part of the epicardium derives from a second proepicardial organ located at the arterial pole (aPEO) and covers the intrapericardial part of the aorta and pulmonary trunk. In avian and mouse embryos, disturbance of epicardium differentiation causes a spectrum of cardiac anomalies including coronary artery abnormalities, deficient annulus fibrosis with rhythm disturbances, valve malformations, and non-compaction cardiomyopathies. Late in prenatal life the epicardium becomes dormant, losing the activity of many genes.

In human cardiac diseases, both arterial and venous epicardium can be activated again into EPDCs. The epicardial reactivation observed after experimental

A.C. Gittenberger-de Groot (✉)
Departments Cardiology, Leiden University Medical Center, Leiden, The Netherlands

Anatomy and Embryology, Leiden University Medical Center, Leiden, The Netherlands
e-mail: a.c.gittenberger_de-groot@lumc.nl

E.M. Winter • M.M. Bartelings
Anatomy and Embryology, Leiden University Medical Center, Leiden, The Netherlands

M.J. Goumans
Molecular Cell Biology, Leiden University Medical Center, Leiden, The Netherlands

R.E. Poelmann
Departments Cardiology, Leiden University Medical Center, Leiden, The Netherlands

Institute of Biology, Leiden University, Leiden, The Netherlands

© The Author(s) 2016
T. Nakanishi et al. (eds.), *Etiology and Morphogenesis of Congenital Heart Disease*,
DOI 10.1007/978-4-431-54628-3_2

myocardial infarction and during aneurysm formation of the ascending aorta provides clinical relevance. EPDCs applied for cell therapy demonstrate repair processes synergistic with the resident cardiac progenitor stem cells that probably share an embryonic origin with EPDCs. Future therapeutic strategies might be possible addressing cell autonomous-based and signaling capacities of the adult epicardium.

Keywords
Epicardium • Venous pole • Arterial epicardium • Cardiac disease • Congenital anomlies

2.1   Origin of the Epicardium

The cardiogenic mesoderm is referred to as first heart field, flanked medially by second heart field (SHF) mesoderm. The addition of SHF-derived cardiac mesoderm enables the formation of all cardiac components. A secondary layer will cover the complete myocardial tube (Fig. 2.1) and the developing roots of the great arteries. At the venous pole, the vPEO develops from which the epicardium (cEP) spreads over the cardiac tube up to the ventriculo-arterial junction [1]. Here, the cEP meets the PEO located at the arterial pole (Fig. 2.2) [2] forming arterial epicardium (aEP) that is continuous with the pericardium. Both PEO structures and the spreading cEP and aEP express Wilms' tumor-1 suppressor gene (WT-1) among others and harbor endothelial and mesenchymal cells.

2.2   Epicardium-Derived Cells (EPDCs)

Epicardial cells lose epithelial contacts by epithelial–mesenchymal transition (EMT) and EPDCs relocate subepicardially [3, 4]. Early EPDCs invade the thin myocardial wall. Proliferation of the myocardium depends on both endocardial- and epicardial-derived signals, including Raldh2 [5] with a spatiotemporal differ-ence between right and left ventricle (Fig. 2.1). The aEP starts EMT slightly later than cEP, and ensuing EPDCs can be detected in the outer layers of the developing great arteries and at the myocardial–endocardial cushion interface (Fig. 2.2). Epicardial heterogeneity with a subset of cells taking part in the initial EMT wave provides the myocardium with the main number of the future interstitial fibroblasts. The next wave of EMT correlates with the formation of the fibrous atrioventricular annulus [6] and contributes to part of the atrioventricular cushion cells. At the ventriculo-arterial junction aEP migrates into the outflow tract forming the future arterial annuli and partake in the semilunar valves [7].

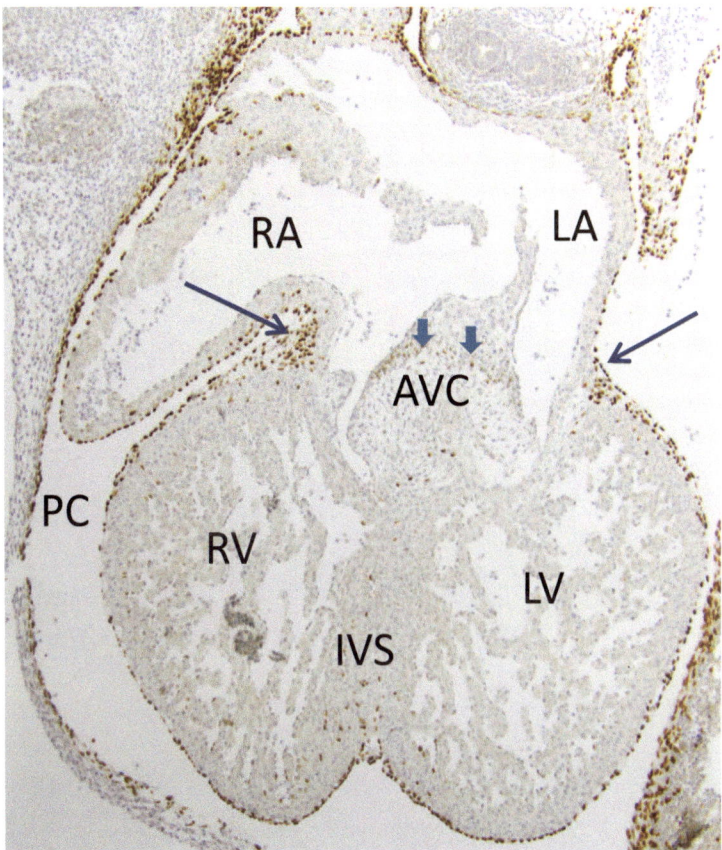

Fig. 2.1 Four-chamber view of a mouse heart (ED12.5) immunostained for WT1. Note brown cells lining the pericardial cavity (PC), including the epicardium. The AV groove is indicated (*arrows*). The left ventricular (LV) wall contains hardly EPDCs, whereas some are present in the RV and in the interventricular septum (IVS). Note the presence of EPDCs at the border of the AV cushion and the interatrial septum (*short arrows*)

2.3   Heterogeneity of Epicardial Cells

2.3.1   The Cardiac Fibroblast

The epicardium is the main source of the interstitial, the fibrous annulus, and the coronary adventitial fibroblasts [7, 8]. The endocardial cells lining the cushions are the other source of the valve fibroblasts [3].

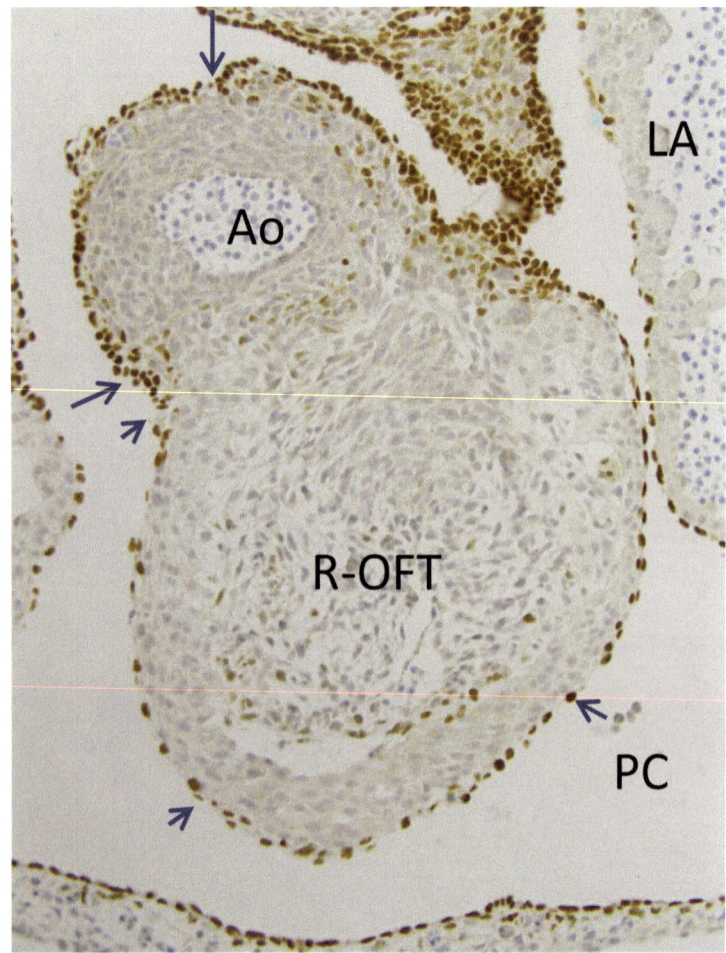

Fig. 2.2 Cross section of aorta (Ao) and right ventricular outflow tract (R-OFT). Aortic epicardium is densely packed and cuboid (*arrows*), whereas cEP is squamous (*short arrows*). Note EPDCs at the border of R-OFT cushions and myocardium and between Ao and R-OFT

2.3.2 Arterial Smooth Muscle Cell

After ingrowth of the peritruncal coronary capillary plexus into the aorta, EPDCs surround the main coronary vessels and differentiate into smooth muscle cells (SMC). Differentiation into SMCs is regulated by many genes including FGF, VEGF, Notch, SRF, and PDGFRb and their ligands [8]. Quail–chicken chimera studies demonstrated the timing of EMT and required cell interactions [4].

2.3.3 Endothelial Cells

The origin of the coronary endothelium is still under debate. Using a quail vPEO transplanted into the isochronous chick as reviewed in [10] pointed out that coronary ECs do not derive from the coelomic lining but from endothelial cells from the sinus venosus sprouting into the stalk of the vPEO. This is also supported by studies using transgenic mice [11]. We have shown that the sinus venosus-derived endothelial cells express the "arterial" Notch1, underlining the plasticity of these cells [9]. The discussion on the origin of the coronary endothelium is kept alive as specific compartments produce restricted numbers of ECs [12, 13].

2.3.4 Cardiomyocytes

Conditional reporter mice studies, using WT-1 and Tbx18 as epicardial marker, suggested that a population of EPDCs might differentiate into a myocardial phenotype. However, based on interaction between BMP and FGF, some of the progenitors of cardiomyocytes in the SHF share the same markers with cells in the underlying mesoderm of the SHF and the epicardial population [14]. Other data, including quail–chicken chimera, also do not support EPDC–cardiomyocyte transition [3].

2.3.5 The Purkinje Fiber

The Purkinje fiber is a specialized cardiomyocyte induced by endothelin produced by endocardium and endothelium. We have postulated an essential interaction between EPDCs and endocardial-/endothelial-derived factors after vPEO tracing and inhibition experiments [15].

2.4 Congenital and Adult Cardiac Disease

In a large-screen microarray [16], no epicardium-specific gene has been encountered. Therefore, it is challenging to attribute specific cardiac malformations and diseases to epicardial malfunctioning. However, in animal models it is possible to link the epicardium to certain cardiac defects and diseases. We are dealing with complex tissues in which epicardial cells are essential.

2.4.1 Non-compaction

The most relevant cardiomyopathy resulting from abnormal EPDC function is the primary left ventricular non-compaction cardiomyopathy [4] demonstrating a spongious myocardium usually including the ventricular septum. Differences in amount and timing of LV and RV invasion by EPDCs might account for predilection of the LV. With respect to congenital heart disease, a spongious ventricular septum can be connected with muscular VSDs.

2.4.2 Conduction System Anomalies

The main components of the conduction system are myocardial in origin. Clinically, it has been hypothesized that the genetically determined long QT syndrome is linked to abnormal Purkinje fiber function. Indirectly, the abnormal formation of the fibrous annulus with persisting accessory pathways can result in reentry tachycardias. PEO inhibition in chick embryos showed defective atrioventricular isolation, delaying the shift from a base-to-apex to an apex-to-base conduction [17].

2.4.3 Valvulopathies

PEO inhibition can lead to deficient AV valve formation [18, 19]. Furthermore, abnormal differentiation including defective undermining of the valve leaflet is similar to Ebstein's anomaly of the tricuspid valve as observed in combination with accessory pathways [4]. EPDCs of aEP origin are found in the outflow tract cushions (Fig. 2.2) probably acting via Notch signaling and hence influencing bicuspid aortic valve formation.

2.4.4 Coronary Vascular Anomalies

Experimental studies disturbing normal coronary development result in a number of malformations that associate human congenital pattern variations with abnormal ventriculo-coronary-arterial communications (fistulae). Fistulae found in avian models with absent coronary arterial orifices in the aorta [20] resembling coronary malformations found in pulmonary atresia without VSD are hypothesized to be a primary coronary vascular disease [21].

2.5   Cardiovascular Repair

Myocardial infarction. Different approaches have focused on the potential of the adult epicardial cell after myocardial infarction (MI). A c-kit-positive subepicardial population indicates renewed epicardial activity with acquisition of stem cell characteristics [22]. Using a retrovirally induced fluorescent Katushka labeling of dormant epicardium showed EPDCs that migrated into the myocardium and differentiated into a myofibroblast phenotype. The activated epicardium and EPDCs reexpressed WT1 not only in the MI border zone but also in remote areas. Another approach is represented by epicardial cells cocultured with cardiomyocytes in which EPDCs change myocardial alignment and contraction [23]. A direct approach was provided by grafting human adult atrial epicardial cells [24] cultured in vitro from a cobblestone epithelium into spindle shape, thereby acquiring characteristics of mesenchymal stem cells. Injection of these adult human EPDCs into immune-incompetent mice resulted in a marked improvement of cardiac function [24] indicative of repair. Combined injection with adult human cardiomyocyte progenitors (CMPCs) aimed at induction of cardiomyocyte regeneration [25] which showed an additive effect on remodeling, although no new cardiac cell types (endothelial cells, fibroblasts, SMCs, or cardiomyocytes) could be traced to human origin. The capacities, expressed within the normal embryonic state, seem to be preserved in adult life and in disease states.

2.6   Future Directions and Clinical Applications

Many positive effects of EPDCs either after injection or by stimulation of the dormant native epicardial covering of the heart are due to a paracrine mechanism [22, 24, 25]. These findings bear important potential for drug and cell-based therapeutic approaches to stimulate the native epicardium in repair of the ischemic cardiac wall. An underdeveloped area is the priming of grafts taken from the pericardium for use in cardiac or arterial repair.

References

1. Vrancken Peeters M-FM, Mentink MMT, Poelmann RE, et al. Cytokeratins as a marker for epicardial formation in the quail embryo. Anat Embryol. 1995;191:503–8.
2. Perez-Pomares JM, Phelps A, Sedmerova M, et al. Epicardial-like cells on the distal arterial end of the cardiac outflow tract do not derive from the proepicardium but are derivatives of the cephalic pericardium. Dev Dyn. 2003;227:56–68.
3. Gittenberger-de Groot AC, Vrancken Peeters M-PFM, Mentink MMT, et al. Epicardium-derived cells contribute a novel population to the myocardial wall and the atrioventricular cushions. Circ Res. 1998;82:1043–52.
4. Lie-Venema H, van den Akker NMS, Bax NA, et al. Origin, fate, and function of epicardium-derived cells (EPCDs) in normal and abnormal cardiac development. Sci World J. 2007;7:1777–98.
5. Kikuchi K, Holdway JE, Major RJ, et al. Retinoic acid production by endocardium and epicardium is an injury response essential for zebrafish heart regeneration. Dev Cell. 2011;20:397–404.

6. Gittenberger de Groot AC, Winter EM, Bartelings MM, et al. The arterial and cardiac epicardium in development, disease and repair. Differentiation. 2012;84:41–53.
7. Zhou B, von Gise GA, Ma Q, et al. Genetic fate mapping demonstrates contribution of epicardium-derived cells to the annulus fibrosis of the mammalian heart. Dev Biol. 2010;338:251–61.
8. Smith CL, Baek ST, Sung CY, et al. Epicardial-derived cell epithelial-to-mesenchymal transition and fate specification require PDGF receptor signaling. Circ Res. 2011;108:e15–26.
9. Van Den Akker NM, Winkel LC, Nisancioglu MH, et al. PDGF-B signaling is important for murine cardiac development: its role in developing atrioventricular valves, coronaries, and cardiac innervation. Dev Dyn. 2008;237:494–503.
10. Winter EM, Gittenberger-de Groot AC. Cardiovascular development: towards biomedical applicability: epicardium-derived cells in cardiogenesis and cardiac regeneration. Cell Mol Life Sci. 2007;64:692–703.
11. Merki E, Zamora M, Raya A, et al. Epicardial retinoid X receptor alpha is required for myocardial growth and coronary artery formation. Proc Natl Acad Sci U S A. 2005;102:18455–60.
12. Katz TC, Singh MK, Degenhardt K, et al. Distinct compartments of the proepicardial organ give rise to coronary vascular endothelial cells. Dev Cell. 2012;22:639–50.
13. Tian X, Hu T, Zhang H, et al. Subepicardial endothelial cells invade the embryonic ventricle wall to form coronary arteries. Cell Res. 2013;23:1075–90.
14. Kruithof BP, van Wijk B, Somi S, et al. BMP and FGF regulate the differentiation of multipotential pericardial mesoderm into the myocardial or epicardial lineage. Dev Biol. 2006;295:507–22.
15. Eralp I, Lie-Venema H, Bax NAM, et al. Epicardium-derived cells are important for correct development of the Purkinje fibers in the avian heart. Anat Rec. 2006;288A:1272–80.
16. Bochmann L, Sarathchandra P, Mori F, et al. Revealing new mouse epicardial cell markers through transcriptomics. PLoS One. 2010;5(e11429):1–13.
17. Kolditz DP, Wijffels MCEF, Blom NA, et al. Persistence of functional atrioventricular accessory pathways in post-septated embryonic avian hearts: implications for morphogenesis and functional maturation of the cardiac conduction system. Circulation. 2007;115:17–26.
18. Gittenberger-de Groot AC, Vrancken Peeters M-PFM, Bergwerff M, et al. Epicardial outgrowth inhibition leads to compensatory mesothelial outflow tract collar and abnormal cardiac septation and coronary formation. Circ Res. 2000;87:969–71.
19. Briggs LE, Kakarla J, Wessels A. The pathogenesis of atrial and atrioventricular septal defects with special emphasis on the role of the dorsal mesenchymal protrusion. Differentiation. 2012;84:117–30.
20. Eralp I, Lie-Venema H, DeRuiter MC, et al. Coronary artery and orifice development is associated with proper timing of epicardial outgrowth and correlated Fas ligand associated apoptosis patterns. Circ Res. 2005;96:526–34.
21. Gittenberger-de Groot AC, Tennstedt C, Chaoui R, et al. Ventriculo coronary arterial communications (VCAC) and myocardial sinusoids in hearts with pulmonary atresia with intact ventricular septum: two different diseases. Prog Pediatr Cardiol. 2001;13:157–64.
22. Limana F, Bertolami C, Mangoni A, et al. Myocardial infarction induces embryonic reprogramming of epicardial c-kit(þ) cells: role of the pericardial fluid. J Mol Cell Cardiol. 2010;48:609–18.
23. Weeke-Klimp A, Bax NA, Bellu AR, et al. Epicardium-derived cells enhance proliferation, cellular maturation and alignment of cardiomyocytes. J Mol Cell Cardiol. 2010;49:606–16.
24. Winter EM, Grauss RW, Hogers B, et al. Preservation of left ventricular function and attenuation of remodeling after transplantation of human epicardium-derived cells into the infarcted mouse heart. Circulation. 2007;116:917–27.
25. Winter EM, Van Oorschot AA, Hogers B, et al. A new direction for cardiac regeneration therapy: application of synergistically acting epicardium-derived cells and cardiomyocyte progenitor cells. Circ Heart Fail. 2009;2:643–53.

Cell Sheet Tissue Engineering for Heart Failure

Hidekazu Sekine, Tatsuya Shimizu, and Teruo Okano

Abstract

In recent years, regenerative medicine using cells for treating tissue defects has been in the spotlight as a new treatment for severe heart failure. Direct injection of bone marrow-derived cells and isolated skeletal myoblasts has already been used clinically as a method to improve cardiac function by regenerating cardiac muscle cells and blood vessels. The research on reconstructing functional three-dimensional (3D) cardiac grafts using tissue engineering methods has also now been addressed as a treatment for the next generation. Our laboratory has proposed an original tissue engineering technology called "cell sheet engineering" that stacks cell sheets to reconstruct functional 3D tissues. Transplantation of cell sheets has already shown it can cure damaged hearts, and it seems clear that the field of cell sheet tissue engineering can offer realistic treatment for patients with severe cardiac disorders.

Keywords

Cardiac tissue engineering • Cell sheet • Regenerative medicine

3.1 Introduction

Heart transplantation is the last hope for treatment of patients with severe heart failure due to ischemia-related disease and dilated cardiomyopathy. However, the lack of donor organs for transplantation continues to be a serious problem around the world. Although there have been many developments in artificial heart systems such as mechanical temporary assist devices or left ventricular assist devices (LVADs), there are also problems in conjunction with thromboembolism, infection, and finite

H. Sekine • T. Shimizu • T. Okano (✉)
Institute of Advanced Biomedical Engineering and Science, Tokyo Women's Medical University, 8-1 Kawada-cho, Shinjuku-ku, 162-8666 Tokyo, Japan
e-mail: tokano@twmu.ac.jp

© The Author(s) 2016

T. Nakanishi et al. (eds.), *Etiology and Morphogenesis of Congenital Heart Disease*, DOI 10.1007/978-4-431-54628-3_3

durability. Given these challenges with current technologies, regenerative therapies are being investigated as an alternative approach and present new possibilities for the repair of a damaged heart. Recently, the direct injection of either autologous skeletal myoblasts or bone marrow-derived cells has been examined in clinical studies as an alternative cell source to cardiac muscle cells [1–3]. The direct injection of the dissociated cells has shown to be slightly effective, but it is often difficult to control the form, dimensions, or the position of implanted cells. In an attempt to solve these problems, research on advanced therapies using functional tissue produced by engineered cardiac grafts has started. Over the past decade, several studies have proved that bioengineered cardiac tissues could improve cardiac function in animal models of impaired heart [4]. In this review, we discuss the progress of research on myocardial regeneration with a focus on our own original approach using cell sheet engineering.

3.2   Cell Sheet Engineering

We have developed our own "cell sheet engineering" method using temperature-responsive culture dishes created by the covalent grafting of a temperature-responsive polymer, poly(N-isopropylacrylamide) (PIPAAm), to normal cell culture dishes [5]. Under normal culture conditions at 37 °C, the dish surface is relatively hydrophobic, and cells can attach, spread, and proliferate similar to commercially available tissue culture surfaces. However, when the temperature is reduced to below the polymer's lower critical solution temperature of 32 °C, the polymer surface becomes hydrophilic and swells, forming a hydration layer between the dish surface and the cultured cells. This allows the cells to detach as a single sheet without the need of enzymatic treatments such as trypsinization. Since the use of proteolysis measures is unnecessary, critical cell surface proteins and cell-to-cell junction proteins remain intact, so that the cells can be harvested noninvasively as an intact sheet while retaining their extracellular matrix (ECM) (Fig. 3.1). Consequently, we can recreate 3D structures such as cardiac tissue by repeated layering of individual cell sheets [6].

3.3   Cardiac Tissue Reconstruction

Harvested cell sheets consist of only confluently cultured cells with their biological ECM on the basal side of the cell sheets, which acts as an adhesive agent for promoting an intimate attachment between each layered cardiac cell sheet. Within layered cardiac constructs, gap junctions are formed which rapidly establish an electrical connection between the cell sheets, leading to 3D cardiac tissues that synchronously pulsatile [7]. Additionally, when these tissues were transplanted onto the subcutaneous tissue of nude rats, the grafts were macroscopically observed to beat synchronously [8]. Importantly, these implanted tissues also showed long-term survival up to 1 year and 8 months, and the grafts also contained elongated sarcomeres, gap junctions, and well-organized vascular networks within the bioengineered cardiac tissues [9].

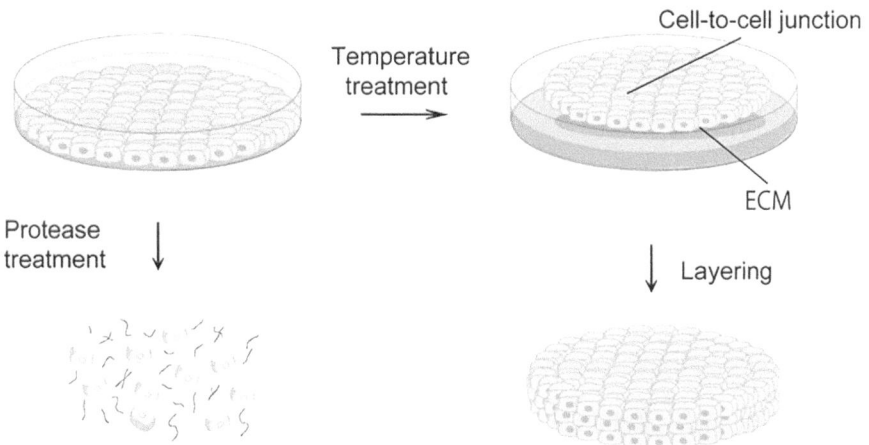

Fig. 3.1 Cell sheet engineering. Using temperature-responsive dishes, cultured cells can be harvested as intact contiguous sheets by simple temperature reduction without proteolytic treatment

3.4 Cell Sheet Transplantation in Small Animal Models

Transplantation of engineered tissue such as cardiac grafts onto infarcted rat hearts (Fig. 3.2) demonstrated morphological and functional connections via bridging cardiomyocytes that migrated from the transplanted grafts to the host heart [10]. Cardiac graft transplantation also improved damaged heart function with significant improvements in the host ejection fraction [11]. We have also demonstrated that control of EC densities in engineered cardiac tissues induces enhanced neovascularization and leads directly to improved function of the ischemic myocardium [12]. Moreover, when compared to direct cell injection, the cardiac graft transplantation exhibited superior cell survival and engraftment [13]. Similarly, skeletal myoblast grafts were able to improve left ventricular contraction, reduce fibrosis, and prevent remodeling via the recruitment of hematopoietic stem cells through the release of various growth factors [14]. The implantation of myoblast grafts also induced the restoration of left ventricular dilatation and prolonged life expectancy in dilated cardiomyopathic hamster [15]. Additionally, mesenchymal stem cell grafts demonstrated improved cardiac function in impaired rat hearts, with the reversal of cardiac wall thinning and prolonged survival after myocardial infarction. This recovery after myocardial infarction suggests that the improvement in cardiac function may be primarily due to the effects of growth factor-mediated paracrine and/or a decrease in left ventricle wall stress, which in turn result from the relatively thick mesenchymal stem cell sheets [16].

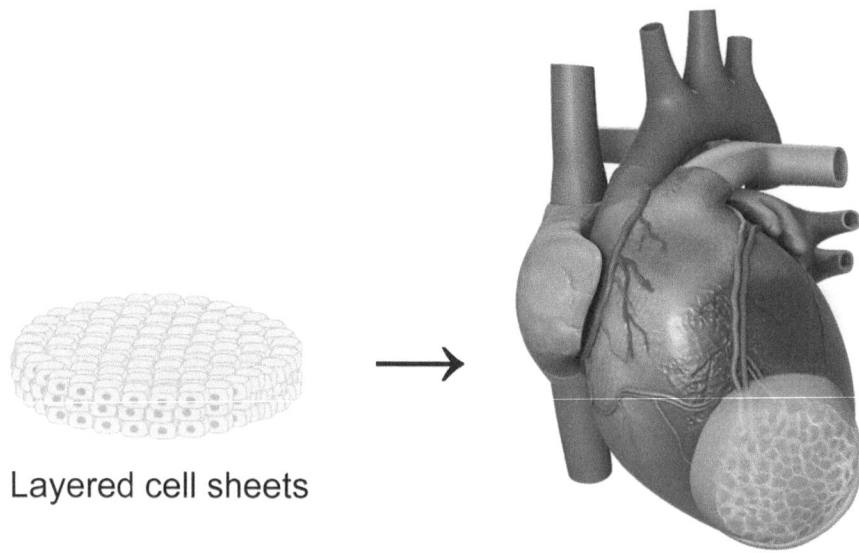

Layered cell sheets

Fig. 3.2 Transplantation of the cardiac grafts onto infarcted rat hearts

3.5   Cell Sheet Transplantation in Preclinical and Clinical Studies

Our latest work has been to transplant engineered tissue in clinically relevant large-animal models. In a pacing-induced canine dilated cardiomyopathy model, transplantation of skeletal myoblast grafts has shown improved cardiac function with reduction of fibrosis and apoptosis [17]. In a porcine cardiac infarction model, skeletal myoblast graft transplantation provided improvement of cardiac function with attenuation of cardiac remodeling [18]. Transplantation of skeletal myoblast grafts has also demonstrated that it is an appropriate and safe treatment for chronic myocardial infarction without increasing the risk of ventricular arrhythmias. Most recently, the transplantation of cardiac progenitor cell grafts derived from embryonic stem cells improved cardiac function without teratoma formation and induced cardiomyogenic differentiation in a simian impaired heart model. Our latest work provides evidence of the safety and efficacy of using embryonic stem cells for myocardial regeneration [19].

Based on the promising results in various animal models, there is a clinical study using cell sheet transplantation currently underway. Autologous skeletal myoblast sheet therapy has demonstrated that it is possible to improve cardiac function to such a degree that LVADs are no longer required for patients with dilated cardiomyopathy [20].

3.6   Conclusions

The field of tissue engineering presents an exciting approach to regenerative therapies. The future solutions scaling up give more powerful construct creation, resulting in the developments of remarkable tissue-engineered cardiac assist devices or organ replacement. Overall, cell sheet tissue engineering is a novel approach for cardiac treatment that promises efficient and effective alternative therapies in regenerative medicine.

References

1. Menasche P, Hagege AA, Scorsin M, Pouzet B, Desnos M, Duboc D, Schwartz K, Vilquin JT, Marolleau JP. Myoblast transplantation for heart failure. Lancet. 2001;357:279–80.
2. Orlic D, Kajstura J, Chimenti S, Jakoniuk I, Anderson SM, Li B, Pickel J, McKay R, Nadal-Ginard B, Bodine DM, Leri A, Anversa P. Bone marrow cells regenerate infarcted myocardium. Nature. 2001;410:701–5.
3. Wollert KC. Cell therapy for acute myocardial infarction. Curr Opin Pharmacol. 2008;8:202–10.
4. Laflamme MA, Murry CE. Regenerating the heart. Nat Biotechnol. 2005;23:845–56.
5. Yamada N, Okano T, Sakai H, Karikusa F, Sawasaki Y, Sakurai Y. Thermo-responsive polymeric surfaces; control of attachment and detachment of cultured cells. Die Makromolekulare Chemie, Rapid Commun. 1990;11:571–6.
6. Shimizu T, Yamato M, Kikuchi A, Okano T. Cell sheet engineering for myocardial tissue reconstruction. Biomaterials. 2003;24:2309–16.
7. Haraguchi Y, Shimizu T, Yamato M, Kikuchi A, Okano T. Electrical coupling of cardiomyocyte sheets occurs rapidly via functional gap junction formation. Biomaterials. 2006;27:4765–74.
8. Shimizu T, Yamato M, Isoi Y, Akutsu T, Setomaru T, Abe K, Kikuchi A, Umezu M, Okano T. Fabrication of pulsatile cardiac tissue grafts using a novel 3-dimensional cell sheet manipulation technique and temperature-responsive cell culture surfaces. Circ Res. 2002;90:e40.
9. Shimizu T, Sekine H, Isoi Y, Yamato M, Kikuchi A, Okano T. Long-term survival and growth of pulsatile myocardial tissue grafts engineered by the layering of cardiomyocyte sheets. Tissue Eng. 2006;12:499–507.
10. Sekine H, Shimizu T, Kosaka S, Kobayashi E, Okano T. Cardiomyocyte bridging between hearts and bioengineered myocardial tissues with mesenchymal transition of mesothelial cells. J Heart Lung Transplant. 2006;25:324–32.
11. Miyagawa S, Sawa Y, Sakakida S, Taketani S, Kondoh H, Memon IA, Imanishi Y, Shimizu T, Okano T, Matsuda H. Tissue cardiomyoplasty using bioengineered contractile cardiomyocyte sheets to repair damaged myocardium: their integration with recipient myocardium. Transplantation. 2005;80:1586–95.

24 H. Sekine et al.

12. Sekine H, Shimizu T, Hobo K, Sekiya S, Yang J, Yamato M, Kurosawa H, Kobayashi E, Okano T. Endothelial cell coculture within tissue-engineered cardiomyocyte sheets enhances neovascularization and improves cardiac function of ischemic hearts. Circulation. 2008;118: S145–52.
13. Sekine H, Shimizu T, Dobashi I, Matsuura K, Hagiwara N, Takahashi M, Kobayashi E, Yamato M, Okano T. Cardiac cell sheet transplantation improves damaged heart function via superior cell survival in comparison with dissociated cell injection. Tissue Eng Part A. 2011;17:2973–80.
14. Memon IA, Sawa Y, Fukushima N, Matsumiya G, Miyagawa S, Taketani S, Sakakida SK, Kondoh H, Aleshin AN, Shimizu T, Okano T, Matsuda H. Repair of impaired myocardium by means of implantation of engineered autologous myoblast sheets. J Thorac Cardiovasc Surg. 2005;130:1333–41.
15. Kondoh H, Sawa Y, Miyagawa S, Sakakida-Kitagawa S, Memon IA, Kawaguchi N, Matsuura N, Shimizu T, Okano T, Matsuda H. Longer preservation of cardiac performance by sheet-shaped myoblast implantation in dilated cardiomyopathic hamsters. Cardiovasc Res. 2006;69:466–75.
16. Miyahara Y, Nagaya N, Kataoka M, Yanagawa B, Tanaka K, Hao H, Ishino K, Ishida H, Shimizu T, Kangawa K, Sano S, Okano T, Kitamura S, Mori H. Monolayered mesenchymal stem cells repair scarred myocardium after myocardial infarction. Nat Med. 2006;12:459–65.
17. Hata H, Matsumiya G, Miyagawa S, Kondoh H, Kawaguchi N, Matsuura N, Shimizu T, Okano T, Matsuda H, Sawa Y. Grafted skeletal myoblast sheets attenuate myocardial remodeling in pacing-induced canine heart failure model. J Thorac Cardiovasc Surg. 2006;132:918–24.
18. Miyagawa S, Saito A, Sakaguchi T, Yoshikawa Y, Yamauchi T, Imanishi Y, Kawaguchi N, Teramoto N, Matsuura N, Iida H, Shimizu T, Okano T, Sawa Y. Impaired myocardium regeneration with skeletal cell sheets – a preclinical trial for tissue-engineered regeneration therapy. Transplantation. 2010;90:364–72.
19. Bel A, Planat-Bernard V, Saito A, Bonnevie L, Bellamy V, Sabbah L, Bellabas L, Brinon B, Vanneaux V, Pradeau P, Peyrard S, Larghero J, Pouly J, Binder P, Garcia S, Shimizu T, Sawa Y, Okano T, Bruneval P, Desnos M, Hagege AA, Casteilla L, Puceat M, Menasche P. Composite cell sheets: a further step toward safe and effective myocardial regeneration by cardiac progenitors derived from embryonic stem cells. Circulation. 2010;122:S118–23.
20. Sawa Y, Miyagawa S, Sakaguchi T, Fujita T, Matsuyama A, Saito A, Shimizu T, Okano T. Tissue engineered myoblast sheets improved cardiac function sufficiently to discontinue lvas in a patient with dcm: report of a case. Surg Today. 2012;42:181–4.

Future Treatment of Heart Failure Using Human iPSC-Derived Cardiomyocytes

4

Shugo Tohyama and Keiichi Fukuda

Abstract

Heart transplantation can drastically improve survival in patients with a failing heart; however, the shortage of donor hearts remains a serious problem with this treatment strategy and the successful clinical application of regenerative medicine is eagerly awaited. To this end, we developed a novel method to generate human induced pluripotent stem cells (iPSCs) from circulating human T lymphocytes using Sendai virus containing Yamanaka factors. To establish an efficient cardiac differentiation protocol, we then screened factors expressed in the future heart site of early mouse embryos and identified several growth factors and cytokines that can induce cardiomyocyte differentiation and proliferation. Subsequent transcriptome and metabolome analysis on undifferentiated stem cells and cardiomyocytes to devise a specific metabolic environment for cardiomyocyte selection revealed completely different mechanisms of glucose and lactate metabolism. Based on these findings, we succeeded in metabolically selecting cardiomyocytes using glucose-free and lactate-supplemented medium, with up to 99 % purity and no teratoma formation. Using our aggregation technique, we also showed that >90 % of the transplanted cardiomyocytes survived in the heart and showed physiological growth after transplantation. We expect that combining these techniques will achieve future heart regeneration.

Keywords

Induced pluripotent stem cell • Purification • Cardiomyocyte • Transplantation • Human

S. Tohyama • K. Fukuda (✉)
Department of Cardiology, Keio University School of Medicine, Tokyo, Japan
e-mail: kfukuda@a2.keio.jp

© The Author(s) 2016
T. Nakanishi et al. (eds.), *Etiology and Morphogenesis of Congenital Heart Disease*,
DOI 10.1007/978-4-431-54628-3_4

25

4.1   Introduction

Heart disease remains a leading cause of death despite recent medical advances, and heart transplantation remains the ultimate treatment for severe heart failure. However, limited donor numbers remain an unsolved problem for transplantation therapy, and both patients and clinicians hold great hope for the future success of heart regenerative cell therapies as an alternative strategy [1]. Pluripotent stem cells (PSCs) including embryonic stem cells (ESCs) and induced pluripotent stem cells (iPSCs) can self-renew infinitely and are potential mass production sources for therapeutic cardiomyocytes. In particular, human iPSCs have the huge advantage of avoiding immunological rejection after cell transplantation.

Human iPSCs were first generated from dermal fibroblasts by Takanashi and Yamanaka in 2007 [2], using a retroviral transduction system. Subsequently, the methodology for generating iPSCs has dramatically improved. We previously reported that integration-free iPSCs could be easily and rapidly generated from terminally differentiated circulating T lymphocytes in peripheral blood using Sendai virus [3]. Our method makes it possible to generate iPSCs from any patients including children, girls, and the very elderly by blood sampling alone.

Such cumulative advances in iPSC generation techniques should accelerate the development of applications for iPSCs generated from patients. However, many hurdles remain in realizing such applications in human heart regeneration, due largely to the assumption that one patient will require at least 1×10^9 cardiomyocytes to recover cardiac function (Fig. 4.1). The important steps to overcome are as follows: (1) cardiac differentiation efficiencies should be stably improved regardless of the source cell lines; (2) large-scale cultivation systems will be required to obtain the billions of differentiated cells required; (3) approximately 1×10^9 cardiomyocytes should be efficiently collected and residual undifferentiated stem cells should be eliminated from large-scale, mixed, differentiated cell populations; (4) purified cardiomyocytes should be functionally and electrophysiologically characterized; (5) tissue engineering technologies might need to be used prior to transplantation because dispersed transplanted cardiomyocytes have not yet achieved high survival rates [4]; and (6) confirmation of safety and efficiency with these techniques is essential in large animal models before clinical application. In this chapter, we discuss these hurdles to realizing heart regenerative therapy in more detail.

4.2   Cardiac Differentiation from Human iPSCs

Many approaches using ESCs have been investigated to induce cardiac differentiation. In general, the differentiation of ESCs into any cell lineage is based on the mechanism of normal early development [5]. The visceral endoderm is known to play a key role in the differentiation of cardiac precursors that are present in the adjacent mesoderm during development, and Mummery et al. [6] previously reported that human ESCs effectively differentiate into cardiomyocytes when

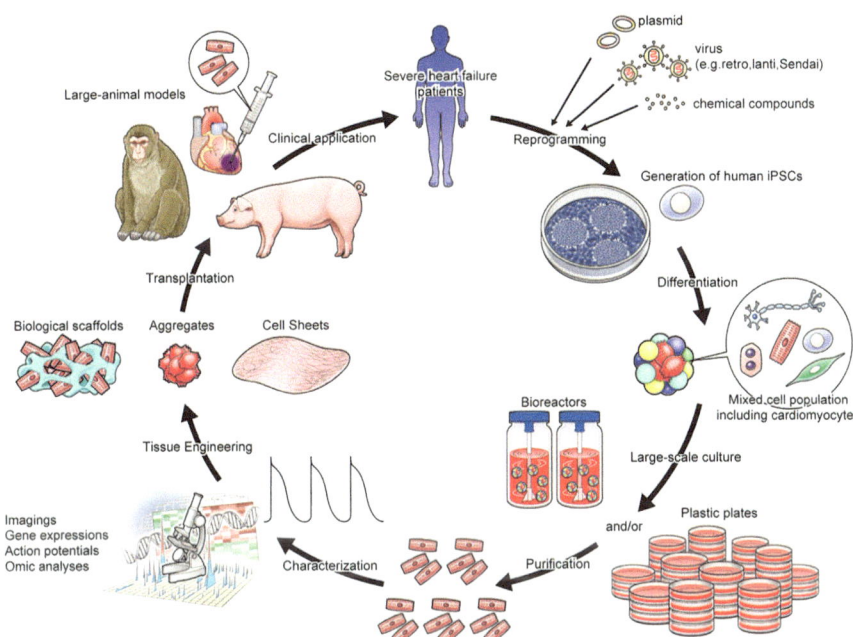

Fig. 4.1 Overview of steps to overcome for realization of heart regenerative therapy. The critical hurdles to overcome are as follows: (1) improvement of cardiac differentiation efficiencies, (2) establishment of large-scale cultivation systems, (3) purification of large-scale cardiomyocytes, (4) electrophysiological characterization, (5) utilization of appropriate tissue engineering technologies, and (6) confirmation of safety and efficiency in large animal models

cocultured with mouse visceral endoderm-like (END-2) cells. In an attempt to improve cardiac differentiation efficiencies inexpensively and easily, Takahashi et al. [6] screened a chemical compound library approved by the United States Food and Drug Administration (FDA) and found that ascorbic acid efficiently induced cardiac differentiation. However, this chemical alone might be not enough to induce cardiac differentiation from iPSCs.

Several studies have shown that various combinations of heart development-related proteins including BMP, activin, Wnt, BMP inhibitor, and Wnt inhibitor induce cardiomyocytes from ESCs [7–10]. We reported that the context-dependent differential action of BMPs in cardiomyocyte induction is explained by the local action of Noggin and other BMP inhibitors and, accordingly, developed a protocol to induce cardiac differentiation of mouse ESCs through transient administration of Noggin [9]. However, to obtain hundreds of millions of cardiomyocytes, it is necessary to establish a cardiac differentiation method that is both efficient and cost-effective due to the many expensive recombinant protein factors used. To address this problem, Minami et al. [11] screened small-molecule compounds to identify those that significantly increase cardiac differentiation induction, and they revealed some inhibitors of canonical Wnt signaling as candidates. In addition, recently novel efficient protocols using small molecules and/or chemically defined media have been reported (Lian et al. Nat Protocols 2013;8:162–175, Burridge et al. Nat Methods. 2014;11:855–60).

Several studies then showed that induction techniques in ESCs could also be applied to iPSCs, although differentiation efficiencies were suggested to be inferior compared to ESCs. However, as such differentiation efficiencies vary greatly with different cell lines [12], further investigation is needed in the future. Furthermore, to efficiently obtain large quantities of cardiomyocytes inexpensively, it is necessary to continue refining efficient cardiac differentiation systems combined with the use of small-molecule compounds. However, despite improved cardiac differentiation efficiencies, it is inevitable that human PSC derivatives will contain not only cardiomyocytes but also undifferentiated stem cells and/or noncardiac cells because all PSCs cannot differentiate into cardiomyocytes. Therefore, to confirm safety after transplantation, it is necessary to remove noncardiac cells and undifferentiated stem cells that could cause tumors.

4.3   Nongenetic Methods for Purifying Cardiomyocytes

One of the biggest risks with in vitro-generated cardiomyocytes for clinical use is teratoma formation due to residual PSC contamination [13]. Current procedures for eliminating such contamination and boosting cardiomyocyte enrichment involve genetic modification [14, 15] and nongenetic methods using a mitochondrial dye [16] or antibodies to specific cell-surface markers [17]. However, none of these methods are ideal for the therapeutic application of PSC-derived cardiomyocytes due to insufficient stability, genotoxicity, and the use of fluorescence-activated cell sorting (FACS). To address this issue, we sought to purify cardiomyocytes efficiently and inexpensively, based on differences among cell-specific nutrition sources.

To create a metabolic environment where "residual undifferentiated stem cells cannot survive and only cardiomyocytes can survive," we performed metabolome and transcriptome analysis in neonatal cardiomyocytes and PSCs. We found that these PSCs mainly depended on activated glycolysis and actively discharged lactate into the extracellular media. In addition, biomass needed for proliferation such as amino acids and nucleic acids were actively synthesized in PSCs compared to cardiomyocytes. On the other hand, cardiomyocytes mainly depended on oxidative phosphorylation in mitochondria to obtain adenosine triphosphate (ATP) efficiently. We also performed metabolome analysis to demonstrate that other noncardiac proliferating cells also depended on glycolysis, like PSCs. Thus, cardiomyocytes and proliferating noncardiac cells including PSCs showed differences in metabolism (Fig. 4.2) that we then successfully exploited to select cardiomyocytes from human PSCs efficiently and inexpensively simply by changing the cell-specific medium to one that is glucose depleted and lactate supplemented [18]. The cardiomyocytes selected by metabolism showed normal electrophysiological properties and did not form teratoma after transplantation.

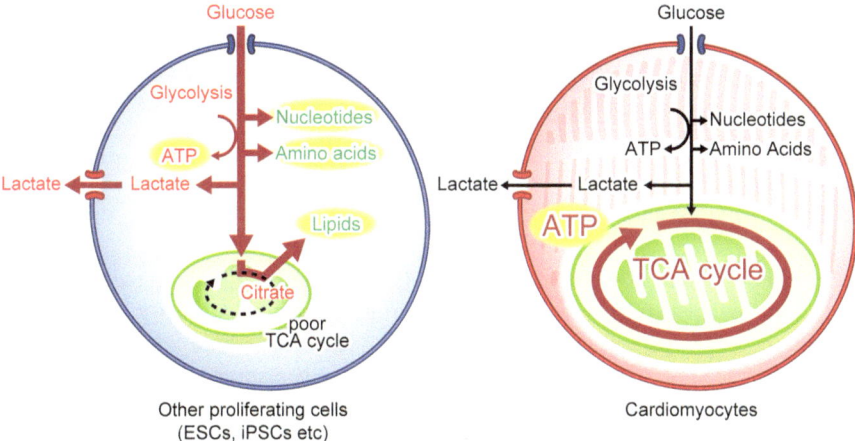

Fig. 4.2 Distinct metabolic differences between cardiomyocytes and other proliferating cells including undifferentiated stem cells. Cardiomyocytes efficiently obtain ATP mainly via oxidative phosphorylation, while other proliferating cells including PSCs obtain ATP, nucleotide, and amino acids via activated glycolysis

Furthermore, we succeeded to establish a practical culture system for generating substantial numbers of purified cardiomyocytes by combining a massive suspension culture system with a metabolic selection medium (Hemmi et al. Stem Cells Transl Med. 2014;3:1473–83).

4.4 Transplantation of Human PSC-Derived Cardiomyocytes

Many studies have been conducted regarding cell transplantation therapies in animal models using human PSC-derived cardiomyocytes. The survival of dispersed transplanted cardiomyocytes is reportedly very low [4], and some ingenuity is required in the method of transplantation. Our group previously reported that transplanted cell survival was dramatically improved by transplantation after formation of cardiomyocyte aggregates [16], while Laflamme et al. [9] showed a similar effect using a prosurvival cocktail that inhibited apoptosis.

With regard to the effectiveness of cell transplantation, Laflamme et al. [9] also demonstrated that transplanted human ESC-derived cardiomyocytes could improve cardiac function after 4 weeks in a rat myocardial infarction model. In addition, Shiba et al. [20] demonstrated that transplanted human ESC-derived cardiomyocytes electrically coupled to the host cardiomyocytes and suppressed arrhythmias in a guinea pig myocardial infarction model. Furthermore, with respect to cell transplantation in large animals, Kawamura et al. [19] recently reported that cardiac cell sheets comprising purified human iPSC-derived cardiomyocytes generated using our method [18] improved cardiac function in a pig myocardial infarction model. Thus, efficacies have been achieved in cell transplantation therapies using human iPSC-derived cardiomyocytes, although many such studies showed only short-term effectiveness. Thus, careful evaluation of the efficacy and safety of human iPSC-derived cardiomyocytes in cell transplantation over the longer term must be ongoing.

4.5 Future Directions

The discovery and refinement of human iPSCs generation is expected to advance not only regenerative medicine but also drug discovery and analyses of genetic disorders using patient-specific iPSCs [19]. The major and common problem remaining in this quest is securing sufficient numbers of mature and functional cardiomyocytes with high purity. To solve this problem, it is essential to develop a stable and efficient mass culture system and to establish a simple system to analyze the electrophysiological function of the generated cardiomyocytes (Fig. 4.1). Furthermore, the future realization of clinical applications using human iPSCs will necessitate a better understanding of the cell biology and techniques involved in tissue engineering generally. Only then will we be able to achieve long-term safety and efficacy in the heart failure models of large animals and finally realize human heart regenerative therapies.

References

1. Passier R, van Laake LW, Mummery CL. Stem-cell-based therapy and lessons from the heart. Nature. 2008;453(7193):322–9.
2. Takahashi K, et al. Induction of pluripotent stem cells from adult human fibroblasts by defined factors. Cell. 2007;131(5):861–72.
3. Seki T, et al. Generation of induced pluripotent stem cells from human terminally differentiated circulating T cells. Cell Stem Cell. 2010;7(1):11–4.
4. Kolossov E, et al. Engraftment of engineered ES cell-derived cardiomyocytes but not BM cells restores contractile function to the infarcted myocardium. J Exp Med. 2006;203(10):2315–27.
5. Srivastava D. Making or breaking the heart: from lineage determination to morphogenesis. Cell. 2006;126(6):1037–48.
6. Takahashi T, et al. Ascorbic acid enhances differentiation of embryonic stem cells into cardiac myocytes. Circulation. 2003;107(14):1912–6.
7. Naito AT, et al. Developmental stage-specific biphasic roles of Wnt/beta-catenin signaling in cardiomyogenesis and hematopoiesis. Proc Natl Acad Sci U S A. 2006;103(52):19812–7.

8. Laflamme MA, et al. Cardiomyocytes derived from human embryonic stem cells in pro-survival factors enhance function of infarcted rat hearts. Nat Biotechnol. 2007;25 (9):1015–24.
9. Yuasa S, et al. Transient inhibition of BMP signaling by Noggin induces cardiomyocyte differentiation of mouse embryonic stem cells. Nat Biotechnol. 2005;23(5):607–11.
10. Kattman SJ, et al. Stage-specific optimization of activin/nodal and BMP signaling promotes cardiac differentiation of mouse and human pluripotent stem cell lines. Cell Stem Cell. 2011;8 (2):228–40.
11. Minami I, et al. A small molecule that promotes cardiac differentiation of human pluripotent stem cells under defined, cytokine- and xeno-free conditions. Cell Rep. 2012;2(5):1448–60.
12. Osafune K, et al. Marked differences in differentiation propensity among human embryonic stem cell lines. Nat Biotechnol. 2008;26(3):313–5.
13. Miura K, et al. Variation in the safety of induced pluripotent stem cell lines. Nat Biotechnol. 2009;27(8):743–5.
14. Hidaka K, et al. Chamber-specific differentiation of Nkx2.5-positive cardiac precursor cells from murine embryonic stem cells. FASEB J. 2003;17(6):740–2.
15. Elliott DA, et al. NKX2-5(eGFP/w) hESCs for isolation of human cardiac progenitors and cardiomyocytes. Nat Methods. 2011;8(12):1037–40.
16. Hattori F, et al. Nongenetic method for purifying stem cell-derived cardiomyocytes. Nat Methods. 2010;7(1):61–6.
17. Dubois NC, et al. SIRPA is a specific cell-surface marker for isolating cardiomyocytes derived from human pluripotent stem cells. Nat Biotech. 2011;29(11):1011–8.
18. Tohyama S, et al. Distinct metabolic flow enables large-scale purification of mouse and human pluripotent stem cell-derived cardiomyocytes. Cell Stem Cell. 2013;12(1):127–37.
19. Egashira T, et al. Disease characterization using LQTS-specific induced pluripotent stem cells. Cardiovasc Res. 2012;95(4):419–29.
20. Shiba Y et al. Human ES-cell-derived cardiomyocytes electrically couple and suppress arrhythmias in injured hearts. Nature. 2012;489(7415):322–5.

Congenital Heart Disease: In Search of Remedial Etiologies

5

Roger R. Markwald, Shibnath Ghatak, Suniti Misra, Ricardo A. Moreno-Rodríguez, Yukiko Sugi, and Russell A. Norris

Abstract

In searching for remedial etiologies for congenital heart disease (CHD), we have focused on identifying interactive signaling pathways or "hubs" in which mutations disrupt fundamental cell biological functions in cardiac progenitor cells in a lineage-specific manner. Based on the frequency of heart defects seen in a clinical setting, we emphasize two signaling hubs – *nodal kinases activated by extracellular ligands* (e.g., *periostin*) *and the cytoskeletal regulatory protein, filamin A (FLNA)*. We discuss them in the context of valve and septal development and the lineages which give origin to their progenitor cells. We also explore developmental windows that are potentially amenable to remedial therapy using homeostatic mechanisms like those revealed by a chimeric mice model, i.e., *irradiated animals whose bone marrow had been reconstituted with GFP+ hematopoietic stem cells*, that shows bone marrow-derived cells track to the heart, engraft, and give rise to bona fide fibroblasts. We propose to use this model to deliver genetic payloads or protein cargos during the neonatal period to override biochemical or structural deficits of CHD associated with valve and septal signaling hubs or fibroblast/myocyte interactions. Preliminary tests of the model indicate remedial potential for cardiac injuries.

Keywords

Heart defects • Valves • Septa • Matricellular proteins • Periostin • Cell signaling • Kinases • Filamin A • Hematopoietic stem cells • Fibroblast • Lineage • Genetic engineering

R.R. Markwald, Ph.D. (✉) • S. Ghatak • S. Misra • R.A. Moreno-Rodríguez • Y. Sugi • R.A. Norris
Department of Regenerative Medicine and Cell Biology, Medical University of South Carolina, 173 Ashley Avenue, Room BSB-648B, P.O. Box 250508, Charleston, SC 29425, USA
e-mail: Markwald@musc.edu

© The Author(s) 2016
T. Nakanishi et al. (eds.), *Etiology and Morphogenesis of Congenital Heart Disease*,
DOI 10.1007/978-4-431-54628-3_5

33

5.1   Introduction

5.1.1   Emerging Concepts

Etiologies of congenital heart defects have been difficult to identify and explain, often invoking discussions of genetics and environment or both. After 40+ years of searching for etiologies, especially ones that might be potentially remedial, a few conceptual observations are emerging (although by no means is there consensus). These include suppositions that:

1. Genetics more than environment is responsible for CHD based on the ability to simulate human heart congenital defects in animal genetic models (*see also Bhattacharya* [1]).
2. The evolution of genetic thought is toward trans-heterozygous, multigenetic interactions vs. single gene hits.
3. Mutations in the downstream intracellular signaling targets of growth factor, transcription factors, or matricellular proteins are more likely to be the root cause of heart defects seen in children by pediatric cardiologists (*as the loss of both alleles for upstream early regulatory genes is usually lethal*).
4. Intersecting regulatory networks or signaling hubs coordinate fundamental biological processes in cardiac progenitor cells in a lineage-restricted manner.
5. Based on the frequency of different types of heart defects, the lineages most likely to be modified by genetic mutations or environmental stressors in CHD are the non-cardiomyocyte lineages, primarily fibroblasts.

5.1.2   Hub Hypothesis

In trying to provide answers to the questions most often asked by parents of children born with heart defects – "why did it happen?" and "what can be done about it?" – we have endeavored to integrate these five emerging concepts around two examples of central signaling hubs or platforms that intersect with multiple gene regulatory networks (like the spokes of a wheel) to regulate or "tune" behavioral changes in the progenitor cells engaged in valvuloseptal morphogenesis (or myocardial remodeling). In so doing, we propose that intersecting signaling hubs explain why so many different genes, if mutated or deleted, can engender similar anatomical dysmorphic phenotypes, e.g., ventricular septal defects. This suggests that there are only so many ways progenitor cells can respond to normal or abnormal signaling inputs: e.g., they can proliferate, activate, or suppress apoptotic pathways; transport ions; secrete, endocytose, adhere/migrate, and generate contractile forces; change polarity or shape; and differentiate. Thus, the conceptual appeal of common final pathways or intersecting signaling hubs/nodes is that they provide potential for exploring "shared" remedial therapies for CHD that do not require an individual approach for correcting each abnormal gene.

5.2   Searching for Candidate Signaling Hubs in Heart Development

5.2.1   Nodal Signaling Kinases

In patients with Down, Marfan, or Noonan Syndromes, there is increasing evidence that genes encoding nodal signaling kinases like FAK/AKTkt/PI3K, RAS, MEKK/ERK1/2, PTPN11, etc., are likely candidates for CHD if they are mutated or overexpressed [2–4]. Such genes are not usually lethal (*as there exist molecular or functional redundancies*), yet as indicated for syndromic heart defects, they have potential to change functional behaviors in progenitor cells that normally mold and remodel the simple tubular heart into a four-chambered organ. As indicated in Fig. 5.1, intracellular signaling kinases (and small regulatory GTPases) are the direct and indirect downstream targets of growth factors (e.g., TGFβ and BMP 2&4) that are normally secreted by the embryonic endocardium or myocardium [5], or in the case of extracellular ligands like matricellular proteins (periostin, the CCN family), they are secreted by the mesenchymal progenitor cells of valves and connective septa which are derived from both endocardium and epicardium [6]. In the case of matricellular and other extracellular proteins (Fig. 5.2), binding to integrin receptors triggers integrin-dependent, downstream signaling kinases/GTPases (FAK/AKT/PI3k) which activate effector mechanisms of growth, survival, and differentiation into the fibrous structures (valves and septa) that assure coordinated and unidirectional blood flow through the right and left sides of the developing heart. Epicardial-derived mesenchymal cells also express periostin as they invade the ventricular myocardium and, like endothelial-derived mesenchyme, secrete collagen and differentiate into ventricular connective tissue but also contribute to the parietal leaflets of the AV valves [7, 8]. Disruption of these signaling pathways by either silencing one or more of the kinases shown in Fig. 5.1, inhibiting β-integrin functions, or deleting the periostin gene itself resulted in septal defects, abnormal (poorly differentiated, hypertrophic) valves, arrhythmias associated with a reduction in the AV fibrous connective tissue, and a reduction in ventricular elastic modulus due to loss of interstitial collagen [9, 10]. These findings are consistent with the relevance of central or nodal signaling kinases to heart development and how different genes, if mutated or inhibited, could produce similar abnormal anatomical phenotypes as a result of either binding to, activating, or encoding a kinase or GTPase component of an interactive signaling pathway. Conceivably, these same interactive signaling pathways could also be used to explore remedial therapies for CHD. Any one or combination of the signaling kinases or effector proteins shown in Fig. 5.1 could be a candidate therapeutic target that could be used to bypass or circumvent a genetic or biochemical block associated with a particular CHD, *if a way could be found to administer drugs or small molecules that silence, simulate, or activate them.* Identifying an in utero approach using orally administered or injected signaling inhibitors, lithium, or retinoids is appealing but lacks target specificity (potentially engendering a broad spectrum of side effects), or it can create a catch-22 in the sense that treatments may

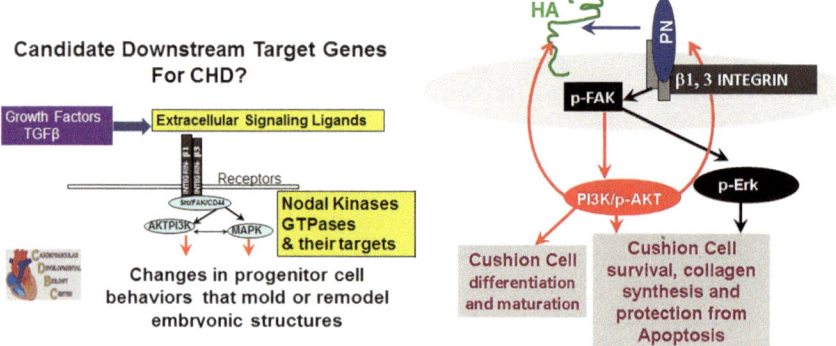

Fig. 5.1 (*Left*): Generalized model proposing the candidate genes for CHD. Based on syndromic mutations or overexpression, the model proposes that the most likely candidates are genes that encode components of a signaling hub such as the one shown for an extracellular ligand and membrane receptors. Their interaction activates a cascade of nodal kinases within valvuloseptal progenitor cells that regulate biological functions in progenitor cells that are normally associated with remodeling the primary heart into a four-chambered organ. (*Right*): Periostin-integrin model depicting downstream targets activated by periostin binding to specific β-integrin heterodimeric receptors expressed in valve and septal progenitor cells. Based on published results [6], periostin binding to integrin induces phosphorylation of p-FAK (a nodal kinase) which, in turn, activates different signaling pathways, p-AKT/PI3K and/or p-Erk, each with distinct biological activities; e.g., p-AKT/PI3K promotes hyaluronan (HA) secretion and enhances periostin expression, whereas Erk activates collagen secretion and survival (anti-apoptotic) mechanisms

have to begin too early, i.e., before it can be actually determined that a development defect is going to happen. Genetic rescues can identify and test candidate target mechanisms (e.g., breeding Noonan mice into an Erk null background) but have little practical remedial potential for preventing CHD.

Another plus for interactive signaling hubs is their potential to reveal a new or unexpected "spoke" whereby genetic, protein, or biochemical connections to signaling pathways are not readily apparent or known, e.g., hyaluronan synthetase 2 (Has2). As shown by Misra et al. [6], Has2 is activated by nodal kinases that are regulated by periostin-integrin-linked signaling [6]. Phosphorylation of Has2 leads to secretion of hyaluronan (HA) (Fig. 5.1, right) which, in addition to its osmotic properties, also binds to CD44, a tyrosine kinase receptor for HA that is expressed by valvular-septal mesenchymal progenitor cells [11]. Binding of HA to CD44 activates some of the same kinases also activated by periostin-integrin signaling, thereby triggering positive feedback loops that can amplify or sustain shared biological effects of periostin including formation, migration, and survival of progenitor valvuloseptal mesenchyme as well as a positive feedback on the secretion of periostin itself (Fig. 5.1) [6, 11]. CD44 signaling also appears to be part of a signaling complex revealed by immunoprecipitation that is required for periostin to activate another candidate signaling hub – filamin A (Fig. 5.2).

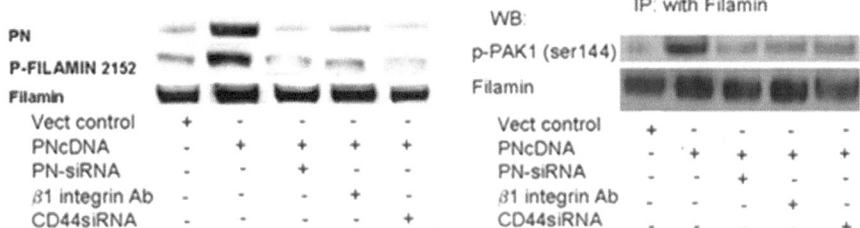

Fig. 5.2 Coordinated activation of Filamin and p21-activated kinase 1 (PAK1) by periostin/β1-integrin signaling in E16.5 mouse AV cells. (**a**) Representative Western blot for periostin (PN), p-filamin A (ser 2152), and filamin A from lysates prepared from isolated mouse E16.5 AV valve cells, transfected with vector controls, PN cDNA, PN-siRNA, and CD44 siRNA for 48 h or treated with 5 μg/ml β1-integrin-antibody for 24 h. (**b**) Same protocol as A but lysates were immunoprecipitated with filamin A and binding proteins in the immunoprecipitate were immunoblotted. The figure shows that upon activation, phospho-PAK1 (ser144) binds to Filamin (and transfers its phosphate to activate the FLNA protein). Because silencing CD44 prevented binding of PAK1 to FLNA, we have suggested that CD44 forms a signaling complex with periostin-integrin-FAK that is required for activating FLNA through PAK1

5.2.2 Filamin A

Filamin A (FLNA) is a homo- or heterodimeric Y-shaped cytoplasmic protein in which each main chain consists of an actin-binding region at the amino terminus, a core of 24 highly homologous Ig-like repeats followed by a carboxyl integrin-binding domain. The most widely studied function of FLNA is its ability, upon activation, to self-assemble into signaling scaffolds or to cross-link cortical actin filaments into a dynamic three-dimensional structure [12]. In the developing heart, FLNA can promote or suppress cell processes important for heart development as evident that when the gene is deleted, lethality ensues. Through its role as an actin-binding protein that anchors various transmembrane proteins to the cytoskeleton, it can generate contractile forces to support migration and fusion of the paired midline AV mesenchymalized "cushions" to form the AV septum [13]. Upon activation, the scaffolds assembled by FLNA form docking sites for second messengers like smad 2/3 which are critical for transmitting TGFβ signaling in heart development [14, 15]. While FLNA is also expressed in the brain and other organs, it is specifically expressed in the heart in valvuloseptal mesenchyme derived from the endocardium and epicardium [16]. The null phenotype, although lethal at ED15.0, is consistent with the pattern of FLNA expression in valvuloseptal primordia, i.e., there are septal and valve defects at *all* axial levels: atrial, atrioventricular, ventricular, and outlet [15–17].

FLNA itself responds to many signaling "inputs" that can activate FLNA or modify its binding to actin [18], including Ras and Rho kinase. Immunoprecipitation studies confirmed that FLNA is also a downstream target of kinases induced by periostin-integrin signaling that promote the binding of PAK-1 to FLNA, resulting in its activation (Fig. 5.2). A large primary atrial septal defect is seen in periostin null mice [19] that is also seen in FLNA nulls, suggesting an interaction between "signaling hubs" that (normally) promote: (1) fusion of paired mesenchymal primordia (often called "cushions") to form septal and valvular primordia and (2) remodeling of the fused cushion primordia during fetal and early postnatal life into mature leaflets or tendon-like septal structures.

Loss-of-function, point mutations in FLNA have also been found in patients with mitral valve disease (e.g., prolapse), all of which occurred in the actin-binding domains of FLNA and similarly effected cell signaling pathways or cytoskeletal organization [20]. For example, a G288R and P637Q point disrupted a signaling network that balanced RhoA and Rac1 GTPases activities and correlated directly with inhibited cell spreading, migration, and contractile force generation [21]. The G288R and P637Q mutation also interfered with FLNA's normal capabilities to promote the expression and transport of beta integrins to the cell surface or bind a variety of intracellular target proteins including vimentin and transglutaminase 2 (TG2) [16, 22]. We found that TG2 homeostatically functions to covalently link serotonin to FLNA enhancing its potential to bind cytoskeletal F-actin and generate contractile forces related to or required for remodeling of valve primordia into compacted, sculpted leaflets [16].

FLNA expression, like periostin, peaks in the postnatal period but is barely detectable in adult heart tissues unless there is an injury ([17, 19, 23]). These findings have three implications: (1) heart development is not over at birth; (2) the postnatal period may still be an open window to explore remedial therapies for CHD, particularly for valve and septal defects [24, 41]; and (3) mutations in genes that cause abnormal structural or functional changes in heart tissues may not always be immediately visible at birth but progressively appear *over time* [25]. Thus, some "adult cardiovascular degenerative diseases" may actually have an embryonic developmental etiology [26].

5.2.3   Relevance of Signaling Hubs to CHD

Shared (interconnected) protein signaling hubs point to a multiplicity of ways by which seemingly unrelated genes (*including ones yet to be identified*) can converge to engender anatomically similar developmental heart defects [27, 28]. Understanding these mechanistic relationships will be important as they have realistic remedial potential for identifying candidate approaches for preventing, modifying, or even reversing some of the clinical consequences of CHD, *particularly if they are recognized early in the neonatal period*. For example, AV valves in which there is diminished function of FLNA are enlarged at birth (and mesenchymal-like) but have elevated phosphorylation of ERK1/2 (16, *unpublished data*) which progressively leads to a more pronounced degenerative, myxomatous-like phenotype. A potential remedial therapy would be to find a way to attenuate ERK signaling and assess whether progression to a degenerative phenotype could be delayed or even reversed. A precedent for this would be the fibrillin-1 knockin model of Marfan syndrome in which the valves become myxomatous or the Loeys-Dietz mice with mutated TGFβ receptors in which the aortic root is dilated and prone to aneurysm [29, 30]. ERK 1/2 pathways are used in both syndromes to transduce elevated, noncanonical TGFβ signaling. Pharmacological approaches (e.g., Losartan) to blunt ERK signaling in both syndromes have improved their cardiovascular function and tissue structural integrity [30]. In a related fashion, understanding that TG2 promotes serotonylation of FLNA (and its binding affinity for actin), altering serotonin uptake, and/or synthesis could also prove to be new remedial approaches for treating valve and septal structural defects during the neonatal period when they normally complete their maturation. *The key would be to identify a means or route for delivering these potential remedial therapies to the "right cell targets" at the "right time".*

5.3 Lineage Is a Key to Remedial Therapy

Finding the "right cell targets" for remedial therapy is a question of lineage. Any hope of efficacious remedial treatments would seemingly require that treatments be directed to those cell populations that are normally involved in valve and septal formation and, if mutated, result in CHD. In heart development, it is the non-myocyte – mostly fibroblast – populations of the heart that appear to be the critical players in valvuloseptal morphogenesis. They are the only heart progenitor cells to express the extracellular signaling ligands like periostin or the cytoskeletal regulatory protein like FLNA. The progenitors of fibroblasts, as noted above, are the mesenchymal stem cells (or "cushion" cells) derived from the transformation of two epithelia: endocardium and epicardium. Cells derived from either lineage have the potential to differentiate into fibroblasts by autonomously secreting periostin [31, 32]. Thus, cardiac fibroblasts are derived from at least two origins which are carried over from intrauterine to postnatal and adult life [32]. This raises two questions: (1) Are new fibroblast progenitor cells added to the heart after birth? (2) If so, are the same signals (e.g., periostin) used to direct their progression into a cardiac fibroblast lineage during embryonic life also used postnatally? The answers to both questions appear to be "yes."

5.3.1 Postnatal Origin of Cardiac Fibroblasts

Recent single cell engraftment experiments indicate that in postnatal and adult life, cardiac fibroblasts are also derived from bone marrow or from pericytes that express the hematopoietic stem cell marker – CD45 [33–38]. In these experiments, adult mice were lethally irradiated and a single (or clone) of a rigorously isolated wild-type CD45+ hematopoietic stem cell (HSC) carrying a green fluorescent protein (GFP) marker was injected into their tail vein. Mice which survived clearly had received a true multipotential stem cell capable of restoring the blood cell lineages. We then asked if any of the original clones of CD45+/GFP+ HSCs left the bone marrow and engrafted elsewhere. They did. We found that CD45+/GFP+ cells migrate and engraft in several organs including the heart. Specifically they engrafted into the inlet and outlet valves, ventricular fibrous interstitium, and as pericytes surrounding coronary microvasculature [36, 37]. In addition to the CD45 marker and GFP, they also expressed fibroblast markers, e.g., collagen-1, HSP47, vimentin, discoid domain receptor 2 (DDR2), and, importantly, periostin [33, 37]. This suggested to us that cardiac fibroblasts are a renewable cell population that can be replenished homeostatically by circulating progenitor cells of bone marrow, HSC origin. Based on marker expression, monocytes are probably the immediate circulating progenitors of the CD45+/collagen I+ cells that engraft into the heart, probably as "blank" cells that differentiate into fibroblasts [38, 39]. Importantly, their numbers increase in the valves and ventricular interstitium significantly if the heart was injured by coronary ligation or cryoablation, indicating that they are also a population of fibrogenic precursors that can respond dynamically to injury or inflammatory signals [38, 40]. In quantitative terms, bone marrow-derived cells accounted for 20–30 % of fibroblasts in normal adult myocardial tissue [36, 37, 39]. In contrast, we found GFP+ label in myocytes at exceedingly low frequency [36] making this approach suitable for a targeted assessment of non-myocyte contributions to cardiac structure and function. It is important to recall, in this context, that not all lineage markers of hematopoietic cell sources (e.g., CD34) are applicable across species.

5.3.2 A Strategy to Use Fibroblast Progenitors to Carry Genetic Payloads

The bone marrow origin of postnatal cardiac fibroblasts does not in any way exclude new fibroblasts arising by proliferation from lineages carried over from intrauterine life. However, those fibroblasts would be expected to continue to carry forward any mutations or other functional or biochemical deficits from embryonic life, whereas those derived from bone marrow provide an opportunity to be isolated and genetically or pharmacologically manipulated (*or reengineered*) and then returned to the marrow. Thus, we propose to use these fibrogenic CD45+ progenitors to carry genetic payloads (or protein cargos) to sites where fibrous valvuloseptal or interstitial tissues are underdeveloped or hyperplastic or where maturation (e.g., valves) has been delayed. Genetic reengineering theoretically could be done to benefit the health of the newborn with a cardiac developmental defect at any time or age, but given its potential for regeneration, earlier in the postnatal period would appear to be the "right time" to implement remedial therapies before "wet cement becomes hardened" [41].

5.3.2.1 This Strategy Calls for a Conceptual Revision in Our Thinking About Fibroblasts

This strategy calls for a conceptual revision in our thinking about fibroblasts that recognizes that they are a renewable cell source that can form a community of active modulators of cell behaviors that can change cardiac function. The full story of cardiac fibroblasts is yet to be realized as to their function in health and disease *including CHD*. Already much is being learned about their potential for repair through homotypic interactions or heterotypic interactions with myocytes. Modes of contact between fibroblasts or fibroblasts and myocytes include connexin-based, *gap junctions* through which ions or small molecules (microRNAs) can be transferred [42], *tunneling nanotubes* [43] through which mitochondria can be exchanged, or by *paracrine secretions* of cytokines, growth factors, or matricellular proteins that facilitate their engraftment, proliferation, and/or differentiation [44].

5.4   Remedial Therapies: Delivering Genetic "Payloads"

Our basic premise is that bone marrow-derived fibroblasts make homeostatic contributions to the non-myocyte cells of adult heart valve and septal connective tissues. We propose to exploit our findings using the single cell engraftment model to deliver relevant payloads to connective tissues of hearts with visible congenital malformations, e.g., valves, septa, hypoplastic ventricles, or non-compacted myocardium in which growth has been compromised. We include myocardial growth malformations because there is emerging evidence that fibroblasts establish heterotypic contacts in vivo with myocytes that can affect myocyte growth or electrophysiology [45]. For example, isolated CD45+ bone marrow cells could be genetically engineered to secrete growth factors like neuregulin which has been shown to reactivate cell cycling in mononucleated cardiomyocytes [46]. For hypoplastic valve or septal tissues, we have already suggested (see above) several genetic or small molecule pharmacological remedial approaches that would be amenable to using CD45 cells for their delivery; for example, viral transduction of CD45+ cells could be used to overexpress or suppress any of kinase or integrin genes known to promote or inhibit fibroblast differentiation [47, 48], focusing on Cre-promoters that allow sharp delineation from myocytes lineages [49, 50]. Antennapedia internalization sequences could also be used to introduce a wide range of "cargo" peptides into isolated CD45 cells.

5.4.1   Preliminary Studies

While we have shown using chimeric mice, i.e., irradiated animals whose bone marrow had been reconstituted with GFP+ hematopoietic stem cells, that it is possible to track bone marrow-derived cells that give rise to bona fide fibroblasts in the heart, it remains to be determined whether they can effectively be transduced to carry a cargo that can have a remedial effect in injured heart tissues. As proof of concept, we prepared a lentivirus periostin-silencing (shRNA) vector directed against the 3′ end of the periostin transcript that was shown to block periostin protein expression (>95 %) in vitro [6, 19]. These silencing or empty control vectors were injected directly into the bone marrow *minutes prior to* an acute myocardial "cryoinjury" administered by a liquid nitrogen cooled microprobe

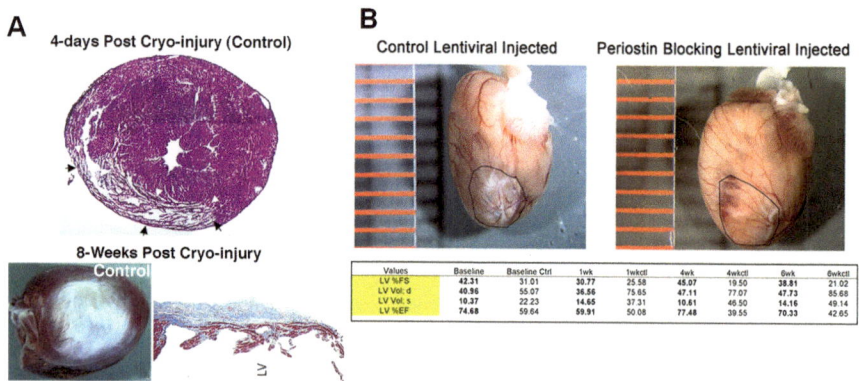

A 4-days Post Cryo-injury (Control)

8-Weeks Post Cryo-injury Control

B Control Lentiviral Injected   Periostin Blocking Lentiviral Injected

Values	Baseline	Baseline Ctrl	1wk	1wkctl	4wk	4wkctl	6wk	8wkctl
LV %FS	42.31	31.01	30.77	25.58	45.07	19.50	38.81	21.02
LV Vol: d	40.96	55.07	36.56	75.65	47.11	77.07	47.73	85.68
LV Vol: s	10.37	22.23	14.65	37.31	10.61	46.50	14.16	49.14
LV %EF	74.68	59.64	59.91	50.08	77.48	39.55	70.33	42.65

Fig. 5.3 Cryoinjured adult hearts treated with lentiviral siRNA vectors to silence periostin injected into the bone marrow. (**a**) *arrows* denote a typical injury site 4 days after cryoablation; after 8 weeks, a large scar (*white color*) is visible on the *left* ventricular surface which when examined histologically reveals extensive fibrosis (*blue* in Masson stain, muscle is *red*). (**b**) *left panel*, empty vector control showing developing scar is turning bluish due to lost vascularity and fibrosis; *right panel* is a heart after bone marrow injection with periostin siRNA vector. The injury site is barely detectable and well vascularized. Performance metrics (*bottom*) indicate that silencing periostin gave results similar to those of baseline controls

directly to the left ventricle of anesthetized mice. This procedure allows for a consistent injury of known size and location within 4 days. As shown in Fig. 5.3, an externally visible scar (blue-white in color) developed within a week in *control* cryoinjured mice, whereas in *experimental* cryoinjured mice in which periostin-silencing vectors were injected into the bone marrow, the scar was barely visible and the tissue appeared healthy and vascularized. Echocardiographic analyses were performed at weekly intervals. Cryoinjured mice that received bone marrow injections of the periostin-silencing vector exhibited performance metrics (e.g., LV vol., ejection fraction) comparable to normal mouse baseline values, whereas those that received the empty control vector exhibited significant reduction of all baseline values. While preliminary, we expect that bone marrow transduction will target cells that potentially can engraft into the heart and modify connective tissue or myocardial remodeling to benefit patient health.

Finally, as developmental biologists searching for remedial etiologies and therapies, we believe the crux of the issue lies in shifting our focus toward understanding that:

1. The later periods of fetal and neonatal cardiovascular development present a window to initiate remedial therapies.
2. The interactions between multiple signaling hubs will become a priority rather than continuing to investigate early events or single genes.
3. The heart is an integrated organ in which lineages can provide insight for improving therapeutic outcomes for CHD through genetic or pharmacological manipulation.

4. The complexity and balance of signals that affect processes controlling the formation of the heart may also come into play after birth or into adult life, where their engineered re-expression can be used to enhance or awaken the ability of heart cells to adapt to pathophysiological stimuli engendered during development.

Acknowledgments Funding for this work was provided by the National Institutes (HL 33756; NIGMS103342, NIGMS 103444-06), the National Science Foundation (EPS0903795), the American Heart Association (11SDG5270006), and the Leducq Foundation, Paris, France (07CVD04).

References

1. Bentham J, Bhattacharya S. Genetic mechanisms controlling cardiovascular development. Ann N Y Acad Sci. 2008;1123:10–9.
2. Wu H, Kao SC, Barrientos T, Baldwin SH, Olson EN, Crabtree GR, Zhou B, Chang CP. Down syndrome critical region-1 is a transcriptional target of nuclear factor of activated T cells-c1 within the endocardium during heart development. J Biol Chem. 2007;282(42):30673–9.
3. Holm TM, Habashi JP, Doyle JJ, et al. Noncanonical TGFbeta signaling contributes to aortic aneurysm progression in Marfan syndrome mice. Science. 2011;332:358–61.
4. Nava C, Hanna N, Michot C, et al. Cardio-facio-cutaneous and Noonan syndromes due to mutations in the RAS/MAPK signalling pathway: genotype-phenotype relationships and overlap with Costello syndrome. J Med Genet. 2007;44:763–71.
5. Person AD, Klewer SE, Runyan RB. Cell biology of cardiac cushion development. Int Rev Cytol. 2005;243:287–335.
6. Ghatak S, Misra S, Norris R, Moreno-Rodriquez RA, Levine RA, Hascall VC, Markwald RR. Periostin induces intracellular cross talk between kinases and hyaluronan in atrioventricular valvulogenesis. J Biol Chem. 2014;289(12):8545–61.
7. Lie-Venema H, Eralp I, Markwald RR, et al. Periostin expression by epicardium-derived cells is involved in the development of the atrioventricular valves and fibrous heart skeleton. Differentiation. 2008;76:809–19.
8. Wessels A, van den Hoff MJ, Adamo RF, Phelps AL, Lockhart MM, Sauls K, Briggs LE, Norris RA, van Wijk B, Perez-Pomares JM, et al. Epicardially derived fibroblasts preferentially contribute to the parietal leaflets of the atrioventricular valves in the murine heart. Dev Biol. 2012;366:111–24.
9. Snider P, Hinton RB, Moreno-Rodriguez RA, et al. Periostin is required for maturation and extracellular matrix stabilization of noncardiomyocyte lineages of the heart. Circ Res. 2008;102:752–60.
10. Norris RA, Moreno-Rodriguez R, Hoffman S, Markwald RR. The many facets of the matricellular protein periostin during cardiac development, remodeling, and pathophysiology. J Cell Commun Signal. 2009;3:275–86.

11. Nakamura T, Colbert M, Krenz M, et al. Mediating ERK 1/2 signaling rescues congenital heart defects in a mouse model of Noonan syndrome. J Clin Invest. 2007;117:2123–32.
12. Camenisch TD, Spicer AP, Brehm-Gibson T, et al. Disruption of hyaluronan synthase-2-abrogates normal cardiac morphogenesis and hyaluronan-mediated transformation of epithelium to mesenchyme. J Clin Invest. 2000;106:349–60.
13. Feng Y, Walsh CA. The many faces of filamin: a versatile molecular scaffold for cell motility and signalling. Nat Cell Biol. 2004;6:1034–8.
14. Sasaki A, Masuda Y, Ohta Y, Ikeda K, Watanabe K. Filamin associates with Smads and regulates transforming growth factor-beta signaling. J Biol Chem. 2001;276:17871–7.
15. Feng Y, Chen MH, Moskowitz IP, et al. Filamin A (FLNA) is required for cell-cell contact in vascular development and cardiac morphogenesis. Proc Natl Acad Sci U S A. 2006;103:19836–41.
16. Sauls K, de Vlaming A, Harris BS, Markwald RR, Norris RA, et al. Developmental basis for filamin-A-associated myxomatous mitral valve disease. Cardiovasc Res. 2012;96:109–19.
17. Norris RA, Moreno-Rodriguez R, Wessels A, et al. Expression of the familial cardiac valvular dystrophy gene, filamin-A, during heart morphogenesis. Dev Dyn. 2010;239:2118–27.
18. Vadlamudi RK, Li F, Adam L, et al. Filamin is essential in actin cytoskeletal assembly mediated by p21-activated kinase 1. Nat Cell Biol. 2002;4:681–90.
19. Norris RA, Moreno-Rodriguez RA, Sugi Y, et al. Periostin regulates atrioventricular valve maturation. Dev Biol. 2008;316:200–13.
20. Kyndt F, Gueffet JP, Probst V, et al. Mutations in the gene encoding filamin A as a cause for familial cardiac valvular dystrophy. Circulation. 2007;115:40–9.
21. Duval D, Lardeux A, Le Tourneau T, et al. Valvular dystrophy associated filamin A mutations reveal a new role of its first repeats in small-GTPase regulation. Biochim Biophys Acta. 2014;1843(2):234–44.
22. MacPherson M, Fagerholm SC. Filamin and filamin-binding proteins in integrin-regulation and adhesion. Focus on: "Filamin A is required for vimentin-mediated cell adhesion and spreading". Am J Physiol: Cell Physiol. 2010;298:C206–8.
23. Norris RA, Borg TK, Butcher JT, Baudino TA, Banerjee I, Markwald RR. Neonatal and adult cardiovascular pathophysiological remodeling and repair: developmental role of periostin. Ann N Y Acad Sci. 2008;1123:30–40.
24. Porrello ER, Mahmoud AI, Simpson E, Hill JA, Richardson JA, Olson EN, Sadek HA. Transient regenerative potential of the neonatal mouse heart. Science. 2011;331 (6020):1078–80.
25. Markwald RR, Norris RA, Moreno-Rodriguez R, Levine RA. Developmental basis of adult cardiovascular diseases: valvular heart diseases. Ann N Y Acad Sci. 2011;1188:177–83.
26. Judge DP, Markwald RR, Hagège AA, Levine RA. Translational research on the mitral valve: from developmental mechanisms to new therapies. J Cardiovasc Transl Res. 2011;4:699–701.
27. Chen PC, Wakimoto H, Conner D, Araki T, Yuan T, Roberts A, Seidman C, Bronson R, Neel B, Seidman JG, Kucherlapati R. Activation of multiple signaling pathways causes developmental defects in mice with a Noonan syndrome–associated Sos1 mutation. J Clin Invest. 2010;120(12):4353–65.
28. Lage K, Greenway SC, Rosenfeld JA, Wakimoto H, Gorham JM, Segrè AV, Roberts AE, Smoot LB, Pu WT, Pereira AC, Mesquita SM, Tommerup N, Brunak S, Ballif BC, Shaffer LG, Donahoe PK, Daly MJ, Seidman JG, Seidman CE, Larsen LA. Genetic and environmental risk factors in congenital heart disease functionally converge in protein networks driving heart development. Proc Natl Acad Sci U S A. 2012;109(35):14035–40.
29. Habashi JP, Doyle JJ, Holm TM, Aziz H, Schoenhoff F, Bedja D, Chen Y, Modiri AN, Judge DP, Dietz HC. Angiotensin II type 2 receptor signaling attenuates aortic aneurysm in mice through ERK antagonism. Science. 2011;332:361–5.
30. Davis F, Rateri DL, Daugherty A. Aortic aneurysms in Loeys-Dietz syndrome – a tale of two pathways? J Clin Invest. 2014;124(1):79–81. See comment in PubMed Commons below.

31. Norris RA, Potts JD, Yost MJ, et al. Periostin promotes a fibroblastic lineage pathway in atrioventricular valve progenitor cells. Dev Dyn. 2009;238:1052–63.
32. Snider P, Standley KN, Wang J, Azhar M, Doetschman T, Conway SJ. Origin of cardiac fibroblasts and the role of periostin. Circ Res. 2009;105:934–47.
33. Ebihara Y, Masuya M, Larue AC, et al. Hematopoietic origins of fibroblasts: II. In vitro studies of fibroblasts, CFU-F, and fibrocytes. Exp Hematol. 2006;34:219–29.
34. Ogawa M, LaRue AC, Mehrotra M. Hematopoietic stem cells are pluripotent and not just "hematopoietic". Blood Cells Mol Dis. 2013;51(1):3–8.
35. Visconti RP, Markwald RR. Recruitment of new cells into the postnatal heart: potential modification of phenotype by periostin. Ann N Y Acad Sci. 2006;1080:19–33.
36. Visconti RP, Ebihara Y, LaRue AC, Markwald R, et al. An in vivo analysis of hematopoietic stem cell potential: hematopoietic origin of cardiac valve interstitial cells. Circ Res. 2006;98:690–6.
37. Hajdu Z, Romeo SJ, Fleming PA, Markwald RR, Visconti RP, Drake CJ. Recruitment of bone marrow-derived valve interstitial cells is a normal homeostatic process. J Mol Cell Cardiol. 2011;51:955–65.
38. Lee R, Perry B, Heywood J, Reese C, Bonner M, Hatfield CM, Silver RM, Visconti RP, Hoffman S, Tourkina E. Caveolin-1 regulates chemokine receptor 5-mediated contribution of bone marrow-derived cells to dermal fibrosis. Front Pharmacol. 2014;5:140.
39. Mollmann H, Nef HM, Kostin S, von Kalle C, Pilz I, Weber M, Schaper J, Hamm CW, Elsasser A. Bone marrow-derived cells contribute to infarct remodelling. Cardiovasc Res. 2006;71:661–71.
40. Haudek SB, Ying X, Huebener P, Lee JM, Signe C, Crawford JR, Pilling D, Gomer RH, Trial JA, Frangogiannis NG, Entman ML. Bone marrow-derived fibroblast precursors mediate ischemic cardiomyopathy in mice. Proc Natl Acad Sci U S A. 2006;103(48):18284–1828.
41. Porrello ER, Olson EN. A neonatal blueprint for cardiac regeneration. Stem Cell Res. 2014;13 (3 Pt B):556–70.
42. Ongstad EL, Gourdie RG. Myocyte-fibroblast electrical coupling: the basis of a stable relationship? Cardiovasc Res. 2012;93:215–7.
43. Ma Z, Yang H, Liu H, Xu M, Runyan RB, Eisenberg CA, Markwald RR, Borg TK, Gao BZ. Mesenchymal stem cell-cardiomyocyte interactions under defined contact modes on laser-patterned biochips. PLoS One. 2013;8(2):e56554.
44. Frangogiannis NG. Matricellular proteins in cardiac adaptation and disease. Physiol Rev. 2012;92:635–88.
45. Kohl P, Gourdie RG. Fibroblast-myocyte electrotonic coupling: does it occur in native cardiac tissue? J Mol Cell Cardiol. 2014;70:37–46.
46. Bersell K, Arab S, Haring B, Kuhn B. ErbB4 signaling induces cardiomyocyte proliferation and repair of heart injury. Cell. 2009;138:257–70.
47. Haudek SB, Gupta D, Dewald O, Schwartz RJ, Wei L, Trial J, Entman ML. Rho kinase-1 mediates cardiac fibrosis by regulating fibroblast precursor cell differentiation. Cardiovasc Res. 2009;83(3):511–8.
48. Ieda M, Tsuchihashi T, Ivey KN, Ross RS, Hong TT, Shaw RM, Srivastava D. Cardiac fibroblasts regulate myocardial proliferation through beta1 integrin signaling. Dev Cell. 2009;16:233–24.
49. Misra S, Hascall VC, Karamanos NK, Markwald RR, Ghatak S. Delivery systems targeting cancer at the level of ECM. In: Karamanos N, editor. Extracellular matrix: pathobiology and signaling. Berlin: DeGruyter; 2012. p. 865–83.
50. Lindsley A, Snider P, Zhou H, et al. Identification and characterization of a novel Schwann and outflow tract endocardial cushion lineage-restricted periostin enhancer. Dev Biol. 2007;307:340–55.

Part II

Left-Right Axis and Heterotaxy Syndrome

Perspective

Bradley B. Keller

The generation of a unique left-right patterning is essential to higher organisms with complex multicellular organs, and errors in patterning represent the origins of some of the most complex forms of congenital heart disease (CHD). Our understanding of these heterotaxy-related syndromes has evolved from initial clinicopathologic correlates over a century ago to the generation and investigation of model organisms (fly, frog, mouse) and the subsequent identification and validation of patterning-related genes and pathways critical for normal human development and responsible for disease states.

Dr. Shiraishi reviews human heterotaxy syndromes associated with CHD, providing a broad overview of the role of cilia, molecular mechanisms involved in left-right patterning, and associated clinical features. Early asymmetric expression of critical morphogens in left-right patterning (Nodal, Lefty2, Pitx2) is required for normal development, and errors in the expression of these morphogens result in patterning errors. Heterotaxy-related CHD is often associated with unbalanced development of the ventricles, resulting in variations of "single ventricle" physiology and requiring staged surgical palliation to separate the venous and systemic circulations. Children and adults with palliated single ventricle physiology, including heterotaxy patients, face a range of medical complications related to cardiac dysfunction and the consequences of increased central venous pressure and reduced cardiac performance. As highlighted by Dr. Shiraishi, large gaps in our

B.B. Keller, M.D. (✉)
Division of Pediatric Heart Research, Cardiovascular Innovation Institute, Louisville, KY, USA

Department of Pediatrics, University of Louisville, 302 East Muhammad Ali Blvd, Louisville 40202, KY, USA
e-mail: Brad.Keller@louisville.edu

understanding of the pathogenesis of heterotaxy syndromes and their optimal medical and surgical management still exist. Hopefully, readers of these chapters will be intrigued and encouraged to pursue solutions.

Dr. Hamada provides an update on the investigation of cilia-mediated flow patterns during early embryogenesis. Motile cilia generate unique extraembryonic flow patterns that mechanically condition non-motile cilia in the ventral node, impact subsequent morphogenesis, and can be explored in fly and mouse model systems. Several key morphoregulatory pathways including agonists and antagonists of noncanonical Wnt signaling and stretch-sensitive Pkd Ca^{2+} channels are clearly involved.

Dr. Shibata and colleagues share some interesting findings on the association of heterotaxy/polysplenia syndrome, single gene mutations in several patterning genes, including *BMPR2*, and pulmonary artery hypertension. These associations require further investigation in model systems where the consequences of CHD including heterotaxy syndrome can be explored in aging animals and in adults with congenital heart diseases.

Thus, the clinical presentation of CHD associated with heterotaxy represents a spectrum of common final pathways and a range of early errors in embryo patterning and morphogenesis, modified during fetal life and then palliated using our current medical and surgical therapies. Investigation of the mechanisms responsible for normal and aberrant patterning and morphogenesis will continue to reveal important genes and pathways that can be used to identify the origins of CHD and may also be important for future targeted therapies, for example, related to pulmonary artery vascular remodeling and hypertension.

Left-Right Asymmetry and Human Heterotaxy Syndrome

6

Isao Shiraishi

Abstract

Heterotaxy syndrome is characterized by a wide variety of cardiac and extra-cardiac congenital malformations that are primarily induced by disorders of the left-right axis determination during early embryonic development. Prognosis of the disease remains unsatisfactory because the syndrome is often associated with complicated congenital heart diseases. Long-term follow-up of heterotaxy patients, particularly those who underwent Fontan procedure, is now one of the most important issues in pediatric and adult congenital heart disease clinics. Collaborative studies between pediatric cardiologists and basic scientists are essential for improving the prognosis of heterotaxy syndrome.

Keywords

Heterotaxy • Left-right axis • Signal transduction • Heart surgery

6.1   Introduction

Heterotaxy syndrome is a rare but serious congenital disease that occurs approximately 1 to 5,000–7,000 of live birth [1]. Patients are generally subdivided into "bilateral right sided" (right isomerism) or "bilateral left sided" (left isomerism) according to the characteristic morphology of atrial appendages of the heart. However, there is a wide spectrum of pathology with considerable overlap of the anatomical features.

I. Shiraishi, M.D., Ph.D. (✉)
Department of Pediatric Cardiology, National Cerebral and Cardiovascular Center, 5-7-1, Fujishirodai, Suita, Osaka 565-8565, Japan
e-mail: shiraishi.isao.hp@ncvc.go.jp

© The Author(s) 2016   49
T. Nakanishi et al. (eds.), *Etiology and Morphogenesis of Congenital Heart Disease*,
DOI 10.1007/978-4-431-54628-3_6

6.2   Molecular and Cellular Mechanisms of Left-Right Determination

The left-right axis determination initiates in the primitive node at E7.5 in mice and develops through the following pathways [2, 3]:

1. Breaking of symmetry as a result of leftward "nodal flow"
2. Transmission of asymmetric signals to the lateral plate mesoderm (LMP)
3. Asymmetric expression of nodal and lefty2 in the LMP
4. Situs-specific morphogenesis mediated by asymmetric expression of Pitx2

6.2.1   Node Cell Monocilia Create Leftward "Nodal Flow" and Activate Asymmetry Signaling Around the Node

The determination of the left-right asymmetry starts as leftward nodal flow generated by rotational movement of monocilia in the primitive node [4, 5]. Clockwise rotation of motile cilia creates unidirectional leftward flow because the rotational axes of cilia tilt caudal direction of the embryos [6, 7].

There are two models why nodal flow is perceived by nodal and perinodal cells. One hypothesis (chemosensory model) is that the nodal flow produces a gradient of left determinant particles (node vesicular parcels) containing hedgehog proteins and nodal [8, 9], which activate downstream signaling of nodal in the left-side perinodal cells.

Alternative hypothesis (mechanosensory model) is that the leftward nodal flow provokes an asymmetrical increase influx of Ca^{2+} ion in the sensory cilia cells through PKD2, a causative gene for human polycystic kidney disease [9, 10]. This Ca^{2+} influx is linked to the activation of nodal in the left-side perinodal cells, which is consequently transferred to the left LPM.

6.2.2   Asymmetry Signaling Transmits to the Left Lateral Plate Mesoderm

Transmission of nodal to the left LMP followed by lefty2 and Pitx2 activation [11] and the consequent heart morphogenesis in the normal subjects, right/left isomerism, and situs inversus is summarized in Fig. 6.1a.

6.2.3   Genes Associated with the Human Heterotaxy Syndrome

Recent human and animal model studies have provided insights into the genetic and developmental etiology of the heterotaxy syndrome. In human, genes that are associated with heterotaxy syndrome are ZIC3, NODAL, CFC1, ACVR2B, LEFTY2, CITED2, and GDF1 [12–14].

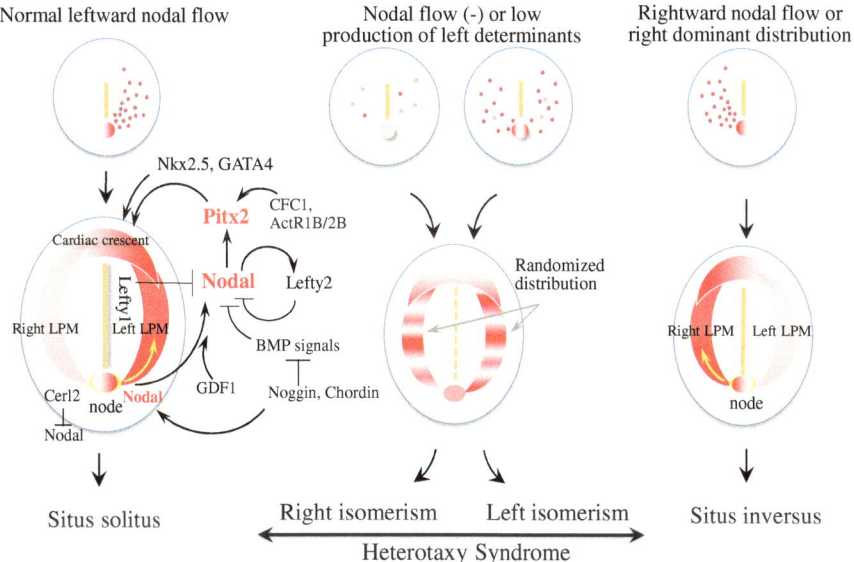

Fig. 6.1 Signal transmission of nodal to the left lateral plate mesoderm followed by Pitx2 activation and the consequent heart morphogenesis in normal and heterotaxy embryos (Adapted and modified from Ref. [1] with permission)

6.3   Clinical Manifestation of the Heterotaxy Syndrome

Factors that deteriorate prognosis of the heterotaxy syndrome are complications with pulmonary venous obstruction, pulmonary arterial distortion, regurgitation of atrioventricular valve, elevated pulmonary vascular resistance, and impaired ventricular function [15].

6.3.1   Right Isomerism

Neonates with right isomerism typically show single atrium, single right ventricle, and univentricular atrioventricular connection often associated with atrioventricular valve regurgitation. First-stage palliation (2–4 weeks after birth) is a control of pulmonary blood flow with pulmonary banding or systemic-pulmonary shunt. Pulmonary venous obstruction due to total anomalous pulmonary venous drainage should be precisely diagnosed and repaired by surgical operation.

The second-stage palliation is the bidirectional Glenn shunt, where the right and/or left superior vena cava is isolated and is connected to the pulmonary artery. This operation is, in general, performed around 6 months after birth.

The third-stage palliation is Fontan procedure. Recently, a modification using extra-cardiac artificial conduit-type total cavo-pulmonary connection (TCPC) is most often employed, because the long-term prognosis of the conventional atrio-pulmonary connection is proved to be unsatisfactory characterized by enlargement of the atrium, intractable atrial tachyarrhythmias, and thromboembolisms.

After successful completion of the TCPC, cyanosis disappears and the general conditions of the patients improve. However, number of patients who underwent successful Fontan procedure is approximately 50 % because right isomerism often accompanied with combination of severe and complicated congenital heart diseases [16].

6.3.2 Left Isomerism

Left isomerism is typically associated with atrioventricular septal defect, persistent left superior vena cava, interrupted hepatic portion of the inferior vena cava, and atrioventricular conduction disturbance. In left isomerism, sinus node and atrioventricular nodes are usually hypoplastic, and sinus bradycardia or complete atrioventricular block is frequently accompanied.

6.4   Long-Term Prognosis of Heterotaxy Patients

Although the medical and surgical treatments of the heterotaxy syndrome have remarkably advanced, long-term prognosis of the patients remains unsatisfactory. Right isomerism has been recognized as one of the worst forms of CHD with overall 5-year survival ranging from 30 to 74 %. The results are better in left isomerism with 5-year survival rates ranging between 65 and 84 %, which is still considerably lower than survival for most other forms of CHD [17]. The main reason is that the nature of the Fontan single ventricle physiology is fundamentally imperfect. Representative long-term complications of the Fontan operation are illustrated in Fig. 6.2.

6.4.1 Protein-Losing Enteropathy

Protein-losing enteropathy (PLE), one of the most severe manifestations of the failing Fontan circulation, occurs in 5–10 % of the total postoperative cases [15]. Chronic loss of serum proteins into the gastrointestinal tracts results in systemic edema, ascites, pleural effusion, diarrhea, gastrointestinal bleeding, susceptibility to infections, and ultimately cachexia. The underlying mechanism of PLE remains uncertain. Elevated inflammatory reactions such as TNF-α or IFN-α, dilatation of intestinal lymphatic vessels, and widening between intestinal epithelial cells may be involved in the protein and fluid losses [18]. Steroids, high molecular weight heparin, sildenafil, surgical interventions, for instance, fenestration of atrial-level communications or conversion of the Fontan circuit, are effective. To date,

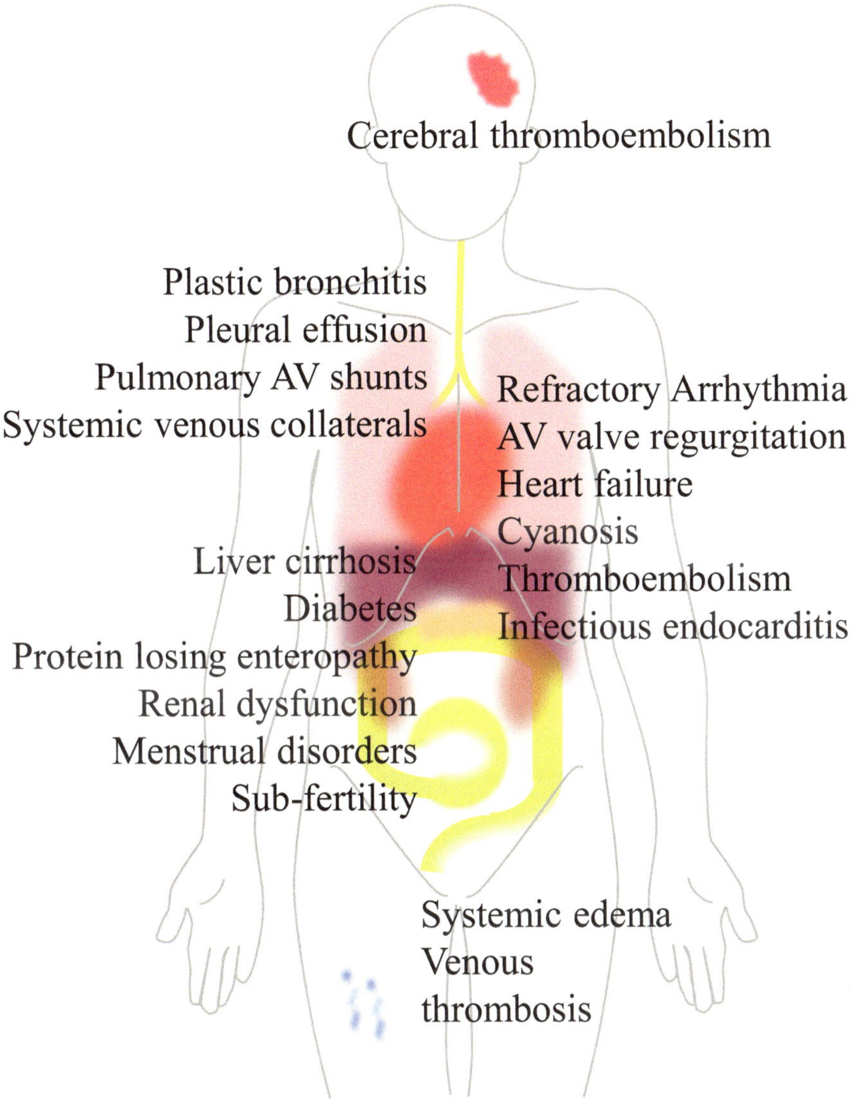

Fig. 6.2 Possible complications of mid- to long term after completion of Fontan procedure for heterotaxy patients (Adapted and modified from Ref. [1] with permission)

cardiac transplantation is considered as the only and complete resolution of PLE pathophysiology.

6.4.2 Arrhythmias

Reentrant atrial tachyarrhythmias are the commonest in patients after Fontan procedure and are often associated with deterioration of hemodynamics, either causally or as a result [15]. Hemodynamic abnormalities such as valve regurgitation or outflow obstruction, if present, should be aggressively treated by surgery.

6.4.3 Heart Failure

Initial feature of heart failure long after the Fontan procedure is characterized by worsening of ventricular relaxation and compliance [15]. These abnormalities may be caused by exposure of hypoxia and volume/pressure overload preceding the Fontan procedure, repetitive surgical operations, and hemodynamic disadvantages of the Fontan circuit. These changes are primarily progressive and consequently lead to failure of the Fontan circuit. Late after the Fontan procedure, systolic dysfunction becomes apparent. Administration of angiotensin converting enzyme inhibitors or β-blockades may be beneficial for particular patients, although the clinical evidence and cellular mechanisms remain to be elucidated.

6.4.4 Hepatic Dysfunction

Hepatic dysfunction, liver fibrosis, and cirrhosis are common complication of patients long after Fontan operation. Recently, cases with hepatocellular carcinoma after Fontan operation have been reported [19, 20]. Careful observation should be necessary to detect the hepatic changes long after Fontan operation.

6.4.5 Management of Failing Fontan Patients

Patients who underwent atrio-pulmonary connection or lateral tunnel procedure are likely to be complicated with thromboembolism or intractable arrhythmias due to enlargement of the right atrium [14]. Surgical intervention with conversion to TCPC is required before such complications become irreversible. Cardiac transplantation may be the only option for patients with severe heart failure, intractable arrhythmias, or recurrent PLE.

6.5   Future Direction and Clinical Implications

In the basic science field, embryonic development of left-right asymmetry has been uncovered by means of mouse genetic engineering. In addition, advanced human genetics have uncovered many responsible genes for heterotaxy syndrome. By means of innovative technologies such as whole genome sequencing or patient-based human inducible pluripotent stem cells, novel genes will be clarified and analyzed. In the clinical field, anatomical and physiological diagnosis from the fetal period, better clinical managements after birth, and tailor-made surgical operations will improve the prognosis. Cell- or tissue-based regeneration therapies and a new ventricular assist device could improve cardiac function of failing Fontan patients. Multiple approaches including basic and clinical science are necessary to improve the prognosis and quality of life of heterotaxy patients.

References

1. Shiraishi I, Ichikawa H. Human heterotaxy syndrome – from molecular genetics to clinical features, managements, and prognosis. Circ J. 2012;76:2066–75.
2. Hamada H. Molecular mechanisms of left-right development. Heart development and regeneration. In: Rosenthal N, Harvey RP, editors. London: Academic Press Inc; 2010. vol 1, p. 297–306.
3. Kawasumi A, Nakamura T, Iwai N, Yashiro K, Saijoh Y, Belo JA, et al. Left-right asymmetry in the level of active Nodal protein produced in the node is translated into left-right asymmetry in the lateral plate of mouse embryos. Dev Biol. 2011;353:321–30.
4. Nonaka S, Tanaka Y, Okada Y, Takeda S, Harada A, Kanai Y, et al. Randomization of left-right asymmetry due to loss of nodal cilia generating leftward flow of extraembryonic fluid in mice lacking KIF3B motor protein. Cell. 1998;95:829–37.
5. Hirokawa N, Tanaka Y, Okada Y. Cilia, KIF3 molecular motor and nodal flow. Curr Opin Cell Biol. 2012;24:31–9.
6. Nonaka S, Yoshiba S, Watanabe D, Ikeuchi S, Goto T, Marshall WF, et al. De novo formation of left-right asymmetry by posterior tilt of nodal cilia. PLoS Biol. 2005;3:e268.
7. Basu B, Brueckner M. Cilia multifunctional organelles at the center of vertebrate left-right asymmetry. Curr Top Dev Biol. 2008;85:151–74.
8. Okada Y, Nonaka S, Tanaka Y, Saijoh Y, Hamada H, Hirokawa N. Abnormal nodal flow precedes situs inversus in iv and inv mice. Mol Cell. 1999;4:459–68.
9. Tanaka Y, Okada Y, Hirokawa N. FGF-induced vesicular release of Sonic hedgehog and retinoic acid in leftward nodal flow is critical for left-right determination. Nature. 2005;435 (7039):172–7.
10. McGrath J, Somlo S, Makova S, Tian X, Brueckner M. Two populations of node monocilia initiate left-right asymmetry in the mouse. Cell. 2003;114:61–73.
11. Martin JF, Amendt BA, Brown NA. Heart development and regeneration. In: Rosenthal N, Harvey RP, editors. London: Academic Press Inc., 2010. vol 1. p. 307–22.
12. Zhu L, Harutyunyan KG, Peng JL, Wang J, Schwartz RJ, Belmont JW. Identification of a novel role of ZIC3 in regulating cardiac development. Hum Mol Genet. 2007;16:1649–60.

13. Bamford RN, Roessler E, Burdine RD, Saplakoğlu U, dela Cruz J, Splitt M, et al. Loss-of-function mutations in the EGF-CFC gene CFC1 are associated with human left-right laterality defects. Nat Genet. 2000;26:365–9.
14. Kaasinen E, Aittomäki K, Eronen M, Vahteristo P, Karhu A, Mecklin JP, et al. Recessively inherited right atrial isomerism caused by mutations in growth/differentiation factor 1 (GDF1). Hum Mol Genet. 2010;19:2747–53.
15. Hutter D, Redington AN. The principles of Management, and outcomes for, patients with functionally univentricular hearts. In: Anderson RH, Baker EJ, Penny DJ, Redington AN, Rigby ML, Wernovsky G, editors. Paediatric cardiology. 3rd ed. Philadelphia: Churchill Livingstone; 2009. p. 687–96.
16. Ota N, Fujimoto Y, Murata M, Tosaka Y, Ide Y, Tachi M, et al. Improving outcomes of the surgical management of right atrial isomerism. Ann Thorac Surg. 2012;93:832–8.
17. Delmo Walter EM, Ewert P, Hetzer R, Hübler M, Alexi-Meskishvili V, et al. Biventricular repair in children with complete atrioventricular septal defect and a small left ventricle. Eur J Cardiothorac Surg. 2008;33:40–7.
18. Serraf A, Bensari N, Houyel L, Capderou A, Roussin R, Lebret E, Ly M, Belli E. Surgical management of congenital heart defects associated with heterotaxy syndrome. Eur J Cardiothorac Surg. 2010;38:721–7.
19. Ostrow AM, Freeze H, Rychik J. Protein-losing enteropathy after fontan operation: investigations into possible pathophysiologic mechanisms. Ann Thorac Surg. 2006;82:695–700.
20. Ghaferi AA, Hutchins GM. Progression of liver pathology in patients undergoing the Fontan procedure: chronic passive congestion, cardiac cirrhosis, hepatic adenoma, and hepatocellular carcinoma. J Thorac Cardiovasc Surg. 2005;129:1348–52.

Roles of Motile and Immotile Cilia in Left-Right Symmetry Breaking

7

Hiroshi Hamada

Abstract

Our body possesses three body axes, anteroposterior, dorsoventral, and left-right (L-R) axes. L-R asymmetry is achieved by three consecutive steps: symmetry breaking at the node, differential patterning of the lateral plate by a signaling molecule Nodal, and finally situs-specific organogenesis. Breaking of L-R symmetry in the mouse embryo takes place in the ventral node, where two types of cilia are found. Whereas centrally located motile cilia generate a leftward fluid flow, peripherally located immotile cilia sense a flow-dependent signal. Although Ca^{2+} signaling is implicated in flow sensing, it is still not clear what triggers Ca^{2+} signaling, a determinant molecule transported by the flow or mechanical force induced by the flow.

Keywords

Cilia • Fluid flow • Laterality • Left-right asymmetry

7.1 Introduction

Most of visceral organs in vertebrates including the human are left-right (L-R) asymmetric in their position or shape. The process by which L-R asymmetry is generated can be divided into three steps (Fig. 7.1):

1. The initial breaking of L-R symmetry, which occurs in or near the node and at the late neural-fold stage

H. Hamada (✉)
Graduate School of Frontier Bioscience, Osaka University, 1-3 Yamada-oka, Suita, Osaka 565-0871, Japan
e-mail: hamada@fbs.osaka-u.ac.jp

© The Author(s) 2016
T. Nakanishi et al. (eds.), *Etiology and Morphogenesis of Congenital Heart Disease*, DOI 10.1007/978-4-431-54628-3_7

57

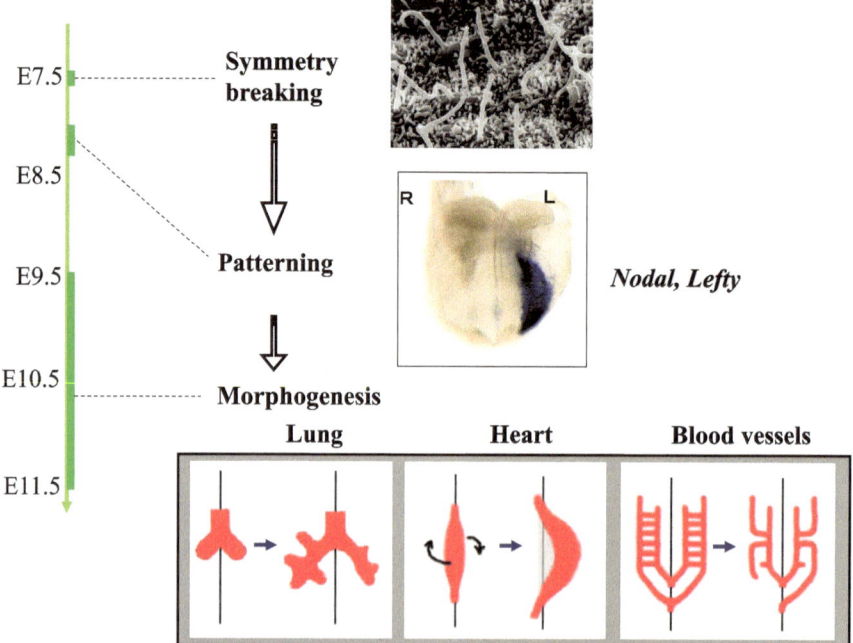

Fig. 7.1 Three steps underlying the generation of L-R asymmetry. Three steps that contribute to the generation of L-R asymmetry are shown: (1) symmetry breaking, (2) molecular patterning of the LPM, and (3) asymmetric organogenesis. The developmental stage (E, embryonic day) corresponding to each step in the mouse is indicated on the *left*

2. Transfer of an L-R-biased signal(s) from the node to the lateral plate mesoderm (LPM), which leads to L-R asymmetric expression of signaling molecules such as the transforming growth factor-β (TGF-β)-related proteins Nodal and Lefty on the left side of the LPM
3. L-R asymmetric morphogenesis of visceral organs induced by these signaling molecules

7.2   Symmetry Breaking by Motile Cilia and Fluid Flow

The breaking of L-R symmetry takes place in the node, an embryonic midline structure located at the anterior tip of the primitive streak in mouse embryos (Fig. 7.2). At the central region of the node, there are about 200 motile cilia that protrude from the ventral side of the node into the node cavity [1] (Fig. 7.2) and rotate in the clockwise direction (when viewed from the ventral side) at a speed of 600 rpm [2]. This rotational movement of the cilia generates the leftward laminar flow of extraembryonic fluid in the node cavity [2], occurs at a speed of ~15–20 μm/ s. This leftward fluid flow in the node, referred to as nodal flow, is responsible for symmetry breaking. Many mutant mice that lack nodal flow because the node cilia

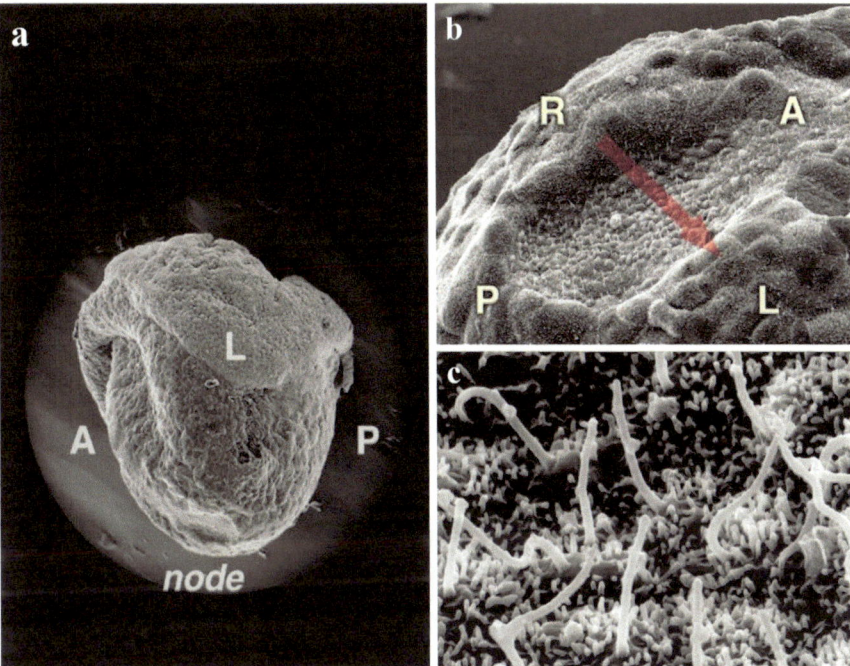

Fig. 7.2 Cilia and fluid flow in the node. A lateral view of the embryonic day 8.0 mouse embryo
(**a**). Note that the node is located at the midline. Left-right (L-R) and anteroposterior (A-P)
orientations are indicated. A ventral view of the mouse node at lower magnification (**b**). The *red
arrow* denotes the leftward flow of extraembryonic fluid. A scanning electron micrograph showing
that each cell on the ventral side of the mouse node has a monocilium (**c**)

are either missing or immotile have been identified, all of which exhibit aberrant
L-R patterning of the LPM. Furthermore, L-R patterning of the embryo can be
reversed when the direction of the flow was experimentally reversed by imposing
the rightward artificial flow [3], establishing that the direction of the flow
determines L-R.

How is the unidirectional fluid flow generated by rotational movement of the
cilia? Hydrodynamic principles predict that the cilia can generate a unidirectional
flow if they are tilted toward a specific direction. When the cilia move closer to the
surface, the movement of fluid near the surface will be restricted as a result of the
"no-slip boundary effect." Conversely, when the cilia move away from the surface,
they move the neighboring fluid more effectively. If cilia are tilted toward the
posterior side, they will be moving toward the right when they come close to the
surface and toward the left when they are far from the surface, thus generating a
leftward flow. Observation of these rotating cilia by high-speed video microscopy
revealed that they are indeed tilted posteriorly at an average angle of 30°
[4, 5]. Recent evidence [6] suggests that, in addition to the "no-slip boundary

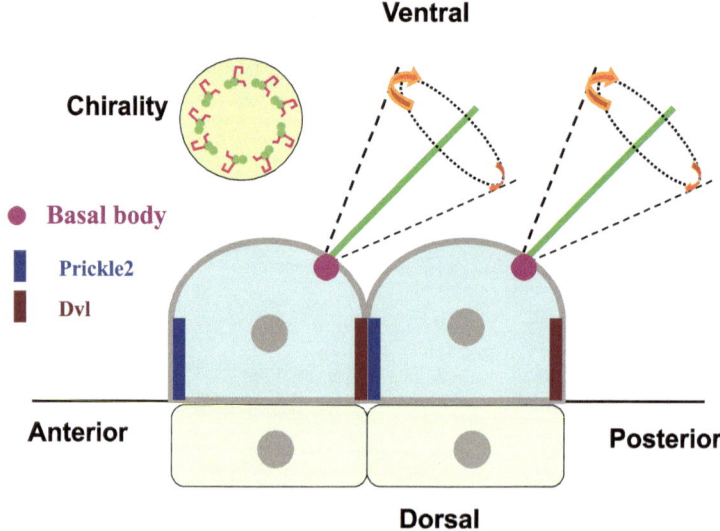

Fig. 7.3 L-R symmetry breaking by preexisting information. Each node cilium (*red bars* on *left*) is posteriorly tilted, likely because the basal body (*green*) is posteriorly shifted within the cell (*blue*). The cilium protrudes from the cell toward the ventral side of the embryo and rotates in a clockwise direction when viewed from the ventral side. Anteroposterior (A-P) and dorsoventral (D-V) orientations are indicated. A schematic representation of a transverse section of a cilium, revealing its chiral structure, is shown on the right. The cilium contains nine pairs of microtubules (*green*) as well as inner and outer arms of dynein (*pink*)

effect," intrinsic asymmetry in rotational stroke may also help generating the unidirectional flow.

Since the L-R axis is the last axis to be determined, symmetry breaking of L-R axis must be achieved by utilizing preexisting positional cues. In fact, two preexisting positional cues are reflected in the cilia: The A-P and D-V axes are thus represented by the posterior tilt and ventral protrusion of the cilia, respectively (Fig. 7.3). The node cilia thus generate the leftward flow by making use both of the preexisting positional cues and their structural chirality.

How is A-P information translated into the posterior tilt of the node cilia? Given its similarity to positioning of the hair in the *Drosophila* wing, a mechanism resembling the planar cell polarity (PCP) pathway involving noncanonical Wnt signaling [7] seems to underlie positioning of the node cilia. Thus, some of the PCP core proteins such as Prickle2 and Vangl1 are localized to the anterior side of node cells [8, 9], whereas Dvl protein is localized to the posterior side [10] (Fig. 7.3). However, it remains unknown what positional cue is responsible for the polarized localization of these PCP core proteins.

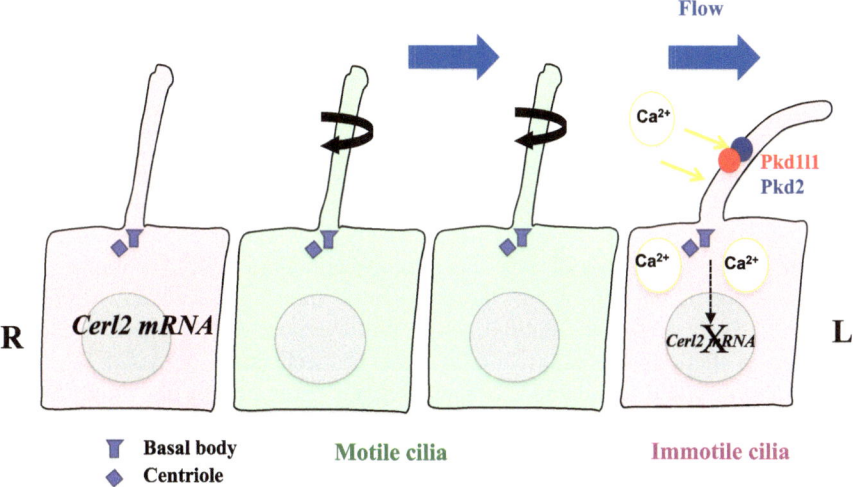

Fig. 7.4 Two types of cilia in the node, motile and immotile. Two types of ciliated cell are present in the node: Those located centrally (*green*) have motile cilia that generate nodal flow, whereas those located peripherally (*pink*) possess immotile cilia that sense the flow. Sensing of the flow requires ciliary localization of a Ca^{2+} channel, the Pkd2-Pkd1l1 complex. The flow-mediated signal results in degradation of *Cerl2* mRNA. In this model, the flow would bend an immotile cilium on the left side

7.3 Sensing of the Fluid Flow by Immotile Cilia

In addition to the motile cilia, there are immotile cilia in the node [11, 12] (Fig. 7.4). Cells located at the central region of the node (pit cells) possess motile cilia, which generate the fluid flow. On the other hand, most cells located at the edge of the node (crown cells) possess immotile cilia [13]. Immotile cilia act as sensors of the fluid flow [13]. Mutant mouse embryo that lack all cilia including those at the node, such as $Kif3a^{-/-}$ mouse embryos, fail to develop nodal flow and show L-R defects [14]. Such cilium-less embryos are also unable to respond to the artificial flow. However, when immotile cilia are restored in crown cells, the resulting embryo can now respond to the artificial flow [13], demonstrating that immotile cilia sense the flow.

Sensing of the fluid flow by immotile cilia requires a Ca^{2+} channel composed of Pkd2 [15] and Pkd1l1 [16, 17]. Indeed, several Ca^{2+} signaling blockers have been shown to disrupt asymmetric gene expression in crown cells [13]. In particular, the effects of $GdCl_3$ [an inhibitor of stretch-sensitive transient receptor potential (TRP) channels], 2-ABP [an inhibitor of the inositol 1,4,5-trisphosphate (IP_3) receptor], and thapsigargin (an inhibitor of Ca^{2+}-dependent ATPase activity in the endoplasmic reticulum) suggest involvement of a TRP-type channel such as Pkd2 and the IP3 receptor in the sensing of nodal flow. A mutation in *Pkd2* that disrupts the ciliary localization of the encoded protein results in L-R defects similar to those of

$Pkd2^{-/-}$embryos [13, 16], suggesting that Pkd2, together with Pkd1l1, functions in the ciliary compartment of crown cells. Whereas *Pkd2* encodes a Ca^{2+} channel with a short extracellular domain, Pkd1l1 possesses a much larger extracellular domain at its amino terminus. Pkd1l1 may be responsible for sensing of the flow signal and regulating Ca^{2+} channel activity of Pkd2. While oscillations of Ca^{2+} signaling with subtle L>R asymmetry were detected in the node [18], direct observation of L-R asymmetric Ca^{2+} signaling in crown cells has not been successful [13].

A long-standing question since the discovery of nodal flow concerns the action of the flow. Two models have been proposed (Fig. 7.5). According to the chemosensor model (Fig. 7.5a), the flow would transport an unknown molecule toward the left side of the embryo, which will eventually act as the L-R determinant. In an alternative model (two-cilia model or mechanosensor model; Fig. 7.5b), the embryo would sense the mechanical force generated by the flow. Several molecules have been proposed to be the determinant transported by the flow. However, none of them fulfill the requirements for the determinant. On the other hand, many lines of circumstantial evidence, including the recent observation that as few as two rotating cilia are sufficient for the breaking of L-R symmetry [19], favor the latter model. However, it is still not clear what exactly the immotile cilia sense during the symmetry-breaking event.

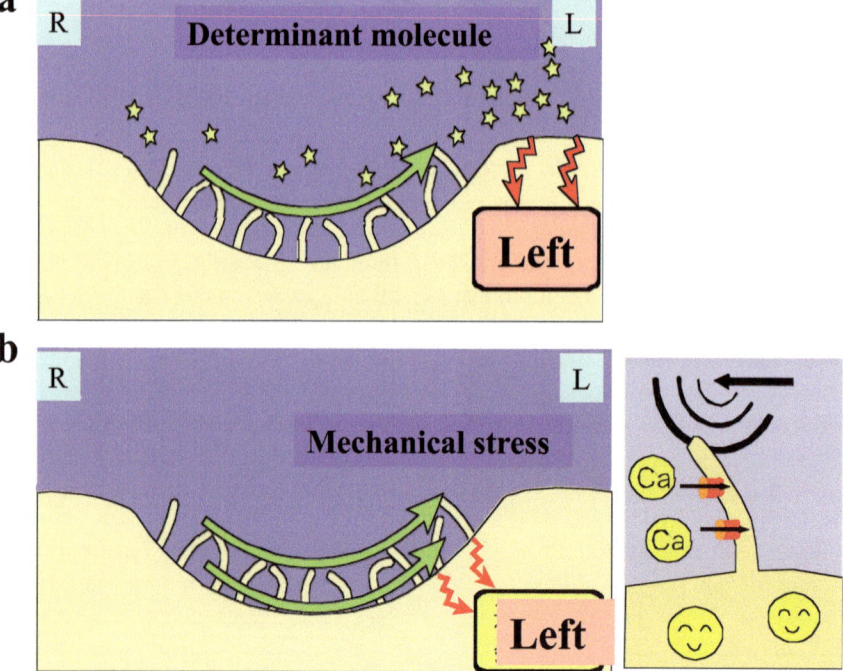

Fig. 7.5 Two models for the mechanism of action of nodal flow. (**a**) Determinant-transporting model. (**b**) Mechanosensory model. *Green arrows* indicate the direction of nodal flow; *yellow stars* denote determinant molecules

7.4   Readouts of the Flow

Cerl2 is the most immediate readout of the flow signal [19, 20]. *Cerl2* encodes a Nodal antagonist, although its precise action is not clear. It is asymmetrically (R > L) expressed in crown cells, and its absence results in randomization of L-R decision making [21]. Whereas expression of *Nodal* is bilateral in crown cells, the R > L expression of *Cerl2* renders Nodal activity in crown cells higher on the left side (Fig. 7.6). The Cerl2-generated asymmetry (R < L) of Nodal activity at the node closely correlates with the asymmetric pattern of *Nodal* expression in LPM [22]. Expression of *Cerl2* is initially symmetric (R = L) at the early headfold stage, but it becomes R > L as the velocity of nodal flow increases, with expression on the left side being downregulated [19, 22]. Finally, $Pkd2^{-/-}Cerl2^{-/-}$ double-mutant embryos manifest randomized *Nodal* expression in LPM, resembling the *Cerl2* single mutant [13]. Therefore, *Cerl2* is the main target of the flow signal.

L-R asymmetry of *Cerl2* expression is generated at a posttranscriptional level [23], by degradation of *Cerl2* mRNA via its $3'$ untranslated region. Preferential decay of *Cerl2* mRNA on the left is initiated by the leftward flow and further enhanced by the operation of *Wnt-Cerl2* interlinked feedback loops, in which Wnt3

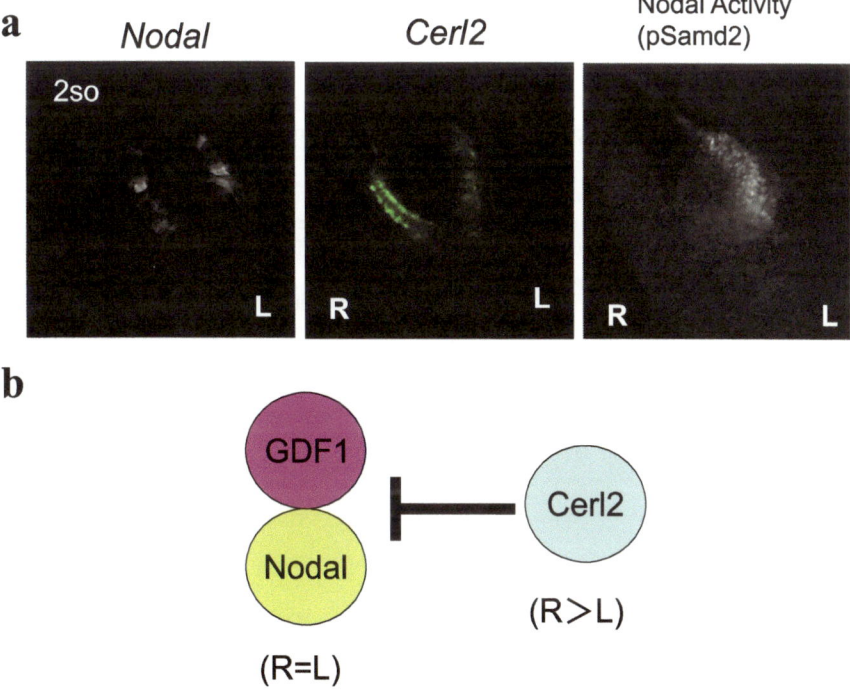

Fig. 7.6 R>L expression of *Cerl2* results in asymmetry in Nodal activity at the node. R>L asymmetric expression of *Cerl2* makes Nodal activity at the node R<L (**a**), Cerl2 is an inhibitor of Nodal (**b**)

upregulates *Wnt3* expression and promotes *Cerl2* mRNA decay, whereas Cerl2 promotes Wnt degradation. Mathematical modeling and experimental data suggest that these feedback loops behave as a bistable switch that is able to amplify in a noise-resistant manner a small bias conferred by fluid flow.

7.5   Future Directions

Although rapid progress has been made in the last 20 years, many important questions remain unanswered. Firstly, how is A-P information translated into the posterior tilt of node cilia? Namely, what is the nature of the A-P information that polarizes node cells along the A-P axis? Secondly, how is the direction of rotation determined for node cilia? Thirdly, how does the nodal flow work? How do motile cilia sense the flow? Do they sense a signaling molecule that is transported by the flow or sense mechanical force? Fourthly, what is the precise role of Ca^{2+} signaling? How does Ca^{2+} signaling induce degradation of Cerl2 mRNA? Finally, to what extent is the mechanism for breaking of L-R symmetry conserved among species? L-R symmetry breaking does not appear to depend on cilia in *Drosophila* and snail [24]. Further development of various approaches (including genetic, cellular, biophysical, and mathematical) will be necessary to answer these questions.

Acknowledgments I thank current and former members of my laboratory for discussion as well as for providing illustrations. The work performed in my laboratory has been supported by CREST, Japan Science and Technology Corporation (JST), and by grants from the Ministry of Education, Culture, Sports, Science, and Technology of Japan.

References

1. Sulik K, et al. Morphogenesis of the murine node and notochordal plate. Dev Dyn. 1994;201 (3):260–78.
2. Nonaka S, et al. Randomization of left-right asymmetry due to loss of nodal cilia generating leftward flow of extraembryonic fluid in mice lacking KIF3B motor protein. Cell. 1998;95 (6):829–37.
3. Nonaka S, et al. Determination of left-right patterning of the mouse embryo by artificial nodal flow. Nature. 2002;418(6893):96–9.
4. Okada Y, et al. Mechanism of nodal flow: a conserved symmetry breaking event in left-right axis determination. Cell. 2005;121(4):633–44.

5. Nonaka S, et al. De novo formation of left-right asymmetry by posterior tilt of nodal cilia. PLoS Biol. 2005;3(8):e268.
6. Takamatsu A, et al. Asymmetric rotational stroke in mouse node cilia during left-right determination. Phys Rev E Stat Nonlin Soft Matter Phys. 2013;87(5):050701.
7. Klein TJ, Mlodzik M. Planar cell polarization: an emerging model points in the right direction. Annu Rev Cell Dev Biol. 2005;21:155–76.
8. Antic D, et al. Planar cell polarity enables posterior localization of nodal cilia and left-right axis determination during mouse and Xenopus embryogenesis. PLoS One. 2010;5(2):e8999.
9. Song H, et al. Planar cell polarity breaks bilateral symmetry by controlling ciliary positioning. Nature. 2010;466(7304):378–82.
10. Hashimoto M, et al. Planar polarization of node cells determines the rotational axis of node cilia. Nat Cell Biol. 2010;12(2):170–6.
11. McGrath J, et al. Two populations of node monocilia initiate left-right asymmetry in the mouse. Cell. 2003;114(1):61–73.
12. Tabin CJ, Vogan KJ. A two-cilia model for vertebrate left-right axis specification. Genes Dev. 2003;17(1):1–6.
13. Yoshiba S, et al. Cilia at the node of mouse embryos sense fluid flow for left-right determination via Pkd2. Science. 2012;338(6104):226–31.
14. Takeda S, et al. Left-right asymmetry and kinesin superfamily protein KIF3A: new insights in determination of laterality and mesoderm induction by kif3A−/− mice analysis. J Cell Biol. 1999;145(4):825–36.
15. Pennekamp P, et al. The ion channel polycystin-2 is required for left-right axis determination in mice. Curr Biol. 2002;12(11):938–43.
16. Field S, et al. Pkd1l1 establishes left-right asymmetry and physically interacts with Pkd2. Development. 2011;138(6):1131–42.
17. Kamura K, et al. Pkd1l1 complexes with Pkd2 on motile cilia and functions to establish the left-right axis. Development. 2011;138(6):1121–9.
18. Takao D, et al. Asymmetric distribution of dynamic calcium signals in the node of mouse embryo during left-right axis formation. Dev Biol. 2013;376(1):23–30.
19. Shinohara K, et al. Two rotating cilia in the node cavity are sufficient to break left-right symmetry in the mouse embryo. Nat Commun. 2012;3:622.
20. Schweickert A, et al. The nodal inhibitor coco is a critical target of leftward flow in Xenopus. Curr Biol. 2010;20(8):738–43.
21. Marques S, et al. The activity of the Nodal antagonist Cerl-2 in the mouse node is required for correct L/R body axis. Genes Dev. 2004;18(19):2342–7.
22. Kawasumi A, et al. Left-right asymmetry in the level of active Nodal protein produced in the node is translated into left-right asymmetry in the lateral plate of mouse embryos. Dev Biol. 2011;353(2):321–30.
23. Nakamura T, et al. Fluid flow and interlinked feedback loops establish left-right asymmetric decay of Cerl2 mRNA. Nat Commun. 2012;3:1322.
24. Okumura T, et al. The development and evolution of left-right asymmetry in invertebrates: lessons from Drosophila and snails. Dev Dyn. 2008;237(12):3497–515.

Role of Cilia and Left-Right Patterning in Congenital Heart Disease

Nikolai Klena, George Gabriel, Xiaoqin Liu, Hisato Yagi, You Li, Yu Chen, Maliha Zahid, Kimimasa Tobita, Linda Leatherbury, Gregory Pazour, and Cecilia W. Lo

Abstract

A central role for cilia in the pathogenesis of congenital heart disease was uncovered by our large-scale mouse mutagenesis screen for mutations causing congenital heart disease. This is supported by human clinical studies, which showed a high prevalence of ciliary dysfunction and respiratory symptoms and disease in patients with congenital heart disease. Our mouse studies indicate this involves essential roles for both primary and motile cilia in the pathogenesis of congenital heart disease. As laterality defects were also observed with high prevalence among the congenital heart disease mutants, this further suggested an important role for left-right patterning in the pathogenesis of congenital heart disease. This finding is reminiscent of the high prevalence of heterotaxy among human fetuses with congenital heart disease, indicating the fetal mouse screen may provide a window into the unborn human fetal population. Clinically, congenital heart disease patients with ciliary dysfunction were found to have more respiratory symptoms and disease, a finding with significant clinical implications, as congenital heart disease patients undergoing surgical palliation often have respiratory complications with high morbidity. While this is usually attributed to the heart disease, we propose this may involve intrinsic airway clearance deficits from ciliary dysfunction. Thus the presurgical screening of congenital heart disease patients for respiratory ciliary dysfunction may provide

N. Klena • G. Gabriel • X. Liu • H. Yagi • Y. Li • Y. Chen • M. Zahid • K. Tobita • C.W. Lo (✉)
Department of Developmental Biology, University of Pittsburgh School of Medicine, Pittsburgh, PA 15201, USA
e-mail: cel36@pitt.edu

L. Leatherbury
Children's National Medical Center, Washington, DC, USA

G. Pazour
Program in Molecular Medicine, University of Massachusetts Medical School, Worcester, MA, USA

© The Author(s) 2016

67

T. Nakanishi et al. (eds.), *Etiology and Morphogenesis of Congenital Heart Disease*, DOI 10.1007/978-4-431-54628-3_8

opportunities to provide perioperative pulmonary therapy to enhance airway clearance for at-risk patients. Such change in the standard of care may improve outcome, especially for those congenital heart disease patients who must endure multiple rounds of cardiac surgeries.

Keywords
Cilia • Ciliary dysfunction • Respiratory symptoms • Laterality • Heterotaxy

8.1   Introduction

Complex congenital heart disease is clinically well described to be highly associated with heterotaxy, a birth defect involving randomization of left-right patterning [1]. The importance of left-right patterning in congenital heart disease is likely a reflection of the fact that the heart is the most left-right asymmetric organ in the body. This asymmetry is critical for establishing systemic and pulmonary circulation required for efficient oxygenation of blood. While heterotaxy is relatively rare, reported at approximately 1 in 10,000 live births, it is clinically of high importance given it is often associated with complex CHD with high morbidity and mortality [2].

8.1.1   Heterotaxy, Primary Ciliary Dyskinesia, and Motile Cilia Defects

Interestingly, heterotaxy and complex CHD have been reported in ~6 % of patients with primary ciliary dyskinesia (PCD), a sinopulmonary disease that arises from airway mucus clearance defects due to immotile or dyskinetic cilia in the respiratory epithelia [3]. Given PCD is also relatively rare at 1 in 16,000 [4], the co-occurrence of heterotaxy and PCD would suggest a mechanistic link for heterotaxy and PCD. This mechanistic link has now been demonstrated to involve a shared disturbance of motile function. Thus animal model studies have shown motile cilia play an important role in embryonic left-right patterning [5].

In the mouse embryo, motile cilia at the node generate nodal flow that helps specify the left-right axis (Fig. 8.1). The nodal flow is sensed by nonmotile or primary cilia at the node periphery, resulting in activation of calcium signaling at the embryo's left, followed by left-sided activation of the nodal signaling cascade in the left lateral plate mesoderm. As motile cilia in the node and airway are constructed in a similar manner with many of the same proteins, it is not surprising that airway ciliary dysfunction might predict nodal cilia dysfunction. This likely accounts for the high prevalence of heterotaxy among PCD patients.

8.1.2   Motile Respiratory Cilia Defects in Other Ciliopathies

Given primary cilia also has been shown to play a role in the embryonic node to establish the left-right axis, this would suggest that mutations affecting primary

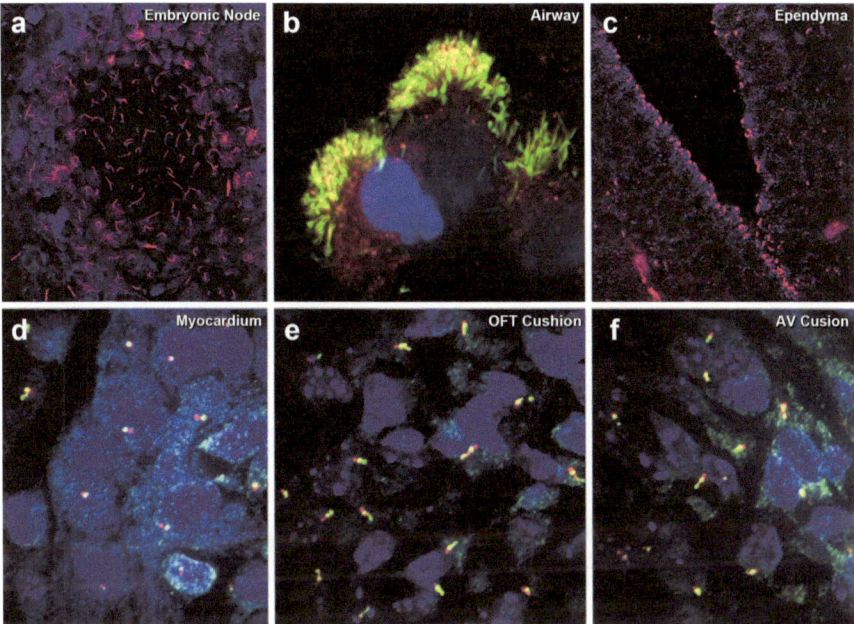

Fig. 8.1 Motile and primary cilia in the mouse embryo. (**a**) Immunostaining with antibodies to acetylated tubulin (*red*) and γ-tubulin (*blue*) was used to visualize motile cilia in the E7.75 mouse embryonic node. (**b–f**) Immunostaining with antibodies to acetylated tubulin and IFT88 visualized cilia in the newborn mouse tracheal airway epithelia (**b**), in E12.5 brain ependyma (**c**), and primary cilia in the myocardium (**d**), outflow tract cushion (**e**), and atrioventricular cushion (**f**) of the E12.5 mouse embryonic heart

cilia function may also contribute to complex CHD associated with heterotaxy. Given the extensive overlap between proteins found in the primary and motile cilia, this would suggest the clinical distinction may be blurred between patients with ciliopathies considered mediated by primary cilia defects vs. those with PCD that have motile cilia defects. Indeed in a recent study, we showed that a patient with cranioectodermal dysplasia, a ciliopathy thought to involve the primary cilia, has presentations consistent with PCD. This includes obstructive airway disease, low nasal NO, and abnormal respiratory ciliary motion [6].

8.1.3   Ciliary Dysfunction in Congenital Heart Disease Patients with Heterotaxy

Even as PCD patients were observed to have a 6 % incidence of heterotaxy, a study of CHD patients with heterotaxy revealed an even higher prevalence of respiratory ciliary dysfunction similar to that seen with PCD [7]. Two tests used for PCD assessment were used to evaluate heterotaxy patients. Ciliary motion analyzed

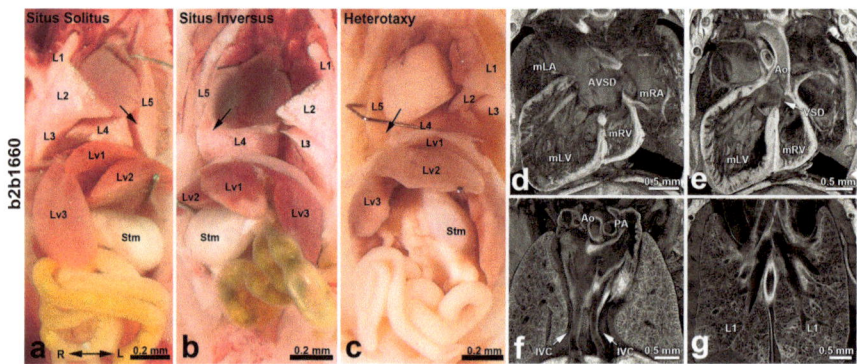

Fig. 8.2 Situs anomalies, congenital heart defects and ciliogenesis defects in laterality mutants. (**a–g**) *Ap1b1^{b2b1660}* mutants exhibit situs solitus (**a**), situs inversus (**b**), or heterotaxy (**c**). Situs solitus, characterized by normal left-right visceral organ positioning, the heart apex (*arrow*) points to the left (levocardia), four lung lobes are on the right and one on the left, stomach is to the left, and the dominant liver lobe is on the right. With situs inversus, there is complete mirror reversal of organ situs, while with heterotaxy, visceral organ situs is randomized, such as dextrocardia with levogastria shown in (**c**). The heterotaxy mutant in (**c**) exhibit complex CHD with AVSD (**d**), ventricular septal defect (VSD) (**e**), duplicated inferior vena cava (IVC) (**f**), and left pulmonary isomerism with bilateral single lung lobes (**g**) (From Li et al. [11])

using videomicroscopy of nasal tissue biopsy and nasal nitric oxide (nNO) was measured, which is typically low in patients with PCD. This analysis showed 42 % of the heterotaxy patients have ciliary dysfunction comprising abnormal ciliary motion and low nNO, presentations typically seen with PCD [7].

Interestingly, a mouse mutant exhibiting complex congenital heart defect associated with heterotaxy was identified to have a pathogenic mutation in *Dnah5*, a cilium outer dynein arm component required for motile cilia function and a gene well described to cause PCD [8]. This mutant exhibited mostly immotile cilia in the airway and in the embryonic node [8], accounting for the laterality disturbance and airway clearance defects seen in PCD patients with *DNAH5* mutations. Interestingly, these *Dnah5* mutant mice exhibited either of three different laterality phenotypes: normal situs solitus, mirror symmetric situs inversus totalis, or randomized visceral organ situs known as heterotaxy (Fig. 8.2). It is only with heterotaxy that complex congenital heart disease was observed, indicating that disturbance of the left-right patterning may play an important role in congenital heart disease. As the mouse *Dnah5* mutants with heterotaxy were mostly inviable to term due to their complex congenital heart disease, this would suggest considerable ascertainment bias in the human population. Consistent with this, a study of PCD patients revealed most had either situs solitus or situs inversus, with only a small fraction exhibiting heterotaxy [4].

8.1.4   Respiratory Complications in Heterotaxy Patients with Ciliary Dysfunction

As the central hallmark of PCD is respiratory disease due to mucociliary clearance defects, the question arises as to whether heterotaxy patients may also have respiratory symptoms and disease. Indeed, heterotaxy patients with ciliary dysfunction are observed to have significantly more respiratory symptoms and disease [7]. Furthermore, those undergoing surgical procedures show increased pulmonary morbidity, including increased use of inhaled β-agonist [9]. β-agonist use is typically avoided in cardiac patients given its arrhythmogenic properties. Hence, the increased use of this medication is a strong indicator of serious respiratory complications.

These findings have important clinical translational ramifications, since respiratory complications in heterotaxy patients are usually attributed to the heart disease, and thus any airway clearance defects are not systematically addressed clinically. In light of these findings, a change in the standard of care may be warranted with the presurgical screening of heterotaxy patients for mucociliary clearance defects and providing airway clearance therapy to help reduce postsurgical respiratory complications in those with airway ciliary dysfunction. This may help improve the prognosis for these patients who typically have to endure multiple high-risk cardiac surgeries to palliate their structural heart defects.

8.1.5   Left-Right Patterning and the Pathogenesis of Congenital Heart Disease

The importance of left-right patterning in the pathogenesis of congenital heart disease has also emerged from a large-scale mouse mutagenesis screen. High-throughput screening of ENU-mutagenized mice using fetal echocardiography allowed the ultrasound phenotyping of greater than 80,000 fetuses (Fig. 8.3) [10, 11]. Fetal echocardiography is ideally suited for recovery of mutants with congenital heart defects, as it is an imaging modality developed in the clinical setting for the assessment of cardiac structure and function (Fig. 8.3). Over 200 mutant mouse lines with a wide spectrum of congenital heart defect were recovered. Surprisingly, this included many mutant lines with laterality defects (~30 %), recovered based on the finding of complex CHD in mutants with heterotaxy. Given our screen was focused on congenital heart defects, not left-right patterning defects, this enrichment of laterality mutants would suggest the disturbance of left-right patterning plays an important role in the pathogenesis of congenital heart disease.

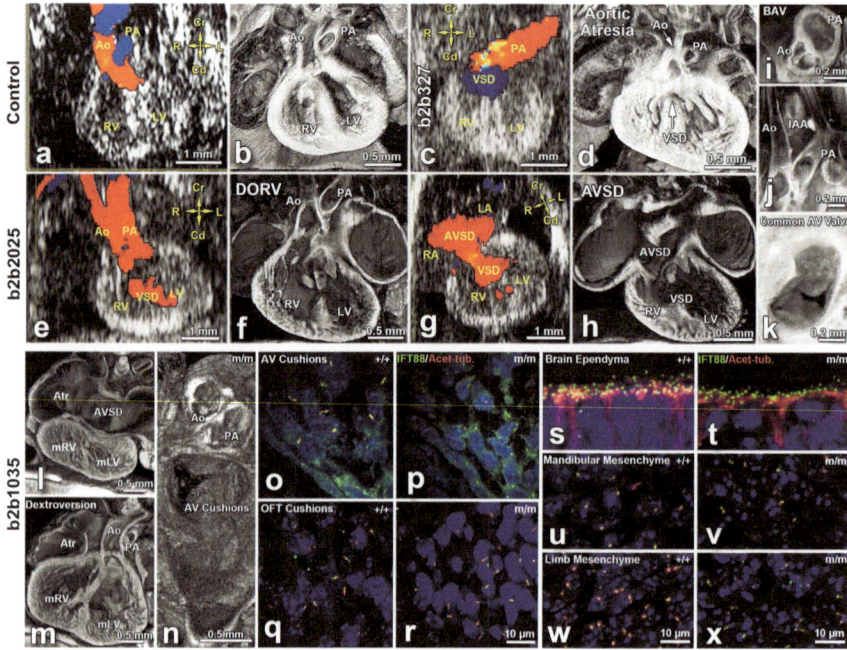

Fig. 8.3 Ultrasound diagnoses of congenital heart disease and cilia defects in mouse mutants with congenital heart disease. Vevo 2100 color flow imaging showed crisscrossing of blood flow, indicating normal aorta (Ao) and pulmonary artery (PA) alignment (**a** Supplemental Movie S1), confirmed by histopathology (**b**). E16.5 mutant (line b2b327) exhibited blood flow pattern indicating single great artery (PA) and ventricular septal defect (VSD) (**c** Supplemental Movie S2), suggesting aortic atresia with VSD, confirmed by histopathology (**d**). Color flow imaging of E15.5 mutant (line b2b2025) with heterotaxy (stomach on right; Supplemental Movie S3c) had side-by-side Ao/PA with Ao emerging from the right ventricle (RV), indicating DORV/VSD (**e**, **f** Supplemental Movie S3a) and presence of AVSD (**g**, **h** Supplemental Movie S3b,S3c). Histopathology also showed bicuspid aortic valve (BAV, **i**), interrupted aortic arch (IAA, **j**), and common AV valve (**k**). (**l–n**) *Cc2d2a* mutant exhibits dextrocardia with ventricular inversion (dextroversion) (**m**) and AVSD (**l**) with malformed AV cushions (**n**) but normal outflow cushions. (**o–x**) Confocal imaging of E12.5 *Cc2d2a* mutant (m/m) vs. wild-type (+/+) embryo sections showed no cilia in AV cushion (**o**, **p**) but normal ciliation in outflow cushion (**q**, **r**). Fewer and shorter cilia were observed in other mutant embryo tissues (**s–x**). *Red*, acetylated tubulin; *green*, IFT88 (From Li et al. [11])

This unexpected finding of a high prevalence of heterotaxy is actually in line with observations in the human fetal population. One clinical study using fetal echocardiography for CHD diagnosis reported that 16 % of human fetuses with congenital heart defects have heterotaxy [12]. This number is likely a minimal estimate, given several clinical studies have shown human fetuses with heterotaxy and congenital heart defect have very high rates of prenatal/intrauterine death (30–60 %) [13–15]. When combined with the fact that only 28 % of CHD is clinically diagnosed prenatally [16], these findings point to the prevalence of congenital heart disease associated with heterotaxy being significantly underestimated clinically. In our mouse screen, we also noted most mouse fetuses with congenital heart defects associated with heterotaxy died in utero. The recovery of these heterotaxy mutants rests entirely on our screen having been conducted prenatally with fetal ultrasound imaging.

Of importance to note is the fact that many of these mutant lines with heterotaxy actually yielded three distinct phenotypes, similar to what had been observed for the *Dnah5* mutants. Thus animals harboring the same mutation can have normal situs solitus, mirror symmetric situs inversus, or heterotaxy (Fig. 8.2). As with the *Dnah5* mutants, congenital heart defects were usually seen only in mutants with heterotaxy [11]. In a subset of these mutants, videomicroscopy of the tracheal epithelia in these mutants also showed immotile or dyskinetic cilia, suggesting they have mutations affecting motile cilia function and may be PCD mouse models [11].

8.1.6   Ciliome Gene Enrichment Among Mutations Causing Congenital Heart Disease

Whole-mouse exome-sequencing analysis was used to recover the pathogenic CHD-causing mutations in mutants recovered from the large-scale mouse muta-genesis screen. This was made possible given the screen was conducted in a C57BL6 inbred strain background. From this analysis, 91 pathogenic mutations were recovered in 61 genes (Fig. 8.4). Of the 61 genes, 35 (58 %) are in cilia-related or ciliome genes (Fig. 8.5); this included 12 genes (34 %) required for motile cilia function (Fig. 8.4). Indeed 8 of these genes are known to cause PCD, including many alleles of *Dnah5* and *Dnah11* (Fig. 8.5). Interestingly, 23 of the cilia genes are actually primary cilia related (66 %). Of these, half are found in mutant lines with laterality defects and half in lines without laterality defects (Fig. 8.4) [11]. These findings suggest the link between cilia and CHD is broader, not merely a reflection of the role of cilia in left-right patterning. This is further supported by the recovery of 15 pathogenic mutations in genes involved in cilia-transduced cell signaling, including mutations in genes involved in Shh, Wnt, Tgfβ, and calcium signaling (Fig. 8.5), all pathways known to play important role in cardiovascular development [11].

Relevant to this, we note cilia is broadly expressed in the embryonic heart, including in the atrial and ventricular myocardium, the atrioventricular and outflow endocardial cushions (Fig. 8.1d–f). Importantly, in the *Cc2d2a* mutant recovered

Fig. 8.4 Distribution of pathogenic congenital heart disease causing mutations recovered from large-scale mouse forward genetic screen. *Top*: Distribution of pathogenic CHD mutations among different mutation types. *Middle*: Ciliome vs. non-ciliome CHD genes found in laterality vs. nonlaterality CHD mutant lines. *Bottom*: Distribution of ciliome CHD genes affecting motile vs. primary cilia among laterality vs. nonlaterality lines (From Li et al. [11])

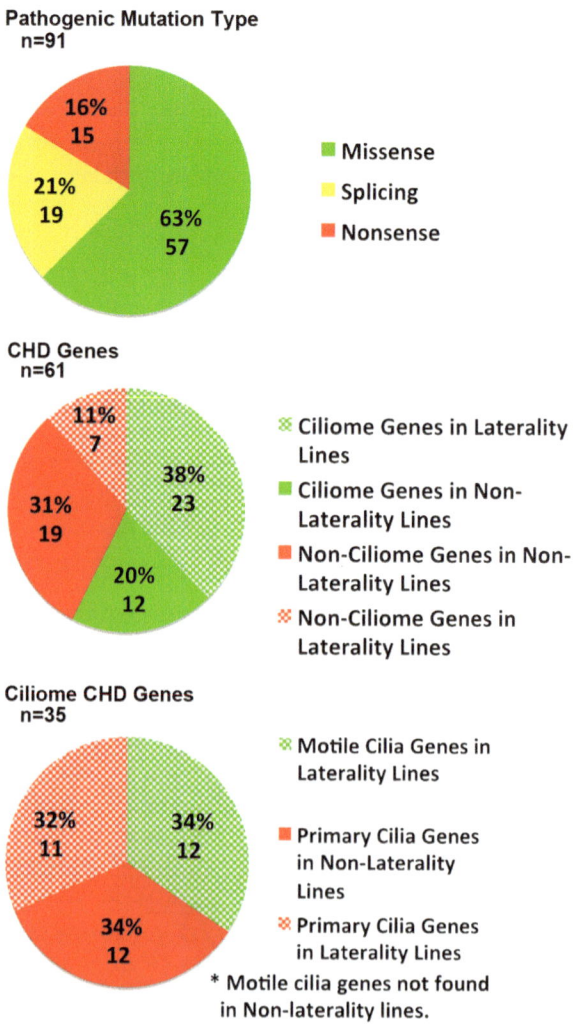

from our screen, atrioventricular septal defects were commonly observed, and this was associated with a selective loss of cilia only in the atrioventricular cushions, while the outflow cushions remain unaffected (Fig. 8.5o–r). The overall marked enrichment observed for cilia mutations in the context of a gene agnostic screen would point to the cilia as playing a central role in the pathogenesis of congenital heart disease, a role that is subserved by both motile and primary cilia and goes beyond the role of cilia in left-right patterning.

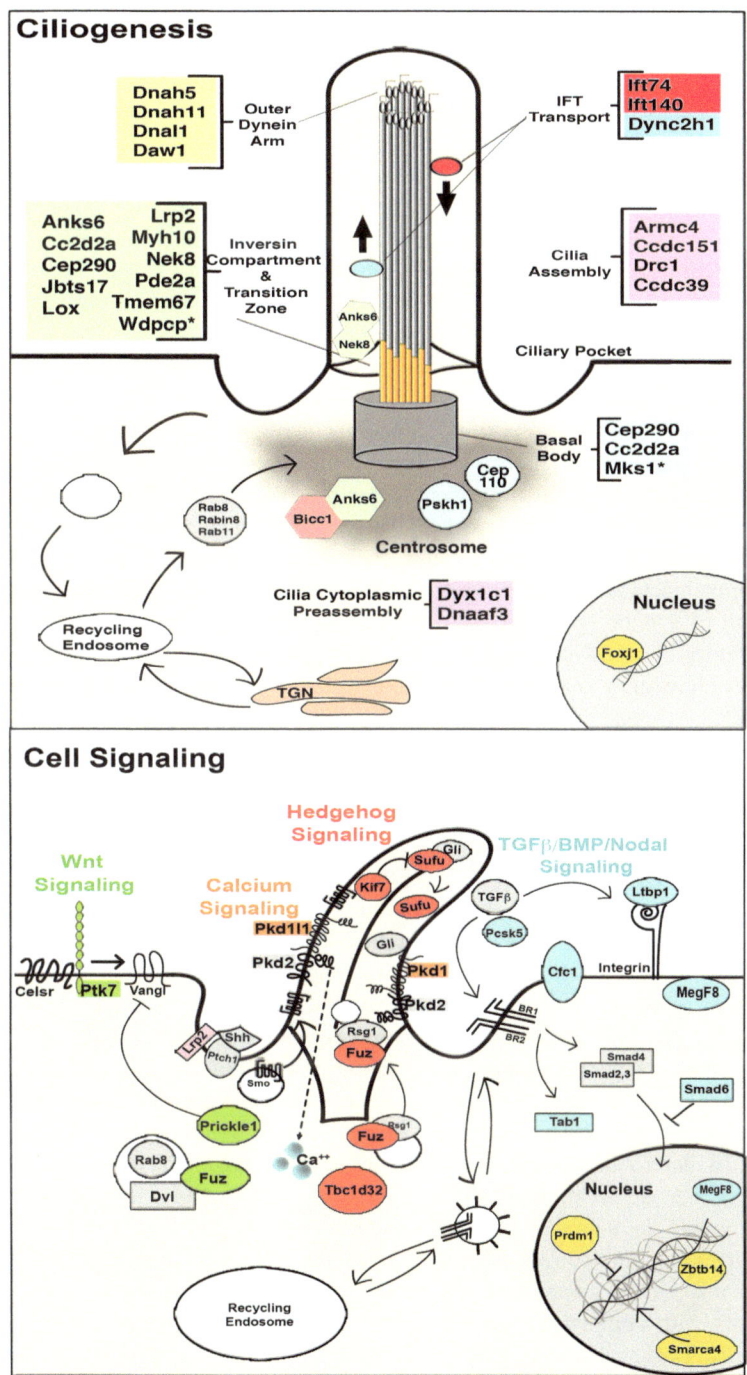

Fig. 8.5 Congenital heart disease genes recovered from mouse mutagenesis screen. Diagrams illustrate biological context of CHD gene function (*color-highlighting* indicates CHD genes recovered; *asterisk* denotes CHD genes recovered from previous screen) (From Li et al. [11])

8.1.7   Ciliary Dysfunction in Congenital Heart Disease Patients Without Heterotaxy

The finding from our mouse studies that cilia defects play a central role in the pathogenesis of CHD would suggest the clinical impact of ciliary dysfunction in congenital heart disease may be broader and have relevance beyond heterotaxy. To examine this question, we conducted a large clinical study of over 200 patients with a broad spectrum of congenital heart disease mostly without heterotaxy to determine the prevalence of ciliary dysfunction [17]. As in our previous study of heterotaxy patients, we assessed for respiratory ciliary dysfunction using videomicroscopy of the nasal epithelia and measured nNO.

This study demonstrated a very high prevalence of ciliary dysfunction, in both the heterotaxy and nonheterotaxy congenital heart disease patients [17]. Moreover, patients with airway ciliary dysfunction had significantly more respiratory symptoms and disease and this did not correlate with heterotaxy status (Table 8.1) [17]. The respiratory symptoms were largely localized to the lower airway and were also significantly associated with PCD symptoms such as chronic otitis media, chronic sinusitis, chronic wet cough, neonatal respiratory distress, pneumonia, and bronchiectasis. Together these findings suggest patients with congenital heart disease of a wide spectrum with or without heterotaxy may have high risk for respiratory ciliary dysfunction. These findings are in agreement with the mouse studies showing a central role for cilia in the pathogenesis of congenital heart disease.

8.1.8   Future Directions and Clinical Implications

We identified a central role for cilia in the pathogenesis of congenital heart disease. This finding uncovered by our mouse mutagenesis screen was unexpected and shows the power of a non-gene-biased phenotype-driven genetic screen to uncover new insights into mechanisms of disease pathogenesis. The important role of cilia in congenital heart disease is supported by the human clinical studies, which also showed a high prevalence of ciliary dysfunction in congenital heart disease patients. Our mouse screen identified primary and motile cilia both playing essential roles in congenital heart disease. The high prevalence of laterality defects among congenital heart disease mutants is reminiscent of the clinicalx observation of a high prevalence of heterotaxy among human fetuses with congenital heart disease. This suggests that observations from our mouse fetal ultrasound screen may provide a window into the unborn human fetal population.

Our finding that congenital heart disease patients with ciliary dysfunction have more respiratory symptoms and disease has important clinical implications. Patients with complex congenital heart disease typically must endure multiple high-risk cardiac surgeries for palliation of their structural heart defects. Not infrequently, respiratory complications with high morbidity are observed postoperatively that can negatively impact outcome. Such respiratory problems are typically

Table 8.1 Multivariate analysis and respiratory symptoms[a]

Covariates	All symptoms		PCD symptoms		Upper symptoms		Lower symptoms	
	exp(b)	P-value	exp(b)	P-value	exp(b)	P-value	exp(b)	P-value
HTX vs. non-HTX	1.103	0.51	1.067	0.75	1.223	0.31	0.973	0.90
Abnormal vs. normal CM	1.426	**0.006**	1.423	**0.043**	1.178	0.4	1.786	**0.003**
Low vs. normal nNO	2.030	**0.002**	2.291	**0.005**	1.520	0.22	2.643	**0.001**

From Garrod et al. [17]

Definition of abbreviations: CM ciliary motion, HTX heterotaxy, nNO nasal nitric oxide, PCD primary ciliary dyskinesia

Significant P values are bolded

[a]In each log-linear regression model, the regression coefficient exp (β) represents one unit increase in the covariate x as a multiplicative effect on the mean of the outcome. When exp (β) >1, the mean of the outcome increases as x increases; if exp (β) <1, the mean of the outcome decreases as x increases

attributed to the heart disease, and the possibility of intrinsic airway clearance defects due to ciliary dysfunction is never considered and hence not tested nor treated. The observations from the human and mouse studies combined would strongly suggest congenital heart disease patients should be presurgically screened for respiratory ciliary dysfunction, and those with ciliary dysfunction should be provided perioperative pulmonary therapy to enhance airway clearance function. Instituting such change in the standard of care may have significant benefit in improving outcome, especially cumulatively through the multiple rounds of surgery that patients with critical congenital heart disease must endure.

Acknowledgments This work was supported by NIH grant HL-U01098180 and funding from the Pennsylvania Department of Health.

References

1. Sutherland MJ, Ware SM. Disorders of left–right asymmetry: heterotaxy and situs inversus. Am J Med Genet C: Semin Med Genet. 2009;151C(4):307–17.
2. Lin AE, Ticho BS, Houde K, Westgate NM, Holmes LB. Heterotaxy: associated conditions and hospital-based prevalence in newborns. Genet Med. 2000;2(3):157–72.
3. Kennedy MP, et al. Congenital heart disease and other heterotaxic defects in a large cohort of patients with primary ciliary dyskinesia. Circulation. 2007;115(22):2814–21.
4. Shapiro AJ, et al. Laterality defects other than situs inversus totalis in primary ciliary dyskinesia: insights into situs ambiguus and heterotaxy. Chest. 2014;146(5):1176–86.
5. Ostrowski LE, Dutcher SK, Lo CW. Cilia and models for studying structure and function. Proc Am Thorac Soc. 2011;8(5):423–9.
6. Li Y, et al. Respiratory motile cilia dysfunction in a patient with cranioectodermal dysplasia. Am J Med Genet A. 2015. doi:10.1002/ajmg.a.37133.
7. Nakleh N, et al. High prevalence of respiratory ciliary dysfunction in congenital heart disease patients with heterotaxy. Circulation. 2012;125:2232–42.
8. Tan SY, et al. Heterotaxy and complex structural heart defects in a mutant mouse model of primary ciliary dyskinesia. J Clin Invest. 2007;117(12):3742–52.
9. Harden B, et al. Increased postoperative respiratory complications in heterotaxy congenital heart disease patients with respiratory ciliary dysfunction. J Thorac Cardiovasc Surg. 2014;147 (4):1291–1298.e2.
10. Liu X, et al. Interrogating congenital heart defects with noninvasive fetal echocardiography in a mouse forward genetic screen. Circ Cardiovasc Imag. 2014;7(1):31–42.
11. Li Y, et al. Global genetic analysis in mice unveils central role for cilia in congenital heart disease. Nature. 2015. doi:10.1038/nature14269(2015).

12. Atkinson DE, Drant S. Diagnosis of heterotaxy syndrome by fetal echocardiography. Am J Cardiol. 1998;82(9):1147–9. A10.
13. Berg C, et al. Prenatal diagnosis of cardiosplenic syndromes: a 10-year experience. Ultrasound Obstet Gynecol. 2013;22(5):451–9.
14. Taketazu M, Lougheed J, Yoo SJ, Lim JS, Hornberger LK. Spectrum of cardiovascular disease, accuracy of diagnosis, and outcome in fetal heterotaxy syndrome. Am J Cardiol. 2006;97(5):720–4.
15. Pepes S, Zidere V, Allan LD. Prenatal diagnosis of left atrial isomerism. Heart. 2009;95 (24):1974–7.
16. Friedberg MK, et al. Prenatal detection of congenital heart disease. J Pediatr. 2009;155 (1):26–31.
17. Garrod AS, et al. Airway ciliary dysfunction and sinopulmonary symptoms in patients with congenital heart disease. Ann Am Thorac Soc. 2014;11(9):1426–32.

Pulmonary Arterial Hypertension in Patients with Heterotaxy/Polysplenia Syndrome

Akimichi Shibata, Keiko Uchida, Jun Maeda, and Hiroyuki Yamagishi

Keywords

Gene • TGF-β • BMPR2

Early progressive pulmonary arterial hypertension (PAH) is often observed in patients with heterotaxy/polysplenia especially who have an intracardiac systemic-to-pulmonary shunt. However, its etiology is uncertain and its management is not well established. There was only a Japanese report about PAH in consecutive patients with heterotaxy/polysplenia syndrome [1]. They seemed to develop pulmonary vascular obstructive disease earlier and more severe than expected, even in cases with only pre-tricuspid systemic-to-pulmonary shunt although more detailed analysis is required.

Improved understanding and studies about the molecular genetics of heterotaxy syndrome indicate that this disease can be caused by single gene mutations. Genes currently implicated in human heterotaxy syndrome include *ZIC3*, *LEFTYA*, *CRYPTIC*, and *ACVR2B* [2]. The establishment of left-right asymmetry is regulated by a number of developmental signaling pathways including the notch, which mediate nodal expression surrounding the node [3]. Nodal, a growth regulator produced by the node, is a signaling molecule belonging to the transforming growth factor (TGF)-β superfamily that plays a variety of roles in the early development [4]. Mouse nodal acts through type I (ALK-4, 5, and 7) and type II (ACVR2B) receptors of the TGF-β superfamily. Ligand activation of the receptors requires one or more co-receptors, Cryptic and Cripto. Lefty-1 (homologue of LEFTYA) is a nodal antagonist that is expressed in medial left lateral plate mesoderm [5].

A. Shibata • K. Uchida • J. Maeda • H. Yamagishi (✉)
Department of Pediatrics, Keio University School of Medicine, Shinanomachi 35, Shinjuku-ku, Tokyo 160-8582, Japan
e-mail: hyamag@keio.jp

The mutation in *BMPR2*, which encodes type II receptor of the bone morphogenic protein (BMP), is a well-known genetic cause of PAH [6]. BMP is also belonging to the TGF-β superfamily. Although there has been no report that describes the association between PAH and mutation of genes implicated in heterotaxy syndrome, they might have some effect on the signaling pathway downstream of BMP and, consequently, be relevant to the pulmonary vascular pathogenesis in progressive PAH.

References

1. Hazama K, Nakanishi T, Momma K, et al. Pulmonary hypertension and pulmonary vascular obstructive disease in patients with polysplenia syndrome. Pediatr Cardiol Cardiac Surg. 1994;10(3):347–53 (in Japanese).
2. Belmont JW, Mohapatra B, Ware SM. Molecular genetics of heterotaxy syndromes. Curr Opin Cardiol. 2004;19:216–20.
3. Krebs LT, Iwai N, Nonaka S, et al. Notch signaling regulates left-right asymmetry determination by inducing nodal expression. Genes Dev. 2003;17:1207–12.
4. Hamada H, Meno C, Watanabe D, et al. Establishment of vertebrate left-right asymmetry. Nat Rev Genet. 2002;3:103–13.
5. Meno C, Shimono A, Saijoh Y, et al. lefty-1 is required for left-right determination as a regulator of lefty-2 and nodal. Cell. 1998;94:287–97.
6. Rabinovitch M. Molecular pathogenesis of pulmonary arterial hypertension. J Clin Invest. 2012;122(12):4306–13.

Part III

Cardiomyocyte and Myocardial Development

Perspective

Adriana C. Gittenberger-de Groot

The heart is the first crucial functioning organ in a developing embryo. The cell essential for the pumping of the blood through the vascular system is the cardiomyocyte. Cardiomyocyte precursors (CPCs) develop in the mesoderm of the embryo in close association with the endoderm. Recent advances in our knowledge on CPC show that during the initial formation of the heart tube, the so-called first heart field (FHF) progenitors are important. These primarily form the left ventricle of the mature heart. Thereafter, second heart field (SHF) CPCs add the right ventricle and outflow tract as well as part of the inflow tract to the looping heart tube. During the process of addition and looping, septation and valve formation take place. The resultant is a four-chambered heart with a unidirectional flow from the right and left atria to the right and left ventricles and respective pulmonary trunk and ascending aorta. This complex remodeling process is highly dependent on instructive and cellular contribution from various surrounding cell populations. These comprise in the early embryo the endoderm as well as later on the endocardium and the epicardium with its epicardium-derived cells (EPDCs). These cell populations are essential in their interaction with the immature myocardial cells for their proper differentiation and the eventual myocardial architecture. Last but not least, the development of a coronary vascular system with endothelial cells and EPDC-derived smooth muscle cells completes the demands of the oxygenation of the cardiomyocytes. Whether the differentiation of a specific population of the

A.C. Gittenberger-de Groot (✉)
Departments Cardiology, Leiden University Medical Center, Leiden, The Netherlands

Anatomy and Embryology, Leiden University Medical Center, Leiden, The Netherlands
e-mail: A.C.Gittenberger-de_Groot@lumc.nl

cardiomyocytes into the cardiac conduction system requires an interaction with surrounding cells is still unresolved.

Interest in cardiomyocyte origin and maturation has been highly augmented by the use of cardiac stem cells and inducible progenitor cells (iPCs) for therapeutic aims in life-threatening myocardial pathology and possibly in the future congenital heart disease.

In this section, many aspects of cardiomyocyte (patho-)biology are addressed. For the possible use in the future of CPCs, it is essential to know whether there are differences between FHF- and SHF-derived CPCs as they have their specific contributions to the right and left ventricle with again their specific cardiac failure problems. Single cell transcriptome analysis might be a novel technique to unravel this question. In early development, the inductive role of the endoderm with several important molecular factors like FGFs including the importance of the related thyroid hormone system should be considered with a possible impact on cardiac disease.

Intriguing is the question of cardiomyocyte regeneration capacities in the mature heart not only by reactivation of the dormant CPC population but also by understanding the molecular mechanism underlying the normal cardiomyocyte cell cycle. Cardiomyocyte proliferation is highly active in the prenatal heart and stops postnatally. Understanding of the mechanism underlying this process and the possibility to extend or to reactivate cardiomyocyte proliferation, in which Meis1 is reported to have a role, might be of great clinical value.

Of similar interest are aspects of formation of the architecture of the myocardial wall consisting of 30 % cardiomyocytes and 70 % cardiac fibroblasts. The myocardial wall develops from a simple two-layered structure into a trabecular layer on the inside and a compact layer on the outside. During development, the relative contributions vary in which the compact layer increases in thickness and the trabecular layer through compaction diminishes. A proper balance of trabecular and compact layer formation seems to be regulated by the Notch pathway-mediated endocardial to myocardial interaction. Disturbance of this pathway in mice may result in a phenotype that resembles a left ventricular non-compaction cardiomyopathy (LVNC) also observed in human familial mutations in this pathway. Likewise, epicardial to myocardial interaction is essential for compact myocardial layer formation as well as ventricular septation. A role of the epicardium in the formation of an anterior-folding ventricular septum is described, shedding new light on the development of ventricular septal defects.

In conclusion, understanding of myocardial pathology and resultant therapeutic and preventive measures, in CHD as well as adult cardiovascular disease, deserves a highly diverse research approach as is elegantly highlighted in this section.

Single-Cell Expression Analyses of Embryonic Cardiac Progenitor Cells

10

Kenta Yashiro and Ken Suzuki

Abstract

For this decade, heart development has been extensively elucidated by the introduction of the concepts of "heart fields" and "cardiac progenitor cells (CPCs)". It is believed that multipotent CPCs are specified among the most anterior part of lateral plate mesoderm as belonging to the two anatomical fields; the first heart field (FHF), which is the future left ventricle and atria, and the second heart field (SHF), which is the future right ventricle and outflow tract of the heart. However, the paradigm of two heart fields dependent on conventional marker genes is still disputed, so the existence of independent CPCs specific to each HF remains an open question. In addition, the molecular mechanism underlying the specification of CPCs remains largely unknown. A single-cell transcriptomic approach, which is realized by the recent advances in molecular biology, can be one of the solutions to bring some breakthrough in this subject.

Keywords

Cardiac progenitor cells • First heart field • Second heart field • Single-cell transcriptomics

10.1 Introduction

Better understanding of the molecular mechanism of heart development is vital to clarify the pathophysiology of congenital heart diseases and will benefit us in terms of prediction, prevention and treatment of such diseases in the future. In addition,

K. Yashiro (✉) • K. Suzuki
Translational Medicine and Therapeutics, William Harvey Research Institute, Barts and The London School of Medicine and Dentistry, Queen Mary University of London, Charterhouse Square, London EC1M 6BQ, UK
e-mail: k.yashiro@qmul.ac.uk

© The Author(s) 2016
T. Nakanishi et al. (eds.), *Etiology and Morphogenesis of Congenital Heart Disease*,
DOI 10.1007/978-4-431-54628-3_10

the information of multipotent cardiac progenitor cells (CPCs) is expected to be valuable for stem cell-based regeneration therapy for the failed heart. However, our knowledge is still insufficient for full translation to the clinical arena.

Here, we review the current knowledge of the earliest phase of cardiac development in mice focusing on CPC specification and discuss the potential of single-cell transcriptomic analyses to elucidate the mechanism [1, 2].

10.2   CPCs of the Two Heart Fields

The heart is one of the first anatomical structures formed during embryogenesis. Fate mapping studies indicated that the nascent mesodermal cells contributing to the heart came through the anterior half of primitive streak on embryonic day (E) 6.5 (Fig. 10.1a) [3]. These mesodermal cells migrate anteriorly acquiring the identity of lateral plate mesoderm. At E7.5, the anterior part of lateral plate mesoderm is identified as the heart fields, the direct source of the heart tube [1, 4]. Then the subsequent morphogenesis to create the heart with the four chambers, including the heart tube formation and looping, occurs.

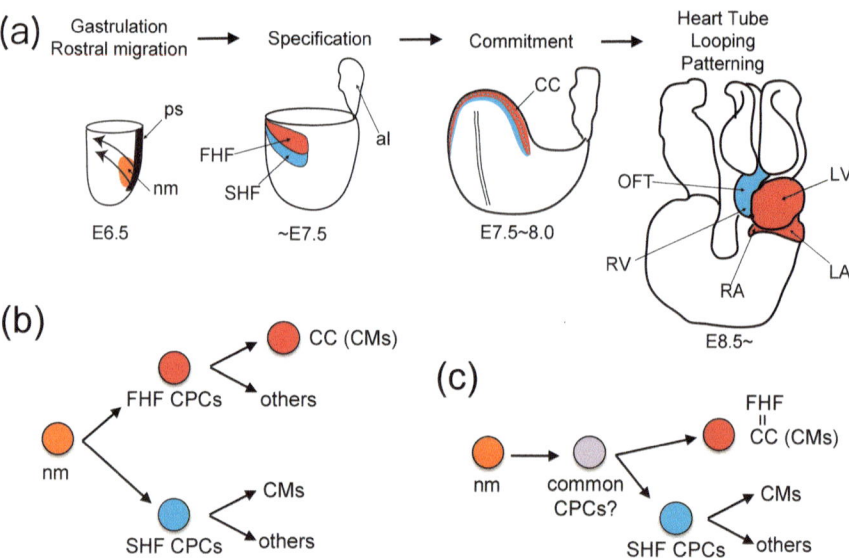

Fig. 10.1 The heart development. (**a**) Schematic illustration of the heart development. The progeny of FHF is indicated by *red* and that of SHF is by *blue*. (**b**) A hypothetical model of the lineage tree of the heart mesoderm. In this model, it is hypothesized that the independent multipotent CPCs of each heart field. (**c**) An alternative model of the heart mesoderm lineage tree. In this model, common CPCs are hypothesized, and only in the case of SHF, the multipotent state might be maintained. *al* allantois, *CC* cardiac crescent, *CMs* cardiomyocytes, *LA* left atrium, *LV* left ventricle, *nm* nascent mesoderm, *OFT* outflow tract, *ps* primitive streak, *RA* right atrium, *RV* right ventricle

According to the fate map, the anatomical heart fields are divided into two major subgroups, the first heart field (FHF) and the second heart field (SHF) (Fig. 10.1a) [1]. FHF is the classical cardiac crescent and is known to contribute to the left ventricle and parts of the atria. SHF contributes to the right ventricle, outflow tract and also parts of atria. These anatomical fields are currently distinguished with the following molecular markers. The genes encoding transcription factors, *Nkx2-5* and *Tbx5*, mark the FHF, whereas *Isl1* marks the SHF at E7.5 in the mouse [1].

CPCs are believed to appear in the period from gastrulation (E6.5) to the heart fields' formation (E7.5). It is thought that CPCs have been already committed toward the heart but are multipotent to give rise to cardiomyocytes, electric conduction system, smooth muscle and endothelium (Fig. 10.1b, c) [1]. Given the paradigm of the anatomical two heart fields, CPCs are classified into FHF CPCs and SHF CPCs (Fig. 10.1b). Unfortunately multipotent FHF CPCs have not been identified (Fig. 10.1c). The cells in the cardiac crescent, the theoretical progeny of FHF CPCs, are unlikely to be "multipotent progenitors", because these already express the terminal differentiation markers of cardiomyocytes, such as *Actc1* and *Myl7* [1, 5, 6]. To add to this, no marker is available to identify FHF before *Nkx2-5* and *Tbx5* expression. By contrast, the presence of multipotent SHF CPCs, which are marked by *Isl1* expression, seems to be validated with clonal tracing experiments [7]. However, the paradigm of FHF and SHF based on *Nkx2-5* and *Isl1* is now disputed [1]. The expression of *Isl1* was shown to be not specific for SHF and was suggested to represent only the developmental stages [8, 9]. Thus, true characteristics of CPCs should be clarified with revalidation of the conventional biomarkers and the concept of two heart fields.

10.3   CPC Specification

The molecular mechanism underlying the specification of CPCs remains largely unknown. In the mouse embryo, specification might occur inside the heart fields [1, 4]. When the nascent mesoderm cells at E6.5 containing the presumptive heart fields were transplanted heterotopically to a location other than the heart fields, the fate was not the heart but the same as the cells of the recipient surrounding the graft [4]. On the other hand, when the same graft was transplanted to the heart fields of E7.5, it was committed almost only to the heart. These suggest that the migrating presumptive heart fields are plastic and that some inductive cue for cardiac specification is present within the heart fields. The surrounding tissue, especially the endoderm, likely provides such an inductive cue [1, 4]. However, we cannot exclude the possibility that some sequential events of specification take place during the gastrulation and migration so that the unfixed cardiac competence might be gradually and sequentially consolidated.

It has been shown that the signalling pathways of WNT, FGF, BMP, NODAL (ACTIVIN), SHH and NOTCH are involved in the specification [1, 10]. However, validating the precise role of each signalling pathway is difficult because their accurate cardiogenic function has been hard to distinguish from their overall effects. Thus, some alternative strategy is needed to bring a breakthrough.

10.4 The Potential of Single-Cell Transcriptomics in the Study of CPC Specification

Accurate quantitative evaluation of genome-wide gene expression provides an essential platform to understand the states of cells of interest. In the case where a specific subset of rare cells plays a vital role, single-cell transcriptomics are a strong tool. Indeed, thus far, this strategy has given fruitful results in the study of the specification of the olfactory receptors, spinal cord neurons and primordial germ cells [11–13]. Currently, several protocols of single-cell transcriptomics have been developed as applicable to next-generation sequencing (NGS) with sufficient coverage, sensitivity and accuracy (Table. 10.1) [2].

The protocol for genome-wide analysis was first developed by Kurimoto and colleagues for microarray and has been recently modified for NGS (Fig. 10.2a) [14–16]. The feature of this protocol is to add artificially a poly A tail to the 3′-end of the first-strand cDNA to generate the appropriate adapter sites for PCR amplification. At the final step, cDNA is obtained as PCR amplicon. Disadvantageously Kurimoto's protocol provides only strongly 3′-biased cDNA, although Tang's protocol likely overcomes this problem [15].

Quartz-seq is also a method applicable to microarray and NGS, based on PCR amplification, and the architecture of the protocol is based on "suppressive PCR" (Fig. 10.2a) [17]. This method looks superior to Kurimoto's in terms of simple operation, coverage and technical reproducibility.

SMART-seq utilizes a unique nature of reverse transcriptase [18]. During reverse transcription, Moloney murine leukaemia virus reverse transcriptase adds a short poly C tail to the 3′-end of the first-strand cDNA. Reverse transcriptase uses a helper oligonucleotide with poly G motif as the template ("template switch") after annealing to the poly C tail, which results in the addition of adaptors to cDNA (Fig. 10.2b). This template switch happens preferentially to 5′-capped mRNA so that finally a full-length cDNA library is obtained as suppressive PCR amplicon if oligo(dT) primer is used for the first-strand synthesis. This enrichment of full length

Table 10.1 Methods of single-cell transcriptomics

Method	Principle	Application	References
Kurimoto et al. 2006	Poly A tailing/PCR	Microarray	[14]
Tang et al. 2009	Poly A tailing/PCR	NGS	[15]
Brouilette et al. 2012	Poly A tailing/PCR	NGS	[16]
Quartz-seq/-chip	Poly A tailing/suppressive PCR	NGS/microarray	[17]
SMART-seq	Template switch/PCR	NGS	[18]
CEL-seq	IVT/PCR	NGS	[19]

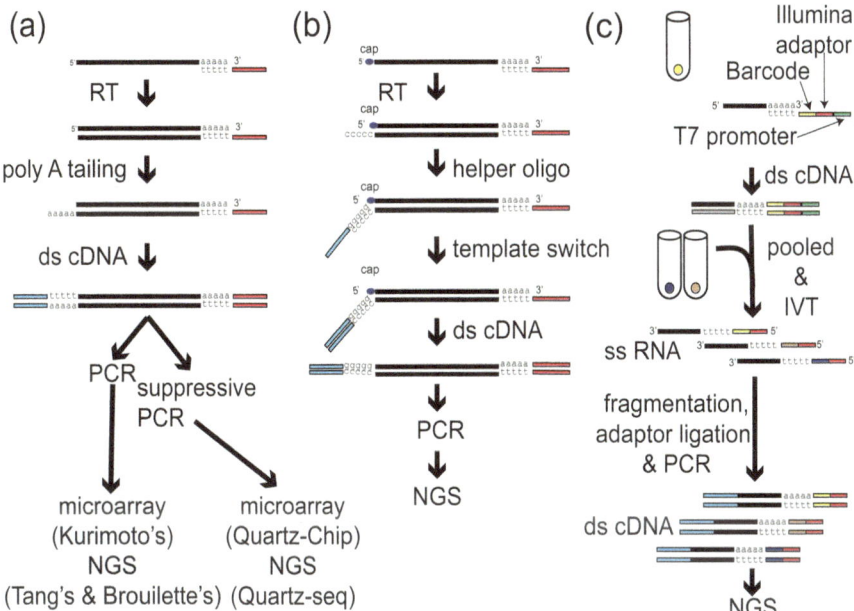

Fig. 10.2 Overview of the methods of single-cell transcriptomics. (**a**) Poly A tailing (Kurimoto's, Tang's, Boulette's protocol and Quartz-seq), (**b**) SMART-seq, (**c**) CEL-seq. Please refer to the main text and the relevant references for details

is a merit, but simultaneously a demerit, because any partially synthesized cDNAs are not amplified so that the yield of cDNAs is inevitably limited.

On the other hand, CEL-seq uses mainly isothermal amplification of in vitro transcription (IVT) by T7 polymerase (Fig. 10.2c) [19]. Theoretically, IVT is superior to PCR in terms of liner and proportional amplification to secure better technical reproducibility. A demerit of this protocol is that reads by NGS are strongly biased toward 3′-end.

Lastly, we should discuss the limitations of single-cell transcriptomics. First, there is always a technical noise due to pippetting error, difference of system (including enzymes, thermal cycler and NGS machine) and amplification bias. Of note, the "Monte Carlo effect", the cause of large quantitative errors of small copy number mRNAs due to the stochastic events within an initial few cycles of PCR reaction, is uncontrollable. Amplification bias is inevitable to some extent in the protocols dependent on PCR, although every protocol has been well validated to secure its sufficient technical reproducibility. Second, the lack of topological information precludes us from referring the obtained expression profile to the cell in the embryo if no molecular markers exist. Thus it is necessary to consider how to fill such lost information.

10.5 Future Direction and Clinical Implication

The strategy of single-cell transcriptomics has great potential to increase understanding of CPC specification, their ground state and differentiation. This knowledge will be utilized in order to control the commitment of pluripotent stem cells including ES cells and iPS cells for future clinical application as stem cell-based regeneration therapy.

Acknowledgements We thank Kazuko Koshiba-Takeuchi, Jun Takeuchi, Shigetoyo Kogaki, Manabu Shirai, Hiroki Kokubo, Yusuke Watanabe and Hiroyuki Yamagishi for helpful discussions and Steven Coppen for editing. Our work is supported by MRC New Investigator Research Grant G0900105, MRC Research Grant MR/J007625/1 and BHF Project Grant PG/11/102/29213 to KY.

References

1. Rana MS, Christoffels VM, Moorman AF. A molecular and genetic outline of cardiac morphogenesis. Acta Physiol (Oxf). 2013;207:588–615.
2. Shapiro E, Biezuner T, Linnarsson S. Single-cell sequencing-based technologies will revolutionize whole-organism science. Nat Rev Genet. 2013;14:618–30.
3. Kinder SJ, Tsang TE, Quinlan GA, et al. The orderly allocation of mesodermal cells to the extraembryonic structures and the anteroposterior axis during gastrulation of the mouse embryo. Development. 1999;126:4691–701.
4. Tam PP, Parameswaran M, Kinder SJ, et al. The allocation of epiblast cells to the embryonic heart and other mesodermal lineages: the role of ingression and tissue movement during gastrulation. Development. 1997;124:1631–42.
5. Takeuchi JK, Bruneau BG. Directed transdifferentiation of mouse mesoderm to heart tissue by defined factors. Nature. 2009;459:708–11.
6. Cai CL, Liang X, Shi Y, et al. Isl1 identifies a cardiac progenitor population that proliferates prior to differentiation and contributes a majority of cells to the heart. Dev Cell. 2003;5:877–89.
7. Moretti A, Caron L, Nakano A, et al. Multipotent embryonic isl1+ progenitor cells lead to cardiac, smooth muscle, and endothelial cell diversification. Cell. 2006;127:1151–65.
8. Prall OW, Menon MK, Solloway MJ, et al. An Nkx2-5/Bmp2/Smad1 negative feedback loop controls heart progenitor specification and proliferation. Cell. 2007;128:947–59.
9. Ma Q, Zhou B, Pu WT. Reassessment of Isl1 and Nkx2-5 cardiac fate maps using a Gata4-based reporter of Cre activity. Dev Biol. 2008;323:98–104.
10. Nemir M, Croquelois A, Pedrazzini T, et al. Induction of cardiogenesis in embryonic stem cells via downregulation of Notch1 signaling. Circ Res. 2006;98:1471–8.

11. Dulac C, Axel R. A novel family of genes encoding putative pheromone receptors in mammals. Cell. 1995;83:195–206.
12. Tanabe Y, William C, Jessell TM. Specification of motor neuron identity by the MNR2 homeodomain protein. Cell. 1998;95:67–80.
13. Saitou M, Barton SC, Surani MA. A molecular programme for the specification of germ cell fate in mice. Nature. 2002;418:293–300.
14. Kurimoto K, Yabuta Y, Ohinata Y, et al. An improved single-cell cDNA amplification method for efficient high-density oligonucleotide microarray analysis. Nucleic Acids Res. 2006;34: e42.
15. Tang F, Barbacioru C, Wang Y, et al. mRNA-Seq whole-transcriptome analysis of a single cell. Nat Methods. 2009;6:377–82.
16. Brouilette S, Kuersten S, Mein C, et al. A simple and novel method for RNA-seq library preparation of single cell cDNA analysis by hyperactive Tn5 transposase. Dev Dyn. 2012;241:1584–90.
17. Sasagawa Y, Nikaido I, Hayashi T, et al. Quartz-Seq: a highly reproducible and sensitive single-cell RNA sequencing method, reveals non-genetic gene-expression heterogeneity. Genome Biol. 2013;14:R31.
18. Ramskold D, Luo S, Wang YC, et al. Full-length mRNA-Seq from single-cell levels of RNA and individual circulating tumor cells. Nat Biotechnol. 2012;30:777–82.
19. Hashimshony T, Wagner F, Sher N, et al. CEL-Seq: single-cell RNA-Seq by multiplexed linear amplification. Cell Rep. 2012;2:666–73.

Meis1 Regulates Postnatal Cardiomyocyte Cell Cycle Arrest

11

Shalini A. Muralidhar and Hesham A. Sadek

Abstract

The neonatal mammalian heart is capable of substantial regeneration following injury through cardiomyocyte proliferation (Porrello et al, Science 331:1078–1080, 2011; Proc Natl Acad Sci U S A 110:187–92, 2013). However, this regenerative capacity is lost by postnatal day 7 and the mechanisms of cardiomyocyte cell cycle arrest remain unclear. The homeodomain transcription factor Meis1 is required for normal cardiac development but its role in cardiomyocytes is unknown (Paige et al, Cell 151:221–232, 2012; Wamstad et al, Cell 151: 206–220, 2012). Here we identify Meis1 as a critical regulator of the cardiomyocyte cell cycle. Meis1 deletion in mouse cardiomyocytes was sufficient for extension of the postnatal proliferative window of cardiomyocytes and for reactivation of cardiomyocyte mitosis in the adult heart with no deleterious effect on cardiac function. In contrast, overexpression of Meis1 in cardiomyocytes decreased neonatal myocyte proliferation and inhibited neonatal heart regeneration. Finally, we show that Meis1 is required for transcriptional activation of the synergistic CDK inhibitors p15, p16, and p21. These results identify Meis1 as a critical transcriptional regulator of cardiomyocyte proliferation and a potential therapeutic target for heart regeneration.

Keywords

Cardiomyocytes • Heart injury • Regeneration • Meis1 • Cell cycle

S.A. Muralidhar, Ph.D. • H.A. Sadek, M.D., Ph.D. (✉)
Department of Internal Medicine, Division of Cardiology, The University of Texas Southwestern Medical Center, Dallas, TX 75390, USA
e-mail: hesham.sadek@utsouthwestern.edu

© The Author(s) 2016
T. Nakanishi et al. (eds.), *Etiology and Morphogenesis of Congenital Heart Disease*,
DOI 10.1007/978-4-431-54628-3_11

93

11.1 Introduction

The past decade witnessed a revolution in our understanding of cardiac biology, with groundbreaking research demonstrating that the adult mammalian heart is capable of limited but measurable cardiomyocyte turnover [5–9]. However, the ultimate goal of complete regeneration of the heart remains elusive. In stark contrast to the adult mammalian heart, we recently demonstrate that the mammalian heart is in fact capable of complete regeneration following apical resection of 15 % of the ventricular myocardium [1]. This remarkable regenerative capacity was associated with robust cardiomyocyte proliferation throughout the myocardium. Moreover, lineage-tracing studies demonstrated that the newly formed myocytes were derived from preexisting cardiomyocytes, rather than from a progenitor population. Finally, we showed that cessation of this regenerative phenomenon occurred at postnatal day 7 (P7), which coincides with the developmental window when cardiomyocytes become binucleate and withdraw from the cell cycle [9, 10]. It is unclear whether the loss of this regenerative potential in the adult heart is due to an intrinsic cell cycle block in adult cardiomyocytes or to loss of mitogenic stimuli as the heart ages (or both). Thus, it is important to determine the mechanisms by which the mammalian heart switches off this regenerative capacity in the week after birth.

In an effort to identify genes involved in postnatal regeneration arrest, we performed several gene arrays following MI at multiple postnatal time points. This allowed us to identify Meis1 as one of the few transcription factors that were dysregulated between injury at P1 and injury at P7 and P14. Meis1 has been studied extensively in the hematopoietic system, is required for normal hematopoiesis, and also plays an important role in leukemogenic transformation [11–14]. What little is known about the role of Meis1 in the heart comes from global KO studies resulting in numerous cardiac defects. However, given that global Meis1 deletion results in embryonic lethality by E14.5, full characterization of the role of Meis1 in cardiomyocytes has been difficult [13, 15, 16]. Despite the role of Meis1 in regulation of hematopoiesis and cardiac development, the mechanism of action of Meis1 remains poorly characterized. Our results indicate that Meis1 expression and nuclear localization in cardiomyocytes coincide with cell cycle arrest. Cardiomyocyte cell cycle exit is associated with downregulation of positive cell cycle regulators (CDK2, CDK3, CDK4, CCND1, and CDK cofactors) and induction of cell cycle inhibitors (CDKI, members of the INK4 and CIP/KIP families) [10, 17–20]. We identified conserved Meis1 domains in only two key CDK inhibitors, namely, p16 and p21, which are known to regulate all three cell cycle checkpoints. We demonstrated that Meis1 regulates the pattern of expression of these two cell cycle inhibitors following Meis1 knockdown. These results provide role and mechanism of cell cycle regulation by Meis1.

A Increased Meis1
expression at P7

B Increased Meis1 expression
Following MI at P7 post 7 days

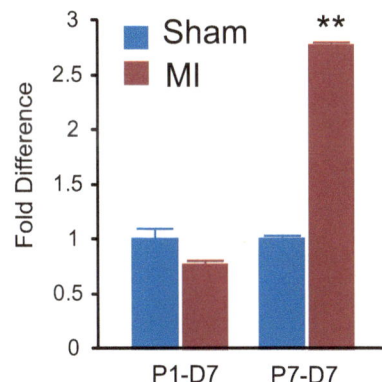

C Expression profile of Meis1 in the neonatal mouse heart

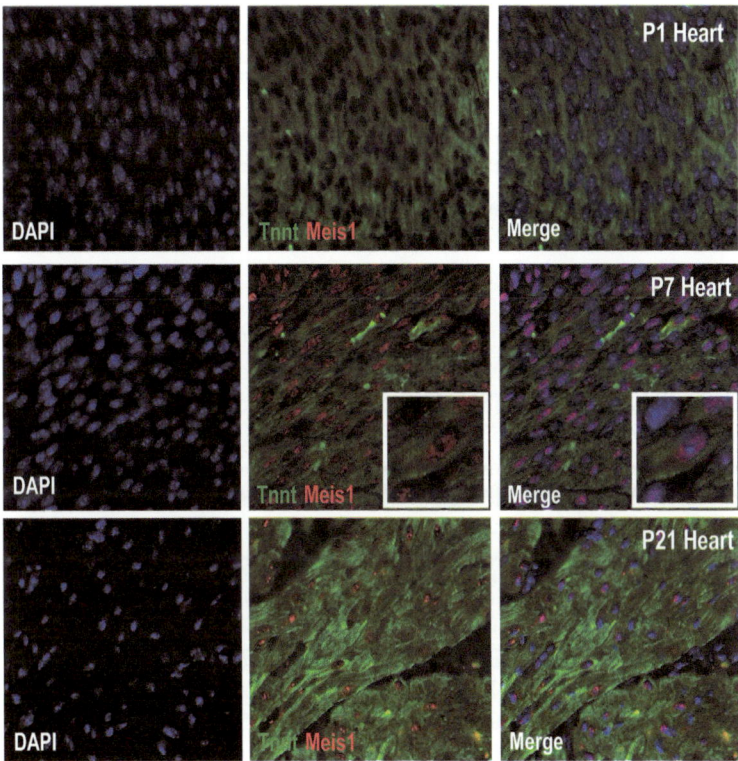

Fig. 11.1 Expression profile of Meis1 in postnatal heart. (**a**) qRT–PCR showing increased expression of Meis1 at postnatal day 7 (P7), a time point that coincides with cell cycle arrest of cardiomyocytes. (**b**) qRT–PCR showing expression levels of Meis1 7 days post Sham or MI at P1 or P7. (**c**) Expression of Meis1 is absent at P1 and nuclear localization of Meis1 in P7 and P21 cardiomyocytes

11.2 Results

11.2.1 Expression of Meis1 During Neonatal Heart Development and Regeneration

Transcriptional regulation of postnatal cardiomyocyte cell cycle arrest is unclear. Our initial screens identified Meis1 as a potential transcriptional regulator of neonatal heart regeneration. Therefore, we conducted this study to determine the role of Meis1 in regulation of cardiomyocyte cell cycle. We first examined the expression pattern of Meis1 during neonatal heart development and regeneration. qPCR showed a modest increase in Meis1 expression between postnatal day 1 (P1) and P7 (Fig. 11.1a). Meis1 was localized to perinuclear regions in neonatal cardiomyocytes at P1 and nuclear localized by P7 and P21 (Fig. 11.1c). Myocardial infarction (MI) at P1, which is associated with an induction of robust cardiomyocyte proliferation at day 7 post-MI, was associated with a modest decrease in the expression of Meis1, whereas Meis1 mRNA expression levels were significantly increased following MI at P7, a time point coinciding with lack of mitotic induction of cardiomyocytes (Fig. 11.1b).

11.2.2 Cardiomyocyte Proliferation Beyond Postnatal Day 7 Following Meis1 Deletion

We generated cardiomyocyte-specific Meis1 knockout (KO) mice by crossing $Meis1^{f/f}$ mice with αMHC-Cre mice (Fig. 11.2a). qRT–PCR of cardiomyocytes from $Meis1^{f/f}$ αMHC-Cre (Meis1KO) compared to αMHC-Cre (control) mouse hearts confirmed a change in gene expression consistent with Meis1 deletion in cardiomyocytes (Fig. 11.2b). Phenotypic characterization of Meis1KO mice at P14 (1 week beyond the normal window of postnatal cardiomyocyte cell cycle arrest) demonstrated that heart size (Fig. 11.2c, d) and cardiac function (Fig. 11.2e) were unaffected by Meis1 deletion. However, Meis1KO cardiomyocytes were smaller compared to control cardiomyocytes (Fig. 11.2f), which may imply that the cardiomyocyte number is increased in Meis1KO (smaller cardiomyocyte size, with no change in heart-to-body weight ratio). Therefore, we examined the Meis1KO hearts for myocyte proliferation using the mitosis marker pH3 (phosphorylated histone H3) and the cytokinesis marker Aurora B kinase. Meis1 deletion resulted in induction of cardiomyocyte proliferation as quantified by an increase in the number of pH3+TnnT2+ (troponin T2) cells (Fig. 11.2g). In addition, we found that Aurora B was markedly expressed in the cleavage furrow between proliferating myocytes in the KO hearts (0.5-fold) (Fig. 11.2h). We also found a significant increase which did not result in an increase in myocyte apoptosis by TdT-mediated dUTP nick end labeling (TUNEL) staining (Fig. 11.2k).

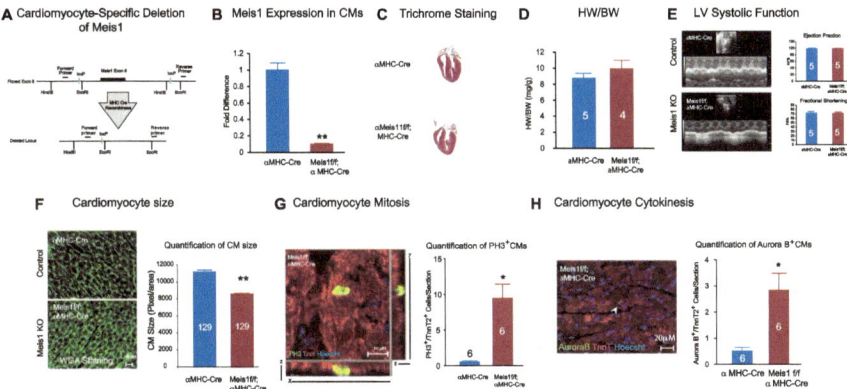

Fig. 11.2 Cardiomyocyte proliferation at P14 following Meis1 deletion. (**a**) Schematic of Meis1 floxed allele. Control mice were αMHC-Cre, Meis1KO mice were Meis1$^{f/f}$αMHC-Cre. (**b**) qRT–PCR demonstrates deletion of Meis1 in cardiomyocytes at P14. (**c**) Trichrome staining of wild-type (WT) and Meis1KO hearts at P14. (**d**) Heart weight (HW)-to-body weight (BW) ratio in WT and Meis1KO hearts. (**e**) Left ventricular (LV) systolic function quantified by ejection fraction (EF) and fractional shortening (FS). (**f**) Wheat germ agglutinin (WGA) staining and cell size quantification. (**g**) Confocal image with z-stacking showing co-localization of pH3, TnnT2, and Hoechst in a Meis1KO heart at P14. Immunostaining showing sarcomere disassembly and normal sarcomeric structure in Meis1KO hearts. Graph shows quantification of the number of pH31+TnnT21+ nuclei. (**h**) Expression of Aurora B in Meis1KO cardiomyocytes at P14 and quantification of the number of Aurora B+TnnT2+ cardiomyocytes

11.2.3 MI in Meis1 Overexpressing Heart Limits Neonatal Heart Regeneration

To determine whether forced Meis1 expression can inhibit neonatal cardiomyocyte proliferation, we generated a cardiac-specific Meis1 overexpressing mouse (Meis1 OE) by crossing pTRE-Meis1 mice with αMHC-tTA mice to allow for specific overexpression of Meis1 in cardiomyocytes (Fig. 11.3a) around birth in the absence of tetracycline. We used a Meis1 line that overexpressed Meis1 by 2.5-fold (Fig. 11.3b). Overexpression of Meis1 did not result in a significant increase in heart-to-body weight ratio (Fig. 11.3c, d), normal systolic function (Fig. 11.3e), and increased cardiomyocyte size (Fig. 11.3f). Despite the increase in cardiomyocyte size in Meis1 (OE) hearts, the lack of decrease in heart-to-body weight ratio most probably reflects a decrease in the number of cardiomyocytes. This is supported by a decrease in the number of mitotic cardiomyocytes in the neonatal Meis1 (OE) hearts (Fig. 11.3g). More notably, Meis1 OE inhibited neonatal heart regeneration following induction of MI at P1 (Fig. 11.3h–j), whereas the wild-type hearts regenerated normally. Finally, Meis1 OE in cardiomyocytes resulted in upregulation of CDK inhibitors, most significantly p21 (Cdkn1a) (Fig. 11.3k). These results indicate that Meis1 overexpression in the neonatal heart results in premature cardiomyocyte cell cycle arrest.

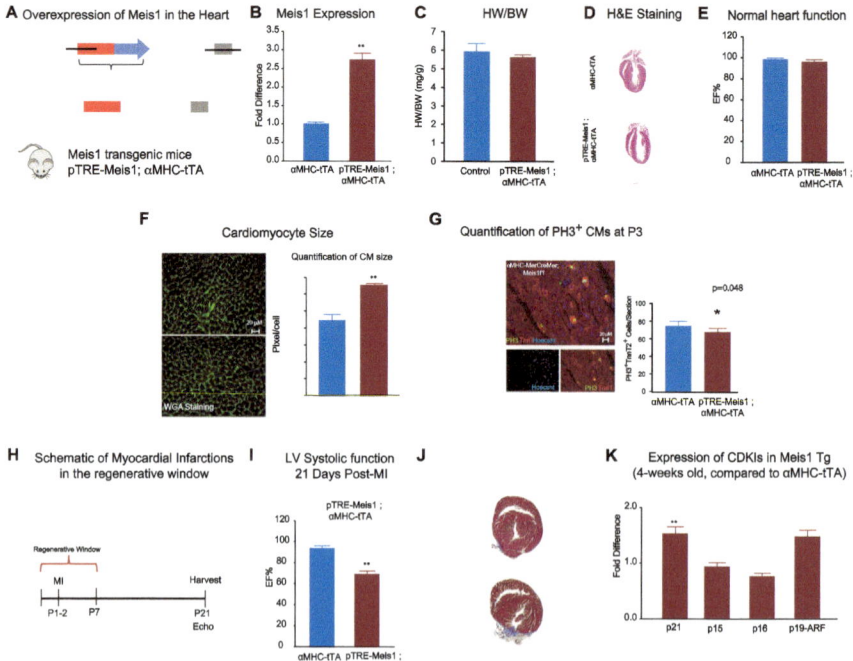

Fig. 11.3 Meis1 overexpression limits neonatal heart regeneration following MI. (**a**) Schematic of Meis1 overexpression (OE) in the heart. Control mice were αMHC-tTA, Meis1 (OE) mice were pTREMeis1-αMHC-tTA. (**b**) qRT–PCR demonstrates overexpression of Meis1. (**c**) HW-to-BW ratio in control and Meis1 (OE) mice. (**d**) H & E staining of WT and Meis1 (OE) hearts. (**e**) LV systolic function quantified by EF. (**f**) WGA staining and cell size quantification. (**g**) Immunostaining image showing co-localization of pH3, TnnT2, and Hoechst in Meis1 (OE) heart at P3. Graph shows quantification of pH3+TnnT2+ nuclei. (**h**) Schematic of neonatal MI during the regenerative window at P1. (**i**) LV systolic function of WT and Meis1 (OE) hearts at 21 days post-MI. (**j**) Trichromes at day 21 post-MI. (**k**) qRT–PCR of CDKIs in hearts of Meis1 (OE) compared to control

11.2.4 Regulation of Cyclin-Dependent Kinase Inhibitors by Meis1

In order to determine the mechanism by which Meis1 regulates cardiomyocyte proliferation, we performed a cell cycle PCR array. We found that Meis1 deletion resulted in downregulation of cyclin-dependent kinase inhibitors in isolated cardiomyocytes, including members of the Ink4b–Arf–Ink4a locus (p16, p15, and p19ARF) and CIP/ KIP family (p21 and p57), as well as upregulation of a number of positive regulators of the cell cycle (Fig. 11.3a, b). qRT–PCR confirmed these results (Fig. 11.4a). Of all the dysregulated cell cycle genes, we identified conserved Meis1 consensus binding sequences in the promoter region of only two loci, namely, the Ink4b–Arf–Ink4a (which includes p16INK4a, p19ARF, and p15INK4b) (Fig. 11.4b) and p21 loci (Fig. 11.4c) using the UCSC genome browser (http://genome.ucsc.edu). To test if Meis1 can transcriptionally activate Ink4b–

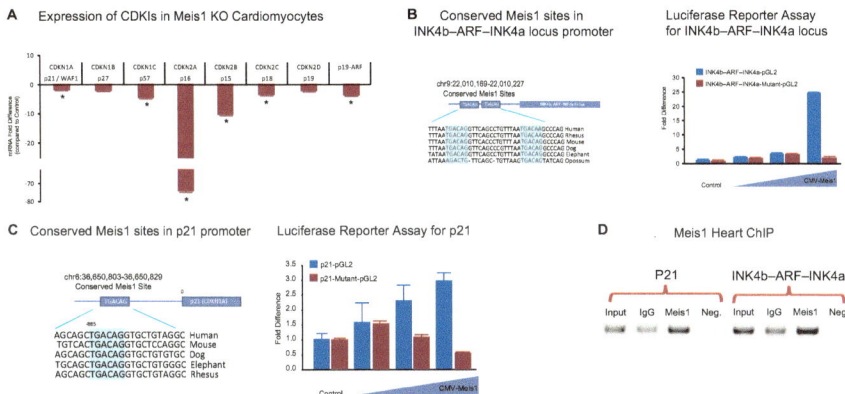

Fig. 11.4 Regulation of cyclin-dependent kinase inhibitors by Meis1. (**a**) qPCR of CDKIs in purified Meis1KO cardiomyocytes demonstrates significant downregulation of CDKN1A (p21), CDKN2B (p15), CDKN2A (p16), p19ARF, CDKN2C (p18), and CDKN1C (p57). (**b**) Highly conserved Meis1 motifs located in p16INK4a/p19ARF/p15INK4b promoter and luciferase reporter assay for INK4b–ARF–INK4a promoter demonstrating induction of INK4b–ARF–INK4a pGL2 reporter by Meis1 and loss of reporter activity following mutation of Meis1 motifs (INK4b–ARF–INK4a-Mutant). (**c**) Highly conserved Meis1 motif located in p21 promoter and luciferase assay for p21 demonstrating induction of p21 pGL2 reporter by Meis1 and loss of reporter activity following mutation of Meis1 motifs (p21 Mutant). (**d**) In vivo confirmation of Meis1 interaction with INK4a/ARF/INK4b and p21 promoters by ChIP assay

Arf–Ink4a and p21, we generated luciferase reporter constructs containing the conserved Meis1 binding motifs. Luciferase reporter assays with INK4b–ARF–INK4a–pGL2 (Fig. 11.4b) and p21–pGL2 reporters (Fig. 11.4c) demonstrated a dose-dependent activation by Meis1. Mutation of the putative Meis1 binding sites abolished the Meis1-dependent activation of the luciferase reporters (Fig. 11.4b, c). Finally, we demonstrated an in vivo interaction between Meis1 and Ink4b–Arf–Ink4a and p21 promoters in the adult mouse heart by ChIP (Figs. 11.4d and 11.5).

11.3   Future Direction and Clinical Implications

The current study identifies Meis1 as a critical transcriptional regulator of cardiomyocyte cell cycle, upstream of two synergistic CDKI inhibitors. Although the mechanism of activation of Meis1 in the postnatal heart is not quite fully understood, results have demonstrated that Meis1 expression coincides with Hox genes that are known to interact with Meis1, stabilize its DNA binding, and enhance its transcriptional activity. Therefore, it would be important for future studies to define the transcriptional network involved in mediating the effect of Meis1 on postnatal cardiomyocytes. Ultimately, we hope to utilize our understanding of the role and mechanism of Meis1 in cardiomyocyte proliferation to uncover new disease mechanisms and therapeutic approaches for cardiovascular diseases.

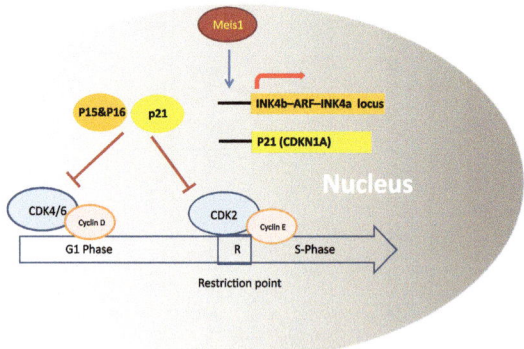

Fig. 11.5 Proposed model of cardiomyocyte cell cycle regulation by Meis1. Meis1 regulates cardiomyocyte cell cycle arrest through transcriptional activation of CDKIs, INK4b–ARF–INK4a locus, and p21, as well as indirectly through a number of other cell cycle regulators. Activation of CDKIs leads to cell cycle arrest by inhibiting CDKs. Statistical significance: Values presented as mean $\pm$ SEM. A Student's t test was used to determine statistical significance. $*P < 0.05$. $**P < 0.01$

Acknowledgments We apologize to whose work was not cited in this chapter and for the omission of some discussion points owing to space constraints. This work was supported by grants from the AHA (Grant in Aid) (Sadek), the Gilead Research Scholars Program in Cardiovascular Disease (Sadek), Foundation for Heart Failure Research, NY, and the NIH (1R01HL115275-01) (Sadek).

References

1. Porrello ER, et al. Transient regenerative potential of the neonatal mouse heart. Science. 2011;331(6020):1078–80.
2. Porrello ER, et al. Regulation of neonatal and adult mammalian heart regeneration by the miR-15 family. Proc Natl Acad Sci U S A. 2013;110(1):187–92.
3. Paige SL, et al. A temporal chromatin signature in human embryonic stem cells identifies regulators of cardiac development. Cell. 2012;151(1):221–32.
4. Wamstad JA, et al. Dynamic and coordinated epigenetic regulation of developmental transitions in the cardiac lineage. Cell. 2012;151(1):206–20.
5. Beltrami AP, et al. Evidence that human cardiac myocytes divide after MI. N Engl J Med. 2001;344(23):1750–7.

6. Bergmann O, et al. Evidence for cardiomyocyte renewal in humans. Science. 2009;324 (5923):98–102.
7. Bergmann O, et al. Cardiomyocyte renewal in humans. Circ Res. 2012;110(1):p. e17–8; author reply e19-21.
8. Soonpaa MH, Field LJ. Assessment of cardiomyocyte DNA synthesis in normal and injured adult mouse hearts. Am J Physiol. 1997;272(1 Pt 2):H220–6.
9. Li F, et al. Rapid transition of cardiac myocytes from hyperplasia to hypertrophy during postnatal development. J Mol Cell Cardiol. 1996;28(8):1737–46.
10. Walsh S, et al. Cardiomyocyte cell cycle control and growth estimation in vivo – an analysis based on cardiomyocyte nuclei. Cardiovasc Res. 2010;86(3):365–73.
11. Moskow JJ, et al. Meis1, a PBX1-related homeobox gene involved in myeloid leukemia in BXH-2 mice. Mol Cell Biol. 1995;15(10):5434–43.
12. Argiropoulos B, Yung E, Humphries RK. Unraveling the crucial roles of Meis1 in leukemogenesis and normal hematopoiesis. Genes Dev. 2007;21(22):2845–9.
13. Imamura T, et al. Frequent co-expression of HoxA9 and Meis1 genes in infant acute lymphoblastic leukaemia with MLL rearrangement. Br J Haematol. 2002;119(1):119–21.
14. Pineault N, et al. Differential expression of Hox, Meis1, and Pbx1 genes in primitive cells throughout murine hematopoietic ontogeny. Exp Hematol. 2002;30(1):49–57.
15. Azcoitia V, et al. The homeodomain protein Meis1 is essential for definitive hematopoiesis and vascular patterning in the mouse embryo. Dev Biol. 2005;280(2):307–20.
16. Hisa T, et al. Hematopoietic, angiogenic and eye defects in Meis1 mutant animals. EMBO J. 2004;23(2):450–9.
17. Poolman RA, Gilchrist R, Brooks G. Cell cycle profiles and expressions of p21CIP1 AND P27KIP1 during myocyte development. Int J Cardiol. 1998;67(2):133–42.
18. Pasumarthi KB, et al. Targeted expression of cyclin D2 results in cardiomyocyte DNA synthesis and infarct regression in transgenic mice. Circ Res. 2005;96(1):110–8.
19. Gude N, et al. Akt promotes increased cardiomyocyte cycling and expansion of the cardiac progenitor cell population. Circ Res. 2006;99(4):381–8.
20. Sdek P, et al. Rb and p130 control cell cycle gene silencing to maintain the postmitotic phenotype in cardiac myocytes. J Cell Biol. 2011;194(3):407–23.

Intercellular Signaling in Cardiac Development and Disease: The NOTCH pathway

12

Guillermo Luxán, Gaetano D'Amato, and José Luis de la Pompa

Abstract

The heart is the first organ to develop in the embryo, and its formation is an exquisitely regulated process. Inherited mutations in genes required for cardiac development may cause congenital heart disease (CHD), manifested in the newborn or in the adult. Notch is an ancient, highly conserved signaling pathway that communicates adjacent cells to regulate cell fate specification, differentiation, and tissue patterning. Mutations in Notch signaling elements result in cardiac abnormalities in mice and humans, demonstrating an essential role for Notch in heart development. Recent work has shown that endocardial Notch activity orchestrates the early events as well as the patterning and morphogenesis of the ventricular chambers in the mouse and that inactivating mutations in the NOTCH pathway regulator MIND BOMB-1 (MIB1) cause left ventricular non-compaction (LVNC), a cardiomyopathy of poorly understood etiology. Here, we review these data that shed some light on the etiology of LVNC that at least in the case of that caused by *MIB1* mutations has a developmental basis.

Keywords

Ventricles • Trabeculation • Compaction • Cardiomyopathy • LVNC • NOTCH

G. Luxán, PhD • G. D'Amato, MSc • J.L. de la Pompa, PhD (✉)
Intercellular Signalling in Cardiovascular Development & Disease Laboratory, Centro Nacional de Investigaciones Cardiovasculares (CNIC), Melchor Fernández Almagro 3, 28029 Madrid, Spain
e-mail: jlpompa@cnic.es

T. Nakanishi et al. (eds.), *Etiology and Morphogenesis of Congenital Heart Disease*,
DOI 10.1007/978-4-431-54628-3_12

12.1   Introduction

Heart development begins in the mouse at embryonic day 7 (E7) when pre-cardiac precursor cells move forward bilaterally into the lateral plate mesoderm to form the cardiac crescent [14]. By E8, a linear heart tube is formed after folding of the mesodermal layers from both sides of the embryo, and the two main tissues of the heart are present, the endocardium inside and the outer myocardium. At E9.0, the heart loops and grows by addition of cells from both cardiac poles [20]. At E9.5, the heart divides afterward into developmental domains that would allow the formation of two regions, the valve and the chamber regions [24] (Fig. 12.1a).

We will describe in some detail the process of ventricular chamber development because it is the focus of this review. At E9.5, the process of trabeculation begins in the primitive ventricles. Trabeculae are highly organized sheets of cardiomyocytes that protrude toward the light of the ventricles [3]. Trabeculae are the first structural landmark of the developing ventricles [31]. They increase the ventricular surface and allow the myocardium to grow in the absence of coronary vessels that will be formed later. Trabeculae are important for contractibility, formation of the conduction system, and blood flow direction in the ventricle. Although trabeculae start to form first in the left ventricle and later in the right ventricle, there are no differences between these structures in the two ventricles. Formed trabeculae have no free ends and they make the ventricle appear like a sponge [31] (Fig. 12.1b). From E10.5 onward, the ventricular myocardium has two well-defined regions, morphologically and molecularly, the outer compact myocardium and the inner trabecular layer. The compact myocardium is less differentiated and shows a higher proliferation rate than the trabecular myocardium [31]. Differentiated cardiomyocytes will give rise to the working myocardium and the conduction system [31]. After completion of ventricular septation around E14.5, further growth and maturation of the ventricular chambers require a complete trabecular remodeling and reorganization into an apico-basal orientation that determines the shape of the ventricles (Fig. 12.1c). Trabeculae become compacted and as a result there is a significant change in the ratio of compact vs. trabecular myocardium in the ventricle. Trabeculae compaction is accompanied by the hypoxia-dependent invasion of the coronary vasculature from the epicardium into the compact layer of the myocardium [33]. As they compact, the intertrabecular recesses are transformed into capillaries of the rising coronary plexus. This is followed by a reorganization of the muscle fibers of the heart that organize in a three-layered spiral around the heart reflecting the twisting pattern of heart contraction [31]. Trabeculae stop proliferating while the proliferation of the compact zone sustains the growth of the chamber and the contribution of the compact layer to the heart becomes more significant than the one of the luminal trabeculations that will eventually disappear. How exactly this happens and what molecular mechanisms drive this process is unknown.

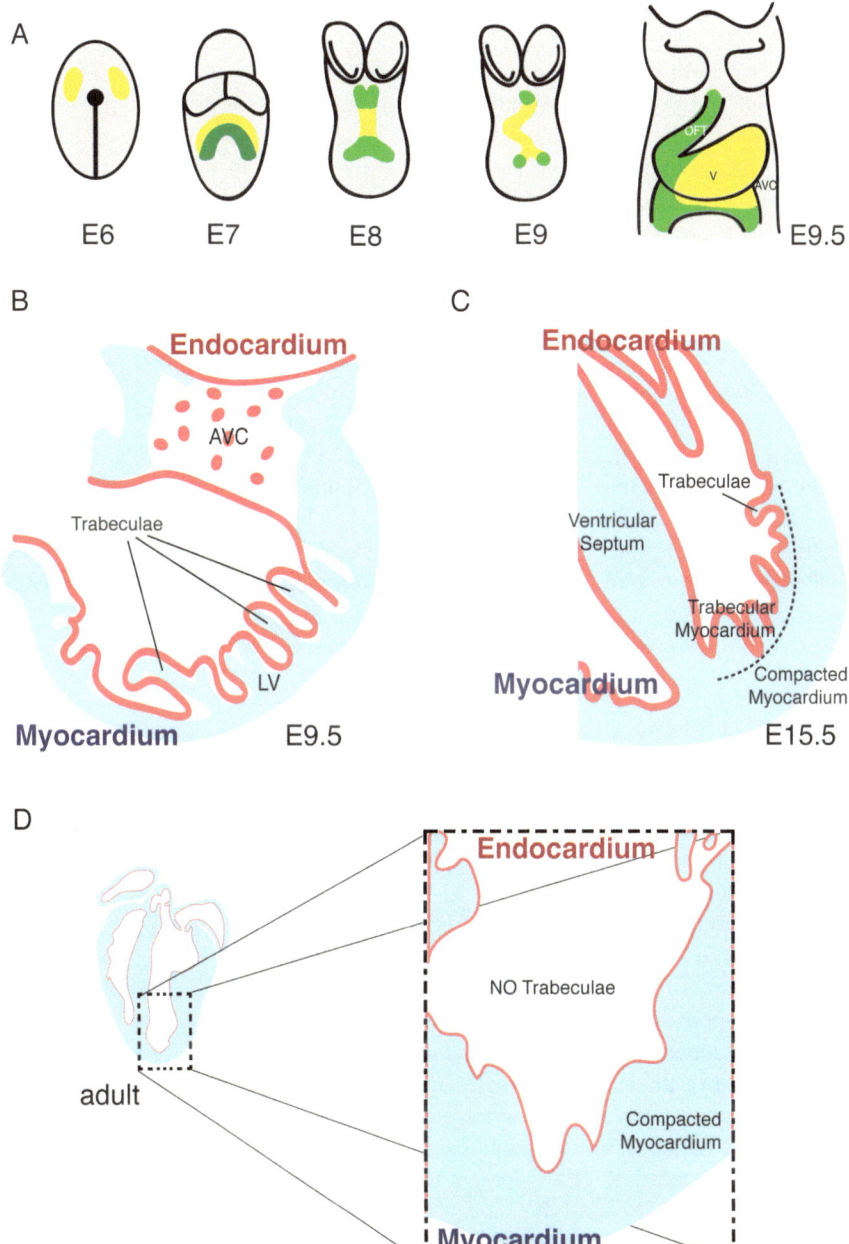

Fig. 12.1 Overview of early heart development. (**a**) Ventral views of the developing mouse embryo. At E7.0, cardiac progenitors (*yellow*) migrate to the center of the embryo to form the cardiac tube and at E7.5, two cardiac fields can be distinguished: the first heart field (FHF; *yellow*) and the second heart field (SHF; *green*). By E8.0, the cardiac tube is formed where endocardium and myocardium are already present. At E9.0 the tube loops and at E9.5, has four anatomically distinct regions: atrium, atrioventricular canal (AVC), ventricle (V), and outflow tract (OFT).

12.2   Left Ventricular Non-compaction (LVNC)

LVNC is a cardiomyopathy of poorly understood etiology that is characterized by prominent and excessive trabeculations with deep recesses in the ventricular wall [18]. LVNC was first diagnosed in 1984, in a 33-year-old woman assessed by echocardiography [10]. Since then, the heterogeneity of its symptoms might have left many LVNC cases undiagnosed [18], giving a lower estimation of its real impact as a cardiomyopathy. The improvement in imaging technologies and the advent of CMRI allow now a better diagnosis of this disease, and 10 years after the first description, LVNC was included in the World Health Organization (WHO) catalogue of cardiomyopathies [30]. In 2006, the American Heart Association included LVNC to its list of genetic cardiomyopathies caused by an arrest of the normal compaction process of the developing myocardium [23]. LVNC can be almost asymptomatic or manifest as depressed systolic function [7], accompanied by systemic embolism, malignant arrhythmias, heart failure, and sudden death [25], although its clinical manifestation and the age of the affected individual at symptom onset are variable. LVNC was shown to be caused by mutations in genes encoding mainly structural proteins of the cardiac muscle, cytoskeleton, nuclear membrane, and chaperone proteins. In general, familial forms of LVNC are transmitted as an autosomal-dominant trait [7]. The underlying molecular mechanisms that produce the structural abnormalities leading to LVNC are still unknown [34].

12.3   The NOTCH Signaling Pathway

Notch is a conserved signaling pathway that regulates cell fate specification, differentiation, and tissue patterning via local cell interactions [2]. There are four Notch receptors in mammals, Notch1–Notch4, which are expressed in different tissues and at different times during embryonic and adult life. Notch is a large type I transmembrane receptor that has two distinct domains, an extracellular domain (NECD) responsible for the interaction with the ligands and an intracellular domain (NICD) responsible for signal transduction (Fig. 12.2). NECD contains a varied number of tandem arrays of epidermal growth factor (EGF)-like repeats [2]. NICD is composed of the RAM23 domain and 6 ankyrin repetitions responsible for the interaction with RBPJK, the transcriptional effector of the pathway; two nuclear localization signals; a transcriptional activation domain only present in Notch1 and Notch2; and a terminal PEST domain that negatively regulates the stability of the receptor [21]. NECD and NICD are synthesized as a single polypeptide, which is directly cleaved after translation by a furin convertase in the Golgi. This first

Fig. 12.1 (continued) Endocardial cushions are formed in valvulogenic regions of the heart (*green*, OFT and *blue*, AVC) and trabeculae appear in the ventricles (**b**). At E15.5, trabeculae are undergoing compaction and there is a thick compact myocardium (**c**) that becomes very prominent in the adult ventricle that has a smooth surface (**d**)

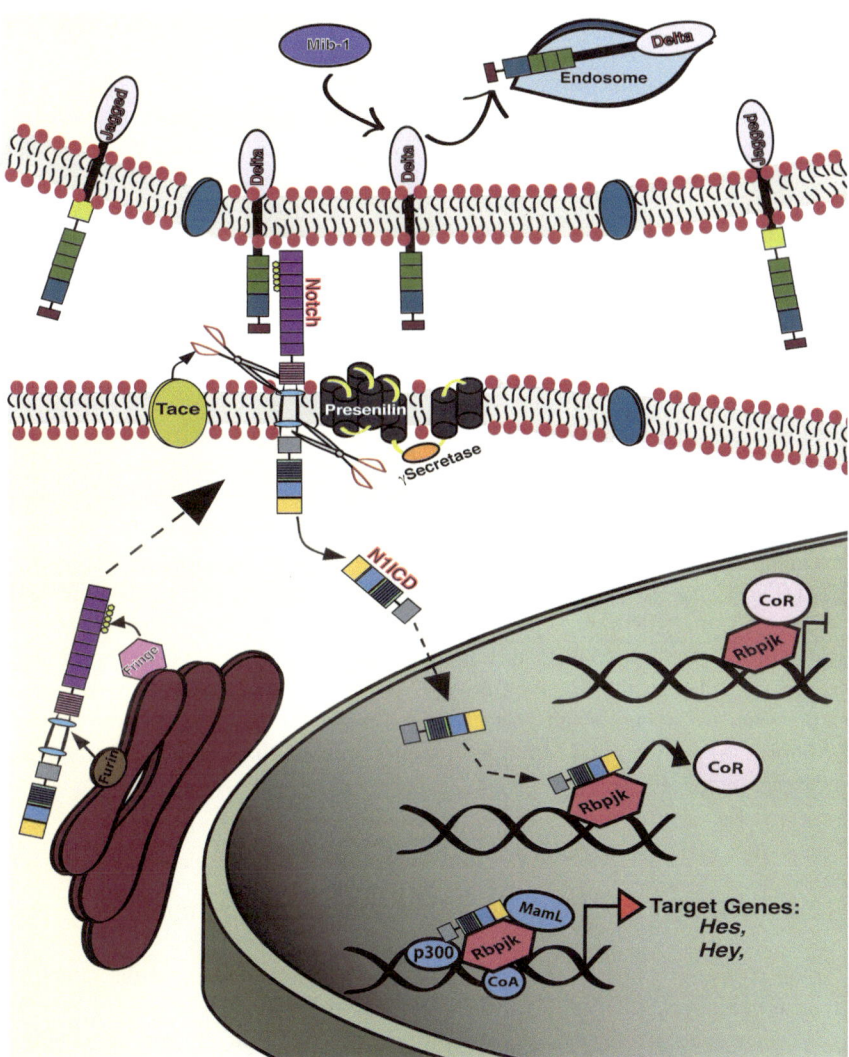

Fig. 12.2 The Notch signaling pathway. The Notch ligands Delta and Jagged are tethered to the membrane of the signaling cell. On their way to the plasmatic membrane, the Notch receptors are modified in their extracellular domains (NECD) by Fringe. Upon ligand-receptor interaction, Mib1 ubiquitylates the ligand, targeting it for endocytosis. Pulling of NECD allows the cleavage and release of NICD by γ-secretase. NICD translocates to the nucleus and forms a transcriptional activation complex together with RBPJK and MAML that activates target genes transcription, including *Hes* and *Hey* that encode transcriptional repressors

cleavage [6] is necessary to form the functional heterodimeric receptor that will be translocated to the membrane where the two domains are still associated by Ca^{2+}-dependent non-covalent bonds [28]. In mammals, the Notch ligands belong to the Delta-like (Dll1, Dll3, and Dll4) and Jagged (Jag1 and Jag2) families. The ligands are type I transmembrane proteins and their extracellular domains are composed by a repetition of EGF-like domains responsible for the interaction with the receptor and the DSL (Delta, Serrate, Lin12) domain (Fig. 12.2) [21]. The glycosyl-transferase Fringe can modify the EGF-like domain in the NECD of the receptor by adding O-fucose glycans [5] (Fig. 12.2). This modification determines which ligand preferentially binds to the receptor. There is evidence that the presence of Fringe shifts the activation balance toward Delta-like ligands at the expense of Jagged ligands when both molecules are expressed at the same time in the same tissue [4]. The E3 ubiquitin ligase Mind bomb1 (Mib1) regulates the endocytosis of the Notch ligands when bound to NECD, a prerequisite for Notch activation [16]. Mib1 binds to the two families of Notch ligands, and it is a point of regulation of the pathway in the signaling cell. Upon interaction of ligand and receptor, Mib1 targets by ubiquitylation the ligand for endocytosis [16]. The ligands are endocytosed, mechanically pulling NECD producing a conformational change that exposes a second cleavage site [27] allowing a metalloprotease of the ADAM family to cleave NECD [13] that is finally endocytosed into the signaling cell together with the ligand. Immediately after, γ-secretase/Presenilin cleaves the receptor in a third site liberating the NICD in the cytoplasm of the receiving cell [26] (Fig. 12.2). NICD translocates to the nucleus where it binds to the transcription factor CSL or RBPJK via by the RAM23 domain [32]. In the absence of Notch signaling activation, RBPJK is bound to nuclear corepressors that inhibit gene expression. Upon the interaction of NICD with RBPJK, the transcriptional repressors are released and transcriptional activators such as Master mind-like (Maml) [35] are recruited forming a protein complex that activates the expression of target genes (Fig. 12.2).

12.4   NOTCH in Ventricular Chamber Development

Notch pathway elements show a complex expression pattern in the developing mouse chambers. At E9.5, Mib1 is expressed in endocardium and myocardium, Dll4 in the endocardium especially at the base of the forming trabeculae, Jag1 in the myocardium, and N1ICD is only active at discrete sites where the trabeculae form [9]. Analysis of standard and endothelial-specific *Notch1* and *RBPJk* mutants shows that trabeculation is severely impaired and only primitive, poorly organized trabeculae can be observed in these mutants [12]. Molecular analysis indicates that during trabeculation, Notch modulates three different signals: EphrinB2 [1], expressed in the endocardium, is a direct Notch target. Nrg1 [15], also expressed in the endocardium, is an indirect Notch target, and its expression depends both of Notch and EphrinB2 [12]. Nrg1 induces the formation of the ventricular conduction system, a feature of ventricular maturation [29]. A third signaling pathway is Bmp10 [8], expressed in trabecular myocardium and whose expression is severely downregulated in Notch pathway mutants, suggesting that Notch is required to produce a signal that activates Bmp10 in the myocardium [12]. During these early

Fig. 12.3 Proposed mechanism of Notch function in trabeculation. (**a**, **b** detail) Dll4 activates Notch1 at the base of the developing trabeculae. N1ICD/RBPJK activates EphrinB2(Efnb2)/ EphB4 signaling in the endocardium, which in turn is required for Nrg1 expression in this tissue. Nrg1 activates ErbB2/ErbB4 signaling in myocardium to promote cardiomyocyte differentiation. In addition, Notch1 signaling in endocardium is required for Bmp10 expression and therefore proliferation of trabecular cardiomyocytes (Modified from [12])

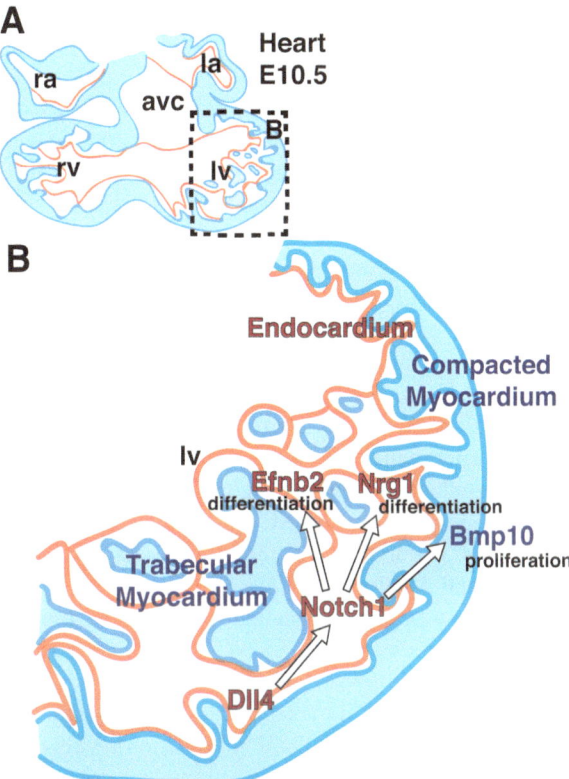

stages, the ligand Dll4 is the main Notch activator in the endocardium [12] (Fig. 12.3).

To precisely determine the contribution of Notch ligands to ventricular chamber development, we bred mice bearing a conditional allele of *Mib1* with myocardium-specific Cre driver line *cTnT-Cre* [19]. Histological examination at E16.5 revealed that *Mib1flox;cTnT-Cre* mutants had a dilated heart with a thin compact myocardium and large, non-compacted trabeculae, protruding toward the ventricular lumen. Morphological and functional analysis using echocardiography and cardiac magnetic resonance imaging (CMRI) showed prominent trabeculations, deep intertrabecular recesses, and a significantly reduced ejection fraction in mutant mice [22]. The hearts of these mice had a ratio of non-compacted to compacted myocardium (non-compaction index, NC) of 2.0, a feature diagnostic of LVNC in humans [11, 17]. These features are all strongly reminiscent of LVNC, establishing *Mib1flox;cTnT-Cre* mice as the first animal model of LVNC [22]. Analysis of chamber markers in E15.5 *Mib1flox;cTnT-Cre* mice revealed expansion of various compact myocardium markers (*Hey2*, *Tbx20* and *n-myc*) to the trabeculae and reduced expression of the trabecular markers *Anf*, *Bmp10*, and *Cx40*. These markers were normally expressed at earlier stages, suggesting that maintenance of trabecular maturation and patterning is impaired in *Mib1^{flox}*;*cTnT-Cre* mutants. Likewise, lost or reduced *Hey1*, *Hey3*, and *EphrinB2* expression in the vessels of the compact myocardium suggested that coronary artery development was defective. Proliferation analysis revealed increased proliferation of trabecular cardiomyocytes in the hearts of E15.5 *Mib1^{flox}*;*cTnT-Cre* embryos, suggesting this as the cause of the enlarged, non-compacted trabeculae in these mutants. RNA sequencing analysis of E14.5 *Mib1^{flox}*;*cTnT-Cre* mutant ventricles identified altered expression of 315 genes (132 upregulated and 183 downregulated). The expression of genes involved in the differentiation of cardiac endothelium/endocardium and cardiomyocytes and coronary vasculogenesis was altered. RNA-Seq data also confirmed the in situ hybridization analysis and demonstrated that *Mib1* inactivation in the myocardium disrupts the differentiation and maturation of cardiac endothelial cells and cardiomyocytes. This may impact in turn the process of ventricular maturation and compaction. Genetic ablation of *Mib1* in the myocardium thus leads to LVNC cardiomyopathy in mice by arresting ventricular trabeculae maturation and compaction and increasing cardiomyocyte proliferation during fetal development [22].

We next examined whether mutations in the human *MIB1* homologue were associated with clinical LVNC and identified two mutations in a cohort of 100 Southern European LVNC cases; one was a heterozygous G to T transversion of nucleotide 2827 in exon 20 (causing a change in amino acid 943 from valine to phenylalanine, the p.Val943Phe mutation). Val943 is located within a region that mediates protein-protein interactions and constitutes the site of MIB1 ubiquitin ligase activity. The second mutation was a heterozygous C to T transition of nucleotide 1587 in *MIB1* exon 11 (causing a premature stop codon instead of arginine at position 530 in the MIB1 ankyrin repeats region, the p.Arg530X mutation). These two mutations were tracked back through three and two generations of LVNC-affected individuals, revealing co-segregation with LVNC, confirming the hereditary nature of the disease [22].

In silico modeling suggested that MIB1 is a homodimer where the p.Val943Phe mutant will alter the alignment of the ring finger domains in heterodimers formed by wild-type and mutant p.Val943Phe MIB1, as well as the angle of interaction with JAG1 of the p.Val943Phe wild-type dimers. This possibility was supported by co-immunoprecipitation experiments with tagged wild-type MIB1 and mutant MIB1 isoforms co-transfected in pair-wise combinations into HEK293 cells. The effect of the *MIB1* mutations on Jag1 ubiquitylation was tested also in HEK293 cells. Jag1 ubiquitylation was strongly reduced in cells co-transfected with wild-type MIB1 plus p.Val934Phe or wild-type MIB1 plus the p.Arg530X mutant [22].

Lastly, to study the effect of these mutant *MIB1* variants in vivo, their mRNAs were microinjected into zebra fish embryos expressing GFP in the developing myocardium. Examination of 72 hpf larvae injected with *p.Val943Phe* or *p.Arg530X* mRNAs revealed severely disrupted embryonic development, disrupted cardiac looping, and kinked tail, typical of defective Notch activity. To complement these findings, we cocultured Jag1-HEK293 cells expressing MIB1 variants with *Notch1*-expressing HEK293 cells co-transfected with *a* Notch luciferase reporter. Jag1-HEK293 cells expressing wild-type *MIB1* increased Notch reporter activity, whereas the p.Val934Phe and p.Arg530X mutant forms reduced reporter activity, indicating that both human mutations disrupt Notch signaling [22].

Accumulated data suggested that both *MIB1* mutations result in loss of MIB1 WT function. In the case of p.Arg530X, this would be due to haploinsufficiency caused by insufficient synthesis of WT MIB1 protein; in the case of p.Val934Phe, this would be due to a dominant-negative effect of the mutant protein, titrating down the amount of functional WT MIB1 dimers through heterotypic or homotypic interactions. In both cases, loss of MIB1 function leads to disease inherited in an autosomal-dominant fashion.

As to the role of Mib1 in ventricular chamber development, we have proposed that myocardial Mib1 activity enables Jag1-mediated activation of Notch1 in the endocardium, to sustain trabeculae patterning, maturation, and compaction [22] and (Fig. 12.4). Abrogation of Mib1-mediated signaling in the myocardium disrupts trabeculae maturation and patterning, arresting the development of chamber myocardium and resulting in LVNC (Fig. 12.4). The expansion of compact zone markers to the trabeculae of E15.5 $Mib1^{flox}$;*cTnT-Cre* mutants further suggests that trabeculae patterning and maturation is impaired (Fig. 12.4). In addition, the disruption of coronary vessel markers in $Mib1^{flox}$;*cTnT-Cre* mutants indicates that abrogation of myocardial-endocardial Notch signaling indirectly impairs coronary vessel formation.

12.5   Future Directions and Clinical Implications

The data reviewed here establish the causal role of NOTCH dysregulation in LVNC, a congenital cardiomyopathy that results from a developmental arrest in ventricular maturation and myocardial compaction. These findings will lead to a better diagnosis and risk stratification, allowing timely intervention in LVNC-associated complications. In addition, since other NOTCH pathway elements may be involved in LVNC, they could serve as novel diagnostic or therapeutic disease targets. Further research is required to determine the molecular mechanisms and regulatory interactions underlying NOTCH function in ventricular chamber development using genetically modified mouse models and the full implication of altered NOTCH signaling in human LVNC.

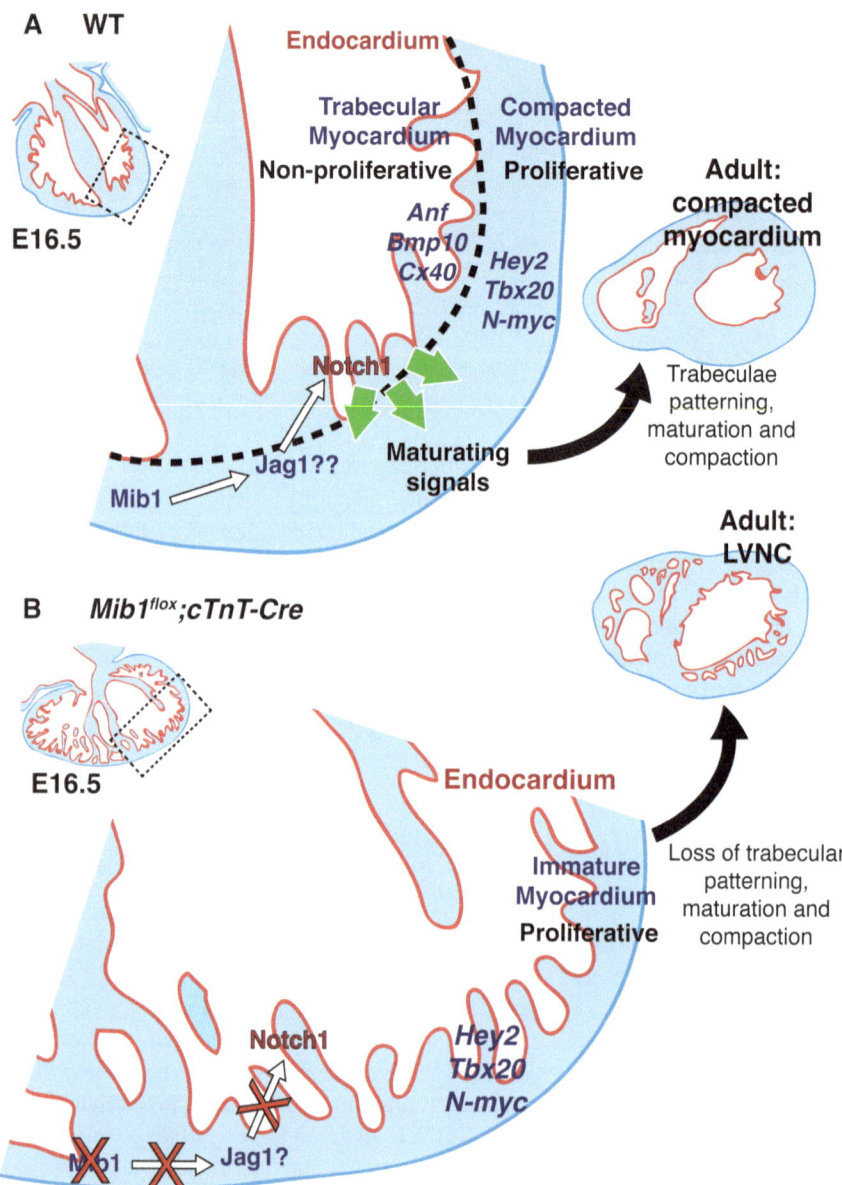

Fig. 12.4 Proposed mechanism of Notch function in trabecular maturation and compaction. (**a**) In wild-type (WT) embryos, ubiquitylation of Jag1 (and another ligand?) by Mib1 in the myocardium allows Notch1 activation in the endocardium. N1ICD is required to sustain trabecular patterning, maturation, and compaction. The compact myocardium (expressing *Hey2*, *Tbx20* and *n-myc*) proliferates actively, unlike the trabecular myocardium (expressing *Anf*, *Bmp10* and *Cx40*). Notch-dependent chamber maturation leads to compacted ventricular myocardium in the adult mouse. (**b**) In *Mib1^flox*; *cTnT-Cre* mutants, Notch1 activity is impaired and compact myocardium markers (*Hey2*, *Tbx20* and *n-myc*) are consequently expanded to the trabeculae, which remain abnormally proliferative. The resulting disruption of trabecular patterning, maturation, and compaction manifests as LVNC (Modified from [22])

Acknowledgments We apologize to those authors whose work was not cited. J.L. de la Pompa is funded by grants SAF2013-45543, RD12/0042/0005 (RIC), and RD12/0019/0003 (TERCEL) from the Spanish Ministry of Economy and Competitiveness (MINECO) and grant FP7-ITN 215761 (NotchIT) from the European Commission. G. Luxán had a PhD fellowship from the MINECO (FPI Program, ref. BES-2008-002904) and G. D'Amato holds a PhD fellowship associated with grant FP7-ITN 215761 (NotchIT). The CNIC is supported by the MINECO and the Pro-CNIC Foundation.

References

1. Adams RH, Wilkinson GA, Weiss C, et al. Roles of ephrinB ligands and EphB receptors in cardiovascular development: demarcation of arterial/venous domains, vascular morphogenesis, and sprouting angiogenesis. Genes Dev. 1999;13:295–306.
2. Artavanis-Tsakonas S, Rand MD, Lake RJ. Notch signaling: cell fate control and signal integration in development. Science. 1999;284:770–6.
3. Ben-Shachar G, Arcilla RA, Lucas RV, et al. Ventricular trabeculations in the chick embryo heart and their contribution to ventricular and muscular septal development. Circ Res. 1985;57:759–66.
4. Benedito R, Roca C, Sorensen I, et al. The notch ligands Dll4 and Jagged1 have opposing effects on angiogenesis. Cell. 2009;137:1124–35.
5. Blair SS. Notch signaling: fringe really is a glycosyltransferase. Curr Biol. 2000;10:R608–12.
6. Blaumueller CM, Qi H, Zagouras P, et al. Intracellular cleavage of Notch leads to a heterodimeric receptor on the plasma membrane. Cell. 1997;90:281–91.
7. Captur G, Nihoyannopoulos P. Left ventricular non-compaction: genetic heterogeneity, diagnosis and clinical course. Int J Cardiol. 2010;140:145–53.
8. Chen H, Shi S, Acosta L, et al. BMP10 is essential for maintaining cardiac growth during murine cardiogenesis. Development. 2004;131:2219–31.
9. Del Monte G, Grego-Bessa J, Gonzalez-Rajal A, et al. Monitoring Notch1 activity in development: evidence for a feedback regulatory loop. Dev Dyn. 2007;236:2594–614.
10. Engberding R, Bender F. Identification of a rare congenital anomaly of the myocardium by two-dimensional echocardiography: persistence of isolated myocardial sinusoids. Am J Cardiol. 1984;53:1733–4.
11. Finsterer J, Stollberger C. Heterogenous myopathic background of left ventricular hypertrabeculation/noncompaction. Am J Med Genet A. 2004;131:221; author reply 222-223.
12. Grego-Bessa J, Luna-Zurita L, Del Monte G, et al. Notch signaling is essential for ventricular chamber development. Dev Cell. 2007;12:415–29.
13. Hartmann D, De Strooper B, Serneels L, et al. The disintegrin/metalloprotease ADAM 10 is essential for Notch signalling but not for alpha-secretase activity in fibroblasts. Hum Mol Genet. 2002;11:2615–24.
14. Harvey RP. Patterning the vertebrate heart. Nat Rev Genet. 2002;3:544–56.

15. Hertig CM, Kubalak SW, Wang Y, et al. Synergistic roles of neuregulin-1 and insulin-like growth factor-I in activation of the phosphatidylinositol 3-kinase pathway and cardiac chamber morphogenesis. J Biol Chem. 1999;274:37362–9.
16. Itoh M, Kim CH, Palardy G, et al. Mind bomb is a ubiquitin ligase that is essential for efficient activation of Notch signaling by Delta. Dev Cell. 2003;4:67–82.
17. Jenni R, Oechslin E, Schneider J, et al. Echocardiographic and pathoanatomical characteristics of isolated left ventricular non-compaction: a step towards classification as a distinct cardiomyopathy. Heart. 2001;86:666–71.
18. Jenni R, Oechslin EN, Van Der Loo B. Isolated ventricular non-compaction of the myocardium in adults. Heart. 2007;93:11–5.
19. Jiao K, Kulessa H, Tompkins K, et al. An essential role of Bmp4 in the atrioventricular septation of the mouse heart. Genes Dev. 2003;17:2362–7.
20. Kelly RG, Buckingham ME. The anterior heart-forming field: voyage to the arterial pole of the heart. Trends Genet. 2002;18:210–6.
21. Kopan R. Notch: a membrane-bound transcription factor. J Cell Sci. 2002;115:1095–7.
22. Luxan G, Casanova JC, Martinez-Poveda B, et al. Mutations in the NOTCH pathway regulator MIB1 cause left ventricular noncompaction cardiomyopathy. Nat Med. 2013;19:193–201.
23. Maron BJ, Towbin JA, Thiene G, et al. Contemporary definitions and classification of the cardiomyopathies: an American Heart Association Scientific Statement from the Council on Clinical Cardiology, Heart Failure and Transplantation Committee; Quality of Care and Outcomes Research and Functional Genomics and Translational Biology Interdisciplinary Working Groups; and Council on Epidemiology and Prevention. Circulation. 2006;113:1807–16.
24. Moorman AF, Christoffels VM. Cardiac chamber formation: development, genes, and evolution. Physiol Rev. 2003;83:1223–67.
25. Oechslin EN, Attenhofer Jost CH, Rojas JR, et al. Long-term follow-up of 34 adults with isolated left ventricular noncompaction: a distinct cardiomyopathy with poor prognosis. J Am Coll Cardiol. 2000;36:493–500.
26. Okochi M, Steiner H, Fukumori A, et al. Presenilins mediate a dual intramembranous gamma-secretase cleavage of Notch-1. EMBO J. 2002;21:5408–16.
27. Parks AL, Klueg KM, Stout JR, et al. Ligand endocytosis drives receptor dissociation and activation in the Notch pathway. Development. 2000;127:1373–85.
28. Rand MD, Grimm LM, Artavanis-Tsakonas S, et al. Calcium depletion dissociates and activates heterodimeric notch receptors. Mol Cell Biol. 2000;20:1825–35.
29. Rentschler S, Zander J, Meyers K, et al. Neuregulin-1 promotes formation of the murine cardiac conduction system. Proc Natl Acad Sci U S A. 2002;99:10464–9.
30. Richardson P, Mckenna W, Bristow M, et al. Report of the 1995 World Health Organization/International Society and Federation of Cardiology Task Force on the Definition and Classification of cardiomyopathies. Circulation. 1996;93:841–2.
31. Sedmera D, Pexieder T, Vuillemin M, et al. Developmental patterning of the myocardium. Anat Rec. 2000;258:319–37.
32. Tamura K, Taniguchi Y, Minoguchi S, et al. Physical interaction between a novel domain of the receptor Notch and the transcription factor RBP-J kappa/Su(H). Curr Biol. 1995;5:1416–23.
33. Tao J, Doughman Y, Yang K, et al. Epicardial HIF signaling regulates vascular precursor cell invasion into the myocardium. Dev Biol. 2013;376:136–49.
34. Towbin JA. Left ventricular noncompaction: a new form of heart failure. Heart Fail Clin. 2010;6:453–69. viii.
35. Wu L, Aster JC, Blacklow SC, et al. MAML1, a human homologue of Drosophila mastermind, is a transcriptional co-activator for NOTCH receptors. Nat Genet. 2000;26:484–9.

The Epicardium in Ventricular Septation During Evolution and Development

13

Robert E. Poelmann, Bjarke Jensen, Margot M. Bartelings, Michael K. Richardson, and Adriana C. Gittenberger-de Groot

Abstract

The epicardium has several essential functions in development of cardiac architecture and differentiation of the myocardium in vertebrates. We uncovered a novel function of the epicardium in species with partial or complete ventricular septation including reptiles, birds and mammals. Most reptiles have a complex ventricle with three cava, partially separated by the horizontal and vertical septa. Crocodilians, birds and mammals, however, have completely separated left and right ventricles, a clear example of convergent evolution. We have investigated the mechanisms underlying epicardial involvement in septum formation in

R.E. Poelmann (✉)
Department of Anatomy and Embryology, Leiden University Medical Center, PO Box 9600, 2300RC Leiden, The Netherlands

Department of Cardiology, Leiden University Medical Center, PO Box 9600, 2300RC Leiden, The Netherlands

Institute of Biology Leiden (IBL), Leiden University, Sylvius Laboratory, Sylviusweg 72, 2333BE Leiden, The Netherlands
e-mail: R.E.Poelmann@lumc.nl

B. Jensen
Department Anatomy, Embryology and Physiology, AMC, Amsterdam, The Netherlands

Department of Bioscience-Zoophysiology, Aarhus University, Aarhus, Denmark

M.M. Bartelings
Department of Anatomy and Embryology, Leiden University Medical Center, PO Box 9600, 2300RC Leiden, The Netherlands

M.K. Richardson
Institute of Biology Leiden (IBL), Leiden University, Sylvius Laboratory, Sylviusweg 72, 2333BE Leiden, The Netherlands

A.C.G.-d. Groot
Department of Cardiology, Leiden University Medical Center, PO Box 9600, 2300RC Leiden, The Netherlands

© The Author(s) 2016
T. Nakanishi et al. (eds.), *Etiology and Morphogenesis of Congenital Heart Disease*,
DOI 10.1007/978-4-431-54628-3_13

115

embryos. We find that the primitive ventricle of early embryos becomes septated by folding and fusion of the anterior ventricular wall, trapping epicardium in its core. This 'folding septum', as we propose to call it, develops in lizards, snakes and turtles into the horizontal septum and, in the other taxa studied, into the folding part of the interventricular septum. The vertical septum, indistinct in most reptiles, arises in crocodilians and pythonids at the posterior ventricular wall. It is homologous to the inlet septum in mammals and birds. Eventually, the various septal components merge to form the completely septated heart. In our attempt to discover homologies between the various septum components, we draw perspectives to the development of ventricular septal defects in humans.

Keywords

Evolution and development • Homology • Ventricular septum • Congenital heart disease • Epicardium

13.1   Introduction

Evolution of full division of the heart into left and right chambers by septa started with the atrium in amphibians, followed by the ventricle in amniotes. Full ventricular septation evolved independently in the lineages of the archosaurs, whose extant representatives are the birds and crocodilians and mammals [1]. Not only is evolution and development of ventricular septation of considerable biological interest, it is also of clinical relevance to the understanding of many types of ventricular septal defects in humans [2].

We demonstrated that part of the ventricular septum depends on interactions between myocardium and the epicardium including the epicardium-derived cells (EPDCs) for its development [3]. The septum will develop abnormally if the epicardium is disturbed [4, 5]. Ventricular development, starting with a primitive common ventricular tube, leading to separation into the left and right ventricle, involves complex mechanisms including the ventricular inflow and outflow compartment [6, 7]. Most reptile groups show a more primitive pattern comprising a horizontal and a vertical septum separating the ventricle into three interconnected cavities. How the primitive reptilian pattern was modified into the complete septum is a matter of debate. Here we report the functional role of the epicardium in ventricular septation in lizard, snake, turtle, chicken, mouse and human using descriptive and experimental approaches including quail-chicken chimeras by transplantation of quail proepicardial organ into the pericardial cavity of chicken [4]. Furthermore, epicardium-deficient animal models such as the podoplanin knockout mouse and epicardial ablation experiments in chicken embryos were investigated [4, 5]. With DiI labelling [8] of the myocardial surface in chicken embryos, the outgrowth of the cardiac compartments was analysed. *Tbx5* expression showed gradients along the cardiac tube and so may be a useful marker for regionalisation of the heart [9]. For experimental details and full material and method description, we refer to [10].

13.2   Septum Components in the Completely Septated Heart

Varying terminology is used for components of the ventricular septum and their respective boundaries; we adopted the following. The posterior component of the interventricular septum, between the left and right atrioventricular junctions, is the *inlet septum*. We propose to use the term 'folding septum' (newly defined here) for the anterior component. The septal band is a muscular ridge on the right ventricular septal surface between the inlet and folding septum, extending as the moderator band towards the right ventricular free wall. In the completely septated heart, the aortopulmonary septum is the last component that closes the interventricular communication, but has not been specifically studied here.

13.3   The Presence of the Epicardium in Amniotes

Among other species we examined embryos of Macklot's python (*Liasis mackloti*), European pond turtle *(Emys orbicularis)* and Chinese soft-shell turtle *(Pelodiscus sinensis)*. The presence of epicardium on the outer face of the myocardial heart tube is confirmed. A pronounced subepicardium in the inner curvature of the looping heart tube at the site of the bulboventricular fold is confirmed, here referred to as epicardial cushion, being almost as elaborate as the adjacent endocardial atrioventricular and outflow tract cushions (Fig. 13.1a). In chicken (Fig. 13.2a, b) and mouse embryos, the presence of the (sub)epicardium is less extensive compared to the turtle. In human embryos at Carnegie stages, 11–15 (3.6–7 mm), the surface of the heart is closely covered by the epicardial epithelium, whereas the inner curvature harbours an extensive epicardial cushion (Fig. 13.1b) comparable to the turtle.

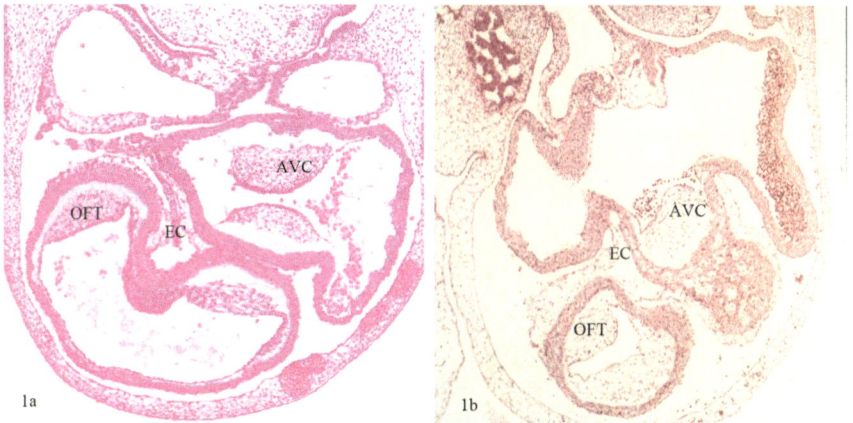

Fig 13.1 (**a, b**) Sections of turtle (*Emys orbicularis*) (**a**) and human embryo (**b**) of comparable developmental stages showing the epicardial cushion (EC) in the inner curvature of the looping heart tube. *AVC* atrioventricular cushion, *OFT* outflow tract cushion

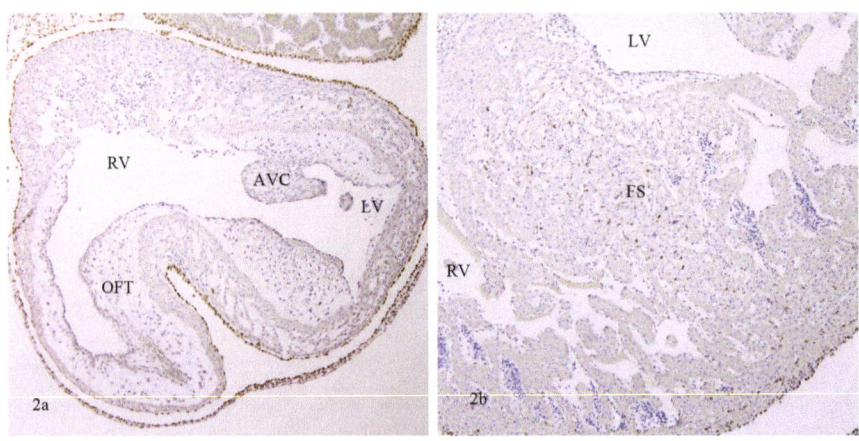

Fig 13.2 (**a**) Chicken embryo stage 27 showing the start of the folding septum between OFT and AVC, stained for the Wilm's tumour antigen (WT1) marking the epicardium (*brown*). (**b**) Chicken embryo stage 31 with an elaborate folding septum (FS) showing dispersed EPDCs (*brown*) between the cardiomyocytes. *LV* left ventricle, *RV* right ventricle

13.4   The Epicardium in the Avian Heart

To investigate the fate of the epicardium during ballooning of the ventricles and subsequent folding of the septum, we constructed quail-chicken chimeras. An isochronic quail proepicardial organ (PEO) including a small piece of adjacent liver tissue to provide endothelial cells was transplanted into the pericardial cavity of HH15–17 chick embryos in an anterior position. Antibody staining revealed that quail EPDCs and endothelial cells were present in the anterior folding septum, but not in the dorsal inlet septum (Fig. 13.3a, b). The quail epicardial sheet dispersed into individual cells that became distributed between the cardiomyocytes in later stages, mostly in the core of the folding septum. In a second set of chimeras, the quail PEO was positioned more dorsally, providing for quail EPDCs and endothelial cells on the posterior ventricular surface, subsequently migrating into the inlet septum including the septal band, without crossing over to the anterior folding septum. This indicates the inlet septum as a separate component. Furthermore, an epicardial sheet, characteristic for the folding septum, was not encountered. As a consequence, expansion of the ventricles results in anterior folding of the ventricular wall, but not of the posterior wall, probably because of the physical constraints imposed by attachment of the heart to the dorsal body wall [10].

To study septum folding during ballooning of the ventricle, fluorescent DiI was tattooed (HH15–17) onto the ventral myocardial surface and embryos were sacrificed between HH22 and 33. DiI was applied on the surface of the future fold and the dye fragmented during further development into a left-sided part, incorporated into the left side of the folding septum and a right-sided part. This

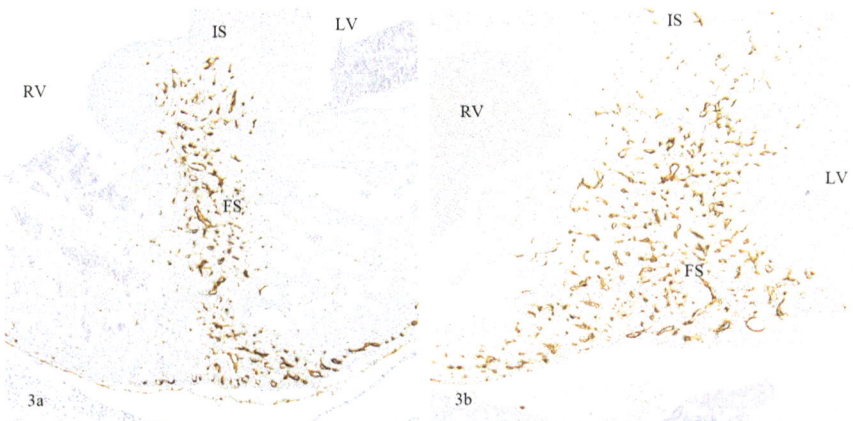

Fig 13.3 (**a**, **b**) Quail-chicken chimeras from the anterior position. The folding septum (FS) harbours quail-derived endothelial cells stained with the QH1 antibody (*brown*), whereas the inlet septum (IS) is virtually negative

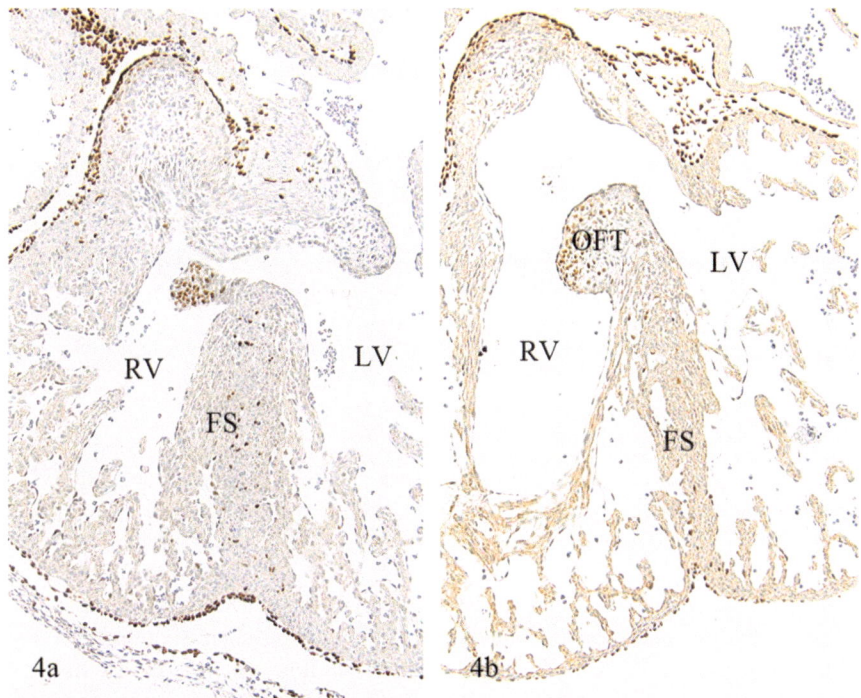

Fig 13.4 Podoplanin wild type (**a**) and –/– mouse (**b**), embryonic day 12.5. Note the diminished number of epicardial cells and EPDCs (*brown* in this WT1 staining) and the thin folding septum in the mutant

Fig 13.5 DiI-labelled
chicken embryo stage 31 with
a thin strip of the dye in the
folding septum adjacent to the
right ventricular surface

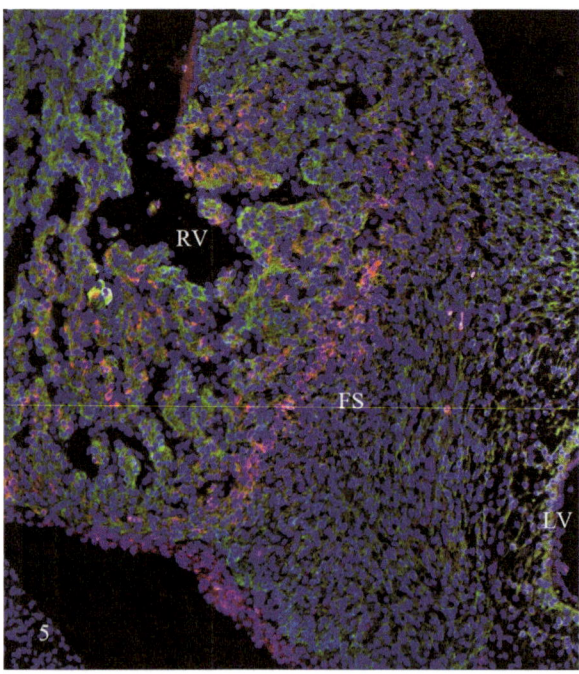

indicates a longitudinally directed morphogenetic expansion of the right ventricular wall compared to a transverse expansion of the left ventricle. At stage 31 the dye was completely embedded in the folding septum as a narrow fluorescent strip (Fig. 13.5) markedly close to the right ventricular face of the septum, indicative of a less elaborate right ventricular contribution to the septum [10].

13.5   Disturbance of the Epicardium

Deficient septation is known in several mouse mutants and we have analysed the podoplanin mutant mouse. Podoplanin has multiple roles in the maintenance of the epithelium and is expressed in the mesothelium of the body cavities including the epicardium. The knockout mouse presents with an underdeveloped PEO and abnormal epicardial covering resulting in only a few EPDCs [5]. The phenotype shows multiple malformations including an atrioventricular septum defect. At embryonic day 12.5, the myocardium and the folding septum are very thin, and the inlet septum is spongy. The diminutive septum is nearly devoid of EPDCs suggestive of an instructive role for these cells in septation (Fig. 13.4a, b).

13.6   Septum Components in Reptilian Hearts

The horizontal septum of developing reptiles is found in comparable stages and location, separating ventricular cavities and harbouring an epicardial sheet similar to the folding septum in mammals and birds. We prefer to address this structure as 'folding septum'. The presence and extent of the vertical septum (homologous to inlet septum) differs among reptiles, being virtually absent in turtles and most non-crocodilian reptiles, but more prominent in varanids and pythonidae [11]. In the python myocardial apical trabeculations traverse the ventricle to partly separate the cavum pulmonale from the cavum arteriosum.

13.7   Tbx5 Expression Patterns

The transcription factor Tbx5 is implicated in full ventricular septation in mammals and is expressed strongly in the left part of the cardiac tube, with expression declining towards the right side [9]; we confirmed this expression pattern also in early stages of the copperhead rat snake, *Coelognathus*. Slightly later in the development, we show that embryonic turtle and python show expression of Tbx5 that declines at the folding septum, but not over the dorsal wall, where a sharp decline was found only near the outflow tract. In the chicken the *Tbx5* mRNA gradient identifies the right ventricle (weak expression) from the outflow tract (negative expression). Furthermore, a *Tbx5* decline is found at the folding septum showing only strong left-sided expression, whereas *Tbx5* expression is present on both sides of the inlet septum including the septal band. Thus, in the chicken the folding and inlet septa are differentially identified by Tbx5 gradients. In the mouse two Tbx5 gradients identify the inlet and folding septum. Protein expression is strongest in the trabeculations of the left ventricle, weaker in the right ventricular inlet including septal band and weakest to negative in the right ventricular outflow.

13.8   Discussion

All amniotes exhibit a folding (horizontal) septum, and here we showed the involvement of the epicardium in fusion of the two opposing myocardial walls. In mammals and birds, the folding septum forms the anterior part of the definitive septum. Reducing the size of the PEO leads to diminished or retarded covering of the myocardium and diminished production of EPDCs [4, 5, 12]. The latter can differentiate into smooth muscle cells of the coronary vessels and into perivascular and interstitial fibroblasts [13]. It is evident that mechanical or genetic interference with the epicardium or EPDCs not only disturbs coronary vascularisation but can also strongly influence cardiomyocyte differentiation and ventricular septation.

Different septal components [6, 7] have been identified in the completely septated hearts which have been the subject of continued debate also involving exploration of the position of central muscular ventricular septal defects and the connection of the tricuspid valve tendinous cords related to either septal components. It was generally agreed that the septal band belonged to the primary septum being the anterior or, as we now refer to it, the folding component of the septum, but we demonstrated that it belongs to the posterior inlet septum. We newly identified the border between inlet and folding component using several criteria: (1) the atrioventricular cushions connect to the inlet component including the septal band; (2) the tip of the septal outflow tract cushion originates where the folding septum merges with the inlet septum; (3) quail cells derived from posterior PEO chimeras do not crossover from inlet septum including septal band to the folding septum, and vice versa in anterior chimeras, there is no crossover from folding to inlet septum; and (4) Tbx5 expression, in contrast to earlier publications [9, 14], is strong on the left and right side of the posteriorly located inlet septum including the septal band, whereas only left-sided positivity is found at the anterior folding septum. From this we postulate that the avian and mammalian embryonic ventricle (homologous to the cavum dorsale of reptiles, which is the cavum arteriosum and venosum combined) give rise to both the left ventricle and the inlet of the right ventricle. This is an indication that the inlet septum originates in its entirety from the wall of the primitive ventricle, the cavum dorsale. In the right ventricle, the boundary of the inlet septum is provided by the septal band, which is not present in the left ventricle, leaving here the boundary between the inlet and folding septum less well determined. Our new model for separating the ventricles implicates more than one component and has consequences for understanding the development of central muscular ventricular septal defects. (Note: The development of the membranous and the outflow tract septum has not been specifically addressed in this study) Interestingly, elephants and some relatives including seacows show a very deep anterior interventricular sulcus [10]. The anatomy of the right ventricular septal band and the attachment of the tricuspid chordae tendineae suggest a diminished folding mechanism resulting in retention of an early embryonic state, also known as neoteny.

In conclusion, we have explored an evo-devo context for the hitherto overlooked role for the epicardium in septation and have clarified complex homologies in amniotes of the ventricular septum to understand clinical disorders of the heart in humans.

Acknowledgements B.J. is supported by the Danish Council for Independent Research I Natural Sciences and M.K.R. is supported by AgentschapNL, Smartmix SSM06010.

References

1. Holmes EB. A reconsideration of the phylogeny of the tetrapod heart. J Morphol. 1975;147:209–28.
2. Jacobs JP, et al. Congenital heart surgery nomenclature and database project: atrial septal defect. Ann Thorac Surg. 2000;69:S18–24.
3. Gittenberger-de Groot AC, et al. Epicardium-derived cells contribute a novel population to the myocardial wall and the atrioventricular cushions. Circ Res. 1998;82:1043–52.
4. Lie-Venema H, et al. Origin, fate, and function of epicardium-derived cells (EPCDs) in normal and abnormal cardiac development. Sci World J. 2007;7:1777–98.
5. Mahtab EAF, et al. Cardiac malformations and myocardial abnormalities in podoplanin knockout mouse embryos: correlation with abnormal epicardial development. Dev Dyn. 2008;237:847–57. doi:10.1002/dvdy.21463.
6. Wenink ACG. Embryology of the ventricular septum. Separate origin of its components. Virchows Arch. 1981;390:71–9.
7. van Mierop LH, Kutsche LM. Development of the ventricular septum of the heart. Heart Vessels. 1985;1:114–9.
8. Darnell DK, et al. Dynamic labeling techniques for fate mapping, testing cell commitment, and following living cells in avian embryos. Methods Mol Biol. 2000;135:305–21.
9. Koshiba-Takeuchi K, et al. Reptilian heart development and the molecular basis of cardiac chamber evolution. Nature. 2009;461:95–8. doi:10.1038/nature08324.
10. Poelmann RE et al. Evolution and development of ventricular septation in the amniote heart. PLoS ONE 2014;9(9):e106569. doi:10.1371/journal.pone.0106569
11. Jensen B, et al. Structure and function of the hearts of lizards and snakes. Biol Rev Camb Philos Soc. 2014;89:302–36.
12. Gittenberger-de Groot AC, et al. Epicardial outgrowth inhibition leads to compensatory mesothelial outflow tract collar and abnormal cardiac septation and coronary formation. Circ Res. 2000;87:969–71.
13. Gittenberger-de-Groot AC, et al. The arterial and cardiac epicardium in development, disease and repair. Differentiation. 2012;84:41–53. doi:10.1016/j.diff.2012.05.002.
14. Greulich F, Rudat C, Kispert A. Mechanisms of T-box function in the developing heart. Cardiovasc Res. 2011;91:212–22. doi:10.1093/cvr/cvr112.

S1P-S1p2 Signaling in Cardiac Precursor Cells Migration

14

Hajime Fukui, Shigetomo Fukuhara, and Naoki Mochizuki

Keywords
Zebra fish • Sphingosine-1-phosphate • Cardia bifida • Endoderm

During embryogenesis, zebra fish cardiac precursor cells (CPCs) originating from anterior lateral plate mesoderm migrate toward the midline between the endoderm and the yolk syncytial layer (YSL) to form cardiac tube. The endoderm functions as a foothold for CPCs as evidenced by the endodermal mutants (*cas/sox32*, *sox17*, *oep*, *fau/gata5*, and *bon*) showing two hearts (cardia bifida) [1]. Furthermore, mutant zebra fish (*toh*) lacking sphingosine-1-phosphate (S1P) transporter which is expressed in the YSL show two hearts [2], indicating the essential role for S1P-mediated signal in cardiac development. This is also supported by a S1p2 receptor mutant (*mil*) which exhibits two hearts [3]. However, it is still unclear how S1P released from YSL regulates CPC migration.

S1p2 is expressed in the endoderm. Thus, we assume that S1P released from the YSL might activate S1p2 expressed in the endoderm, thereby regulating CPC migration. One possibility is that S1p2-mediated signal controls the endoderm formation as a foothold for CPCs. Another possibility is that endodermal cells activated by S1p2 might secrete the chemokines which accelerate CPC migration or secrete the extracellular matrix proteins for guiding CPC movement.

To test these possibilities, we need to delineate the downstream signaling of S1p2. Recently, $G\alpha_{13}$ is reported to inhibit Hippo-mediator Lats1/2 kinase through a RhoGEF/Rho/Rho-kinase signaling [4, 5]. We demonstrate that the inhibition of Hippo signaling in the endoderm by activated S1p2 is essential for endodermal cell

H. Fukui • S. Fukuhara • N. Mochizuki (✉)
Department of Cell Biology, National Cerebral and Cardiovascular Center Research Institute, Fujishirodai 5-7-1, Suita, Osaka 565-8565, Japan
e-mail: nmochizu@ncvc.go.jp

© The Author(s) 2016
T. Nakanishi et al. (eds.), *Etiology and Morphogenesis of Congenital Heart Disease*,
DOI 10.1007/978-4-431-54628-3_14

Fig. 14.1 A model of S1P-S1p2 signaling regulated CPC migration. S1P secreted from the yolk activates S1p2 in the endoderm. Hippo signaling acts downstream of S1P-S1p2 signaling and maintains the endoderm to act as scaffolds for CPC migration

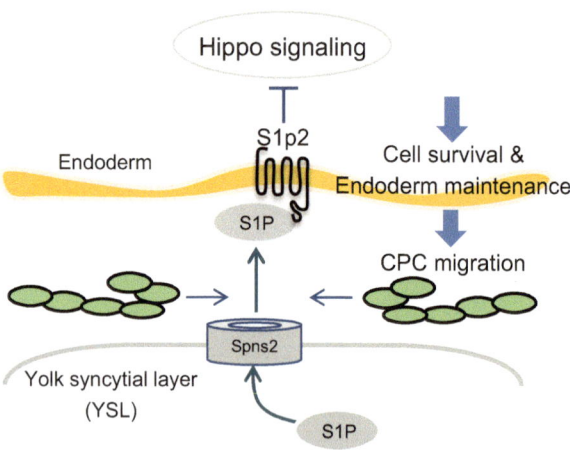

survival and that the endoderm maintained by S1P signaling indeed becomes the foothold for CPC migration (Fig. 14.1).

References

1. Kikuchi Y, et al. casanova encodes a novel Sox-related protein necessary and sufficient for early endoderm formation in zebrafish. Genes Dev. 2001;15(12):1493–505.
2. Kawahara A, et al. The sphingolipid transporter spns2 functions in migration of zebrafish myocardial precursors. Science. 2009;323(5913):524–7.
3. Kupperman E, et al. A sphingosine-1-phosphate receptor regulates cell migration during vertebrate heart development. Nature. 2000;406:192–5.
4. Ye D, Lin F. S1pr2/Gα_{13} signaling controls myocardial migration by regulating endoderm convergence. Development. 2013;140:789–99.
5. Yu FX, et al. Regulation of the Hippo-YAP pathway by G-protein-coupled receptor signaling. Cell. 2012;150(4):780–91.

Myogenic Progenitor Cell Differentiation Is Dependent on Modulation of Mitochondrial Biogenesis through Autophagy

15

Yoshimi Hiraumi, Chengqun Huang, Allen M. Andres, Ying Xiong, Jennifer Ramil, and Roberta A. Gottlieb

Abstract

Over the last decade, stem/progenitor cell therapy has emerged as an innovative approach to promote cardiac repair and regeneration. However, the therapeutic prospects of are currently limited by inadequate means to regulate cell proliferation, homing, engraftment, and differentiation. Autophagy, a lysosome-mediated degradation pathway for recycling organelles and protein aggregates, is recognized as important for facilitating cell differentiation. Studies have shown that induced pluripotent stem cells (iPCs), which exhibit a predominantly glycolytic metabolism, shift toward oxidative mitochondrial metabolism as a prerequisite for the formation of sarcomeres and differentiation into cardiomyocytes. C2C12 myoblasts are a mouse-derived myogenic progenitor cell line which can be induced to differentiate into myotubes. We hypothesize that autophagy is essential in coordinating transcription factor activity and metabolic reprogramming of mitochondria to support myocyte differentiation.

Keywords

Autophagy • Stem cell • Differentiation

C2C12 myoblasts were cultured in DMEM (10 % FBS) and induced to differentiate into myotubes with DMEM (2 % horse serum) for 6 days. To disrupt autophagy, cells were (1) transfected with 50 nM Atg5 siRNA for 8 h twice over a 48 h period

Y. Hiraumi (✉)
Department of Pediatrics Cardiology, Hyogo Amagasaki Prefecture Hospital, Higashidaimotsu-cyo 1-1-1, Amagasaki, Hyogo 660-0828, Japan
e-mail: hirahira@kuhp.kyoto-u.ac.jp

C. Huang • A.M. Andres • Y. Xiong • J. Ramil • R.A. Gottlieb
Donald P. Shiley BioScience Center, San Diego State University, 5500 Campanile Drive, San Diego, CA 92182-4650, USA

© The Author(s) 2016
T. Nakanishi et al. (eds.), *Etiology and Morphogenesis of Congenital Heart Disease*,
DOI 10.1007/978-4-431-54628-3_15

127

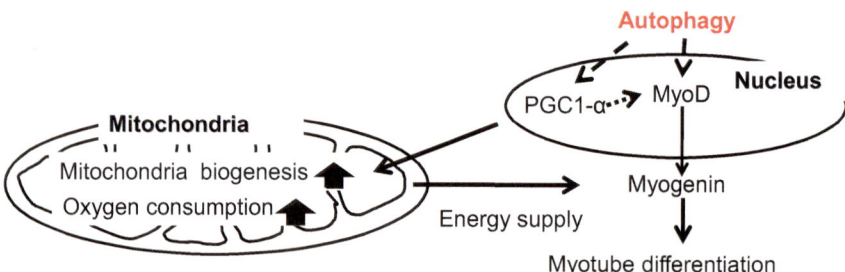

Fig. 15.1 Proposed mechanism for the role of autophagy in C2C12 cell differentiation. Upregulation of PGC1α, MyoD, and myogenin is a hallmark of cell differentiation. Transcription factor regulation by autophagy may affect mitochondrial turnover required for C2C12 cell differentiation. Autophagy is essential for coordinating transcriptional regulation and mitochondrial dynamics to support the progression of cell differentiation

or (2) treated with 10 nM bafilomycin A1 or vehicle control for a 3 h/day for the first 3 days of differentiation. GFP-LC3 adenovirus was employed to visualize autophagy. Western blot and real-time qPCR were used to examine proteins and transcripts of interest. We observed increased LC3-II levels and GFP-LC3 puncta during the differentiation of C2C12 cells, suggesting the involvement of autophagy in this process. Transient inhibition of autophagy during the early stages of differentiation with either Atg5 siRNA or bafilomycin A1 interfered with myotube formation and attenuated the upregulation of myogenic transcription factors MyoD and myogenin. Differentiation was accompanied by an increase in PGC1α mRNA, mitochondrial mass, and oxygen consumption, all of which were blocked by disruption of autophagy. Autophagy coordinates transcription factor expression and mitochondrial turnover essential for cell differentiation (Fig. 15.1).

The Role of the Thyroid in the Developing Heart

16

Kazuhiro Maeda, Sachiko Miyagawa-Tomita, and Toshio Nakanishi

Keywords

Thyroid • Thyroid hormone • Thyroid hormone receptor • Heart • Chick embryo

Congenital hypothyroidism (CH) is one of the most common diseases of the endocrine system among newborns. Infants with CH have been reported to have a high frequency of congenital cardiovascular malformations (CM), such as ventricular and atrial septal defects [1]. Some studies have demonstrated that these cases were due to gene mutations and neural crest abnormality. Infants with CH and CM have been shown to have significantly lower T_4 levels than those with isolated CH. However, the role of thyroid hormone in the developing heart has not been reported. In this study, we show the thyroid anlage in chick embryos by immunohistochemistry and follow the expression of thyroid hormone receptor during heart development.

1. The thyroid anlage appeared close to the aortic sac at H/H 14 of chick embryos, as determined by immunohistochemistry (Fig. 16.1a).
2. Avians have access to thyroid hormone long before the embryonic thyroid gland starts to secrete hormones due to the hormone deposition in the yolk and egg white [2].
3. We found that the expression of thyroid hormone receptors during embryonic heart development was earlier than that reported previously published study using RT-PCR (Fig. 16.1b) [3].

K. Maeda • T. Nakanishi
Department of Pediatric Cardiology, Tokyo Women's Medical University, 8-1 Kawada-cho, Shinjuku-ku, Tokyo 162-8666, Japan

S. Miyagawa-Tomita (✉)
Department of Pediatric Cardiology, Division of Cardiovascular Development and Differentiation, Medical Research Institute, Tokyo Women's Medical University, Tokyo, Japan
e-mail: ptomita@hij.twmu.ac.jp

© The Author(s) 2016
T. Nakanishi et al. (eds.), *Etiology and Morphogenesis of Congenital Heart Disease*,
DOI 10.1007/978-4-431-54628-3_16

129

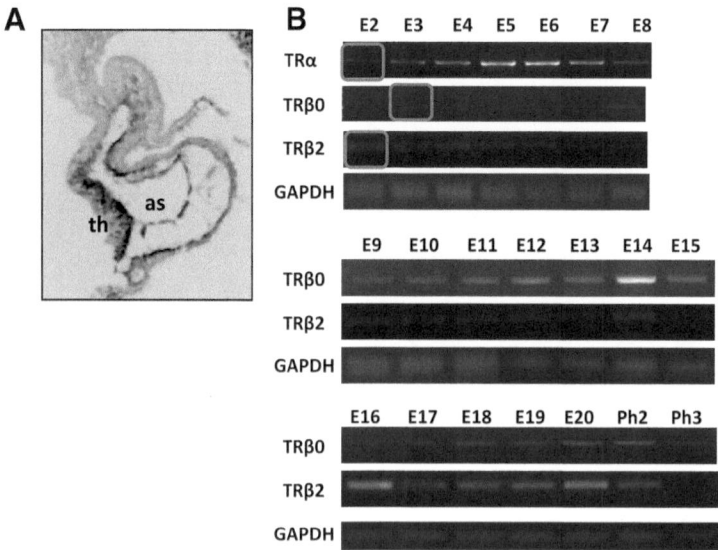

Fig. 16.1 (**a**) The thyroid anlage appeared close to the aortic sac at H/H 14. *as* aortic sac, *th* thyroid anlage. (**b**) The expression of thyroid hormone receptors in the developing chick heart. TRα and TRβ2 were expressed beginning on E2, and TRβ0 was expressed beginning on E3 (*circled*). *E* embryonic day, *Ph* post hatching day

These results suggest that thyroid hormone may contribute to the development of the heart.

References

1. Olivieri A, Stazi MA, Mastroiacovo P, et al. A population-based study on the frequency of additional congenital malformations in infants with congenital hypothyroidism: data from the Italian Registry for Congenital Hypothyroidism (1991–1998). J Clin Endocrinol Metab. 2002;87:557–62.
2. Prati M, Calvo R, Morreale G, Morreale de Escobar G. L-thyroxine and 3,5,3′-triiodothyronine concentrations in the chicken egg and in the embryo before and after the onset of thyroid function. Endocrinology. 1992;130:2651–9.
3. Forrest D, Sjöberg M, Vennström B. Contrasting developmental and tissue-specific expression of alpha and beta thyroid hormone receptor genes. EMBO J. 1990;9:1519–28.

Perspective

Roger R. Markwald

Cardiac valve diseases are common clinical problems that affect from 4 to 10 % of the human population, increasing with age. For many valve problems, there are few options other than surgery, adding to the more than 16,000 surgical cases occurring each year. While surgical techniques continue to improve, the number of surgical cases and associated mortality rates still continue to increase. Secondary complications such as arrhythmias, heart failure, aortic dissection, myocardial hypertrophy, and sudden cardiac death exacerbate primary effects upon blood flow and hemodynamics, indicating the importance of identifying remedial etiologies which remain poorly understood.

Discovery of mutations in patients with congenital heart defects or valve diseases which display progressive (age related) forms of matrix degeneration (e.g., a myxomatous or a calcification phenotype) is currently providing a useful approach for uncovering mechanisms that impact heart valve development and provide candidate targets for therapy. One example of this was the discovery in 2007 by Profs. Herve Le Marec and Jean-Jacques Schott (Nantes, France) of a point mutation in patients with a non-syndromic form of mitral valve prolapse (MVP) that is characterized by progressive, degenerative changes in the extracellular matrix. The mutation occurred in the actin-binding region of a multifunctional cytoskeletal protein, filamin A (FLNA). In mice models, loss of cytoskeletal FLNA function revealed a similar phenotype to MVP resulting in hypertrophied valves with reduced mechanical properties. Further studies indicated that

R.R. Markwald (✉)
Cardiovascular Developmental Biology Center, Department of Regenerative Medicine, Medical University of South Carolina, Charleston, SC 29425, USA
e-mail: markwald@musc.edu

downstream kinase pathways associated with increased canonical and noncanonical TGFβ signaling were associated with the FLNA mutation. How this information might be used to help patients with valve diseases like MVP is an opportunity for basic scientists and clinicians to come together to achieve remedial therapies. In this instance, the finding of altered TGFβ signaling in MVP patients evokes the question of whether similar pharmacological approaches currently being used to treat Marfan patients who also have elevated TGFβ signaling could be used to treat patients with a myxomatous valve phenotype.

The FLNA gene is normally expressed in the mesenchymal progenitors of valve or ventricular interstitial fibroblasts and is downregulated during neonatal life as shown by Russell Norris and his colleagues (Developmental Dynamics 2010). This has two important implications: (i) the roots of adult valve defects and diseases can extend back into embryonic life and (ii) heart valve development does not end at birth but is completed postnatally in neonatal life. The expression of the FLNA gene only in mesenchyme progenitors of valve or ventricular fibroblasts points to yet another important implication for valve development: lineage.

In addition to endothelial-derived, progenitor cells, extracardiac mesenchymal cells of neural crest origin or epicardial origin migrate into the heart adding to the mesenchymal population of future valve cells in the outflow track and AV inlets, respectively. Why multiple lineages of progenitor cells are needed for valvulogenesis, especially those derived from outside the heart, is an important unresolved question in heart development and a subject touched upon by three papers presented in Part IV. Dr. Andy Wessels and his collaborators indicate in their chapter that epicardial-derived cells specifically migrate into the lateral (future parietal) AV cushions and uniquely contribute to the posterior leaflets of the left and right AV valve. He then discusses the role of Bmp signaling, through the Bmp receptor BmpR1A/Alk3, in regulating the migration of epicardial-derived cells or EPDCs initially as an epithelial sheet under the endocardium of the AV mitral valve and later as free cells within the matrix. In another paper presented in Part IV, Dr. Scott Baldwin utilizes tissue-specific Cre lineage drivers in mice to differentially assess whether Tie1 is expressed in the endocardium vs. the prevalvular mesenchyme of the AV junction or outflow tract. Dr. Baldwin's lineage-tracing studies revealed that a non-cell autonomous form of cross talk occurs between the endocardial epithelium and subjacent valve interstitial cells that affects the secretion of extracellular matrix and the normal formation and the differential remodeling of the valves of the AV junction and outflow tract in late gestation and postnatal life. His work also provides in vivo support for the in vitro observations presented in the chapter by Drs. Kei Inai and Yukiko Sugi and their collaborators that valvular endocardial cells can regulate the migration and differentiation of valve interstitial cells. Similarly, in another chapter in Part IV, Mizuta et al. demonstrate that if Tmem100, a novel, endothelial-specific, *membrane* protein that is a downstream of BMP9/BMP10-ALK1 signaling, is genetically deleted, the resulting Tmem100 null embryos exhibit an atrioventricular defect. The authors suggest that the atrioventricular septal defect is the result of disrupting an active dialogue or cross talk between the AV endocardial endothelial cells and the subjacent AV prevalvular mesenchyme.

Whether this putative cross talk between endocardium and prevalvular mesenchyme involves the secretion of a paracrine signal into the extracellular matrix, transport of developmental cues within membrane-enclosed exosomes (or adherons) or direct cell contact (e.g., gap junctions) remains another important developmental heart question to be resolved. The answer to this question could provide more candidate targets for developing or engineering remedial therapies, especially if the signal requires extracellular processing or specific receptors and downstream signaling pathways.

Cross talk does not have to be unidirectional from the endocardium to the mesenchyme but may also originate from the mesenchyme and induce changes in the endocardium. One frequently suggested function for the EPDCs is that they somehow signal the termination of the transformation of AV endothelial cells into new valve progenitor mesenchyme. If correct, this could serve as a mechanism for regulating the normal size or shape of lateral AV cushions and their future parietal leaflets. Whether the neural crest plays a similar instructive "cross-talk" role in determining the size, shape, or fate of the semilunar valves derived from the outflow tract cushions also remains to be determined. Understanding how lineage and "cross talk" may intersect to shape and model inlet and outlet valve development is likely to be one of the more exciting and promising future directions for research into the mechanisms of cardiac valvuloseptal morphogenesis. Already, there are emerging clues and new insights that point to primary (non-motile) cilia as specialized structures on the surfaces of embryonic epithelial and/or mesenchymal cells that may potentially act as sites for sending or receiving developmental signals between interactive cells.

Finally, valve development – be it AV or semilunar – is largely a story about the extracellular matrix and the progenitor cells that normally differentiate into valve interstitial fibroblasts. Like all fibroblasts, they can be "friend or foe." During normal development, they secrete a collagenous matrix which they organize and compact into mature cusps and leaflets. In disease states, they may transdifferentiate into myofibroblasts that secrete metalloproteinases and proteoglycans that disrupt the matrix resulting in a myxomatous phenotype in the AV valves, or, in the outflow tract, they may abnormally enter osteogenic lineages resulting in calcified aortic valves. In her chapter, Dr. Katherine Yutzey and her colleagues review positive and negative changes in the extracellular matrix that precede normal or pathological remodeling and dissect some of the candidate transcriptional regulatory mechanisms that regulate lineage progression and organization of valve extracellular matrix.

Atrioventricular Valve Abnormalities: From Molecular Mechanisms Underlying Morphogenesis to Clinical Perspective

Kei Inai

Abstract

Malformation of the atrioventricular (AV) cushion is a common congenital heart defect. Ebstein's anomaly, characterized by a heart defect related to the AV cushion, involves not only a valve defect but also a myocardial abnormality such as Uhl's anomaly. The morphogenetic features of the heart in the case of these diseases can be used as a reference for investigating valvuloseptal and myocardial formations in the human heart.

The AV endocardium transforms into the cushion mesenchyme through epithelial–mesenchymal transition (EMT). After the EMT, distal outgrowth and maturation of endocardial cushions are important morphogenetic steps for AV valvuloseptal morphogenesis. While bone morphogenetic protein (BMP)-2 is known to be critical for AV EMT, little is known about the functional relationship between BMP and ECM and their roles in cushion mesenchymal cell (CMC) migration after EMT. In our previous study, we showed that BMP-2 and BMP signaling induced AV CMC migration. We have been exploring the role of BMP-2 in the regulation of valvulogenic extracellular matrix (ECM) components, periostin, versican, and hyaluronic acid (HA), and cell migration during post-EMT AV cushion expansion and maturation.

We further examined whether BMP-2-promoted cell migration is associated with expression of periostin, versican, and HA. BMP-2-promoted cell migration is significantly impaired by treatment with versican siRNA and HA oligomer. We also found that transcription of *Twist and Id1*, implicated in cell migration in embryogenesis and activation of the periostin promoter, was induced by BMP-2 but repressed by noggin in CMC cultures.

K. Inai (✉)
Department of Pediatric Cardiology, Heart Institute, Tokyo Women's Medical University,
8-1 Kawada-cho, Shinjuku-ku, Tokyo, Japan
e-mail: pinai@hij.twmu.ac.jp

T. Nakanishi et al. (eds.), *Etiology and Morphogenesis of Congenital Heart Disease*,
DOI 10.1007/978-4-431-54628-3_17

Taken together, we provide evidence that BMP-2 induces expression and deposition of three major ECM proteins, periostin, versican, and HA, and that these ECM components contribute to BMP-2-induced CMC migration during post-EMT AV cushion expansion and maturation.

Based on these findings, we discuss the morphogenetic process of AV valve abnormalities and crosstalk between valve and cardiomyocytes morphogenesis.

Keywords

Bone morphogenetic protein (BMP) • Cardiac cushions • Periostin • Extracellular matrix protein • Valvulogenesis

17.1   Introduction

Our institute houses a huge amount of congenital heart disease specimens, of which approximately 4,000 are referred to as Prof. Ando's collections, and the late Dr. Ando, former professor at our department, wrote 30 books describing the anatomical features of the specimens.

In one of the books, I read the following text at the edge of a page: "I think there is a right ventricle (RV)–tricuspid valve (TV) dysplastic syndrome as a new clinical entity, which comprises three abnormalities, Ebstein's anomaly, Uhl's anomaly, and tricuspid absence." This note can be likened to the one written by Fermat, on the basis of which Fermat's son Clément-Samuel deduced Fermat's last theorem (Fig. 17.1).

Fig. 17.1 Tricuspid valve anomalies syndrome or RV-TV dysplastic syndrome: Ebstein's anomaly, Uhl's anomaly and tricuspid valve absence sometimes coexist in a heart

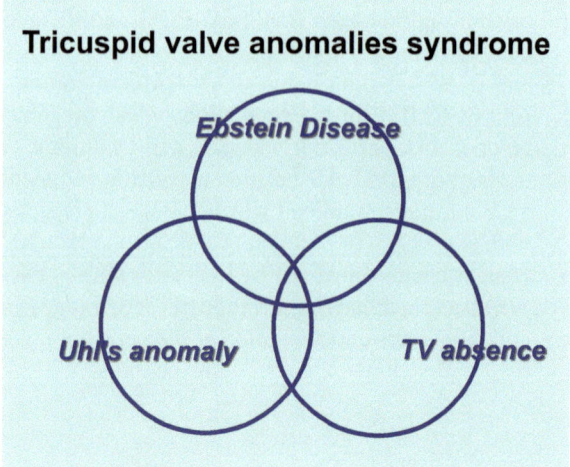

17.2 RV–TV Dysplastic Syndrome

17.2.1 Anatomic Features of the Heart in Ebstein's Anomaly Patients

Ebstein's anomaly is a rare congenital cardiac disease in which the TV leaflets are displaced to the right ventricular cavity, resulting in atrialization of the RV [1]. Marked right atrial enlargement caused by severe tricuspid insufficiency, cardiomegaly, and increased right atrial pressure; cyanosis caused by right–left shunt by way of interatrial communication, if present; and right ventricular volume overload are often observed. Figure 17.2 illustrates Ebstein's anomaly with severe TV insufficiency involving two-thirds of the RV.

17.2.2 Morphogenetic Features of the Heart in Patients with Uhl's Anomaly

In the case of patients with Ebstein's anomaly, the RV of the heart often shows partial or total loss of myocardium; in other words, the features of Ebstein's anomaly overlap those of Uhl's anomaly. Uhl's anomaly, a rare congenital heart disease characterized by the absence of apical trabeculations in the right ventricular cavity with a thin hypokinetic ventricular wall, was first described in 1952 [2]. In Fig. 17.3, almost complete absence of the right ventricular myocardium, preserved septum, and left ventricular myocardium is seen. These clinical and anatomic findings strongly indicate that AV valve formation and ventricular myocardium formation may play a role during cardiac morphogenesis.

Fig. 17.2 Ebstein's anomaly with severe TV plastering down into two thirds of right ventricle (RV). The cushion tissue plastered almost throughout RV with a something like plaster and it looks very smooth surface

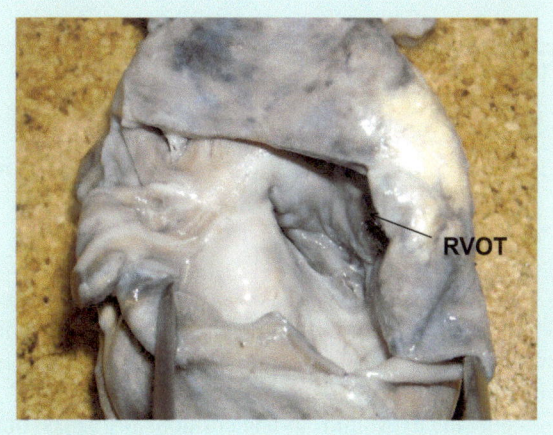

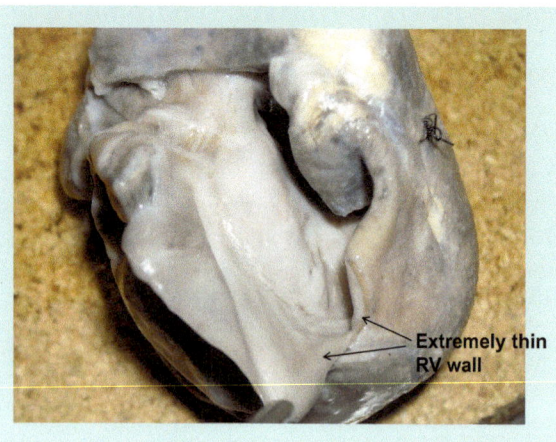

Fig. 17.3 Ebstein's anomaly with Uhl's anomaly. Note the complete absence of the right ventricular myocardium and preserved septal and left ventricular myocardium

17.2.3 Absence of the TV

Unguarded TV orifice is an unusual cardiac malformation that involves partial or total agenesis of the TV tissue (absence of the TV). Kanjuh et al. described the anatomic features of the heart in this disease: the wall of the RV is formed by slightly trabeculated muscle and lined by thickened endocardium [3]; the free wall of the RV is only 2 mm thick, and the ventricular septum is intact. In Fig. 17.4, the specimen shows TV absence and partial loss of the right ventricular myocardium.

TV–RV dysplastic syndrome revealed the important link between TV morphogenesis and right ventricular myocardium formation during the biological process of cardiac morphogenesis. Therefore, it is vital to determine the biological key factors that link the myocardium and cushion tissue.

17.3 Bone Morphogenetic Proteins (BMPs) and Their Important Role in Cushion Formation

17.3.1 Role of BMP2 in Cushion Mesenchymal Cell (CMC) Migration

Anomalies related to valvuloseptal formation are some of the most common congenital heart anomalies. Two segments of the endocardium, the atrioventricular (AV) and outflow tract (OT) endocardium, transform into the cushion mesenchyme—the primordia of the valves and membranous septa—through epithelial–mesenchymal transformation (EMT). Transformed endocardial cells subsequently migrate to the underlying extracellular matrix (ECM), called "cardiac jelly," and remodel the ECM into cardiac cushions [4]. Distal outgrowth and maturation of the cardiac cushion are the initial and critical morphogenetic steps in post-EMT valve formation.

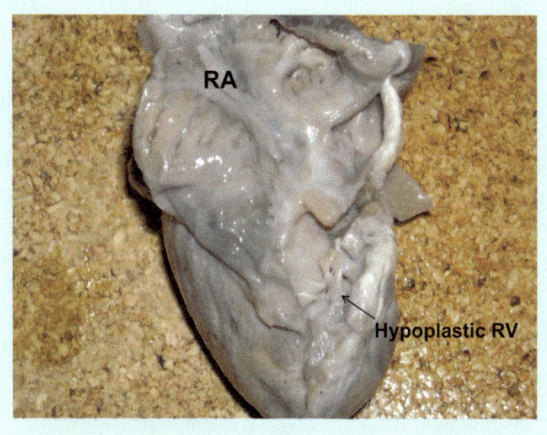

Fig. 17.4 Ebstein's anomaly with tricuspid valve absence and Uhl's anomaly. Unguarded tricuspid valve orifice and partial absence of the right ventricular myocardium concomitant with Ebstein's anomaly. Note the absence of the rudimentum of the tricuspid valve and tension apparatus

BMP is a member of the TGF-ß superfamily proteins, one of many molecules related to AV EMT. BMP signaling was found to be crucial for AV EMT in studies involving explant cultures in mice and chicks and BMP-2 conditional knockout (CKO) experiments in mice [5]. BMP-2 conditional KO at the EMT stage, however, causes subsequent lethality, thereby hindering the examination of the role of BMP-2 in post-EMT valve formation. BMPs bind to the cell surface of BMP receptors (BMPRs)—Type I and Type II receptors. Type II receptors transphosphorylate the glycine–serine-rich domain of Type I receptors and transduce intracellular signals. The Type II BMP receptor, BMPRII, is reported to be expressed ubiquitously throughout the embryonic period [6].

We previously showed the expression patterns of Type I BMPRs by localizing *BMPR-1A (ALK3)*, *BMPR-1B (ALK6)*, and *ALK2* in the post-EMT AV cushion mesenchyme in a chick. These findings indicate that BMP signaling regulates the biological processes necessary for distal outgrowth and maturation of post-EMT cushion mesenchyme during early valvulogenesis [4].

17.3.2   BMP2 Induces CMC Migration and Id and Twist Expression

BMP-2 is localized in the AV cushion mesenchyme during post-EMT valve formation, suggesting autocrine signaling by BMP-2 in CMCs during post-EMT valve formation. Therefore, to determine the roles of BMP-2 and BMP signaling in post-EMT AV CMCs, we established a hanging-drop culture system and spatiotemporal viral gene-transfer technique using chick AV CMC cultures both in vitro and whole embryo cultures in vivo (*ovo*) [5]. Although most BMP receptors have ligand-binding affinity with other TGF-ß superfamily proteins, BMPR-1A and BMPR-1B bind specifically with BMP-2 and BMP-4 (bind with BMP-7 at low affinity) but do not bind with other TGF-ß superfamily proteins. We showed that *dnBMPR-1B* infection as well as treatment with a BMP antagonist, noggin, significantly inhibited endogenous phospho-Smad 1/5/8 expression in AV CMCs, indicating that intracellular BMP signaling in the CMCs is effectively inhibited by treatment with dnBMPR-1B or noggin. On the other hand, BMP-2 or caBMPR-1B-virus treatment induced expression of phospho-Smad 1/5/8, indicating that intracellular BMP signaling was

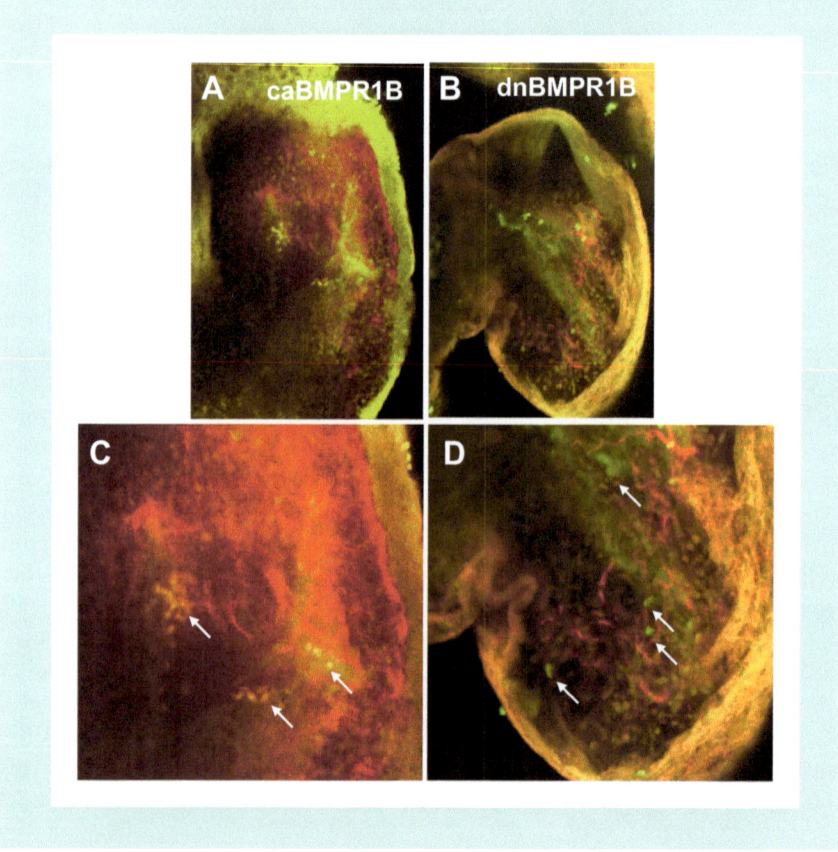

Fig. 17.5 AV cushions were microinjected with caBMPR-1B virus (**a**) or dnBMPR-1B virus. Cushion mesenchyme microinjected with dnBMPR-1B virus showed a significant decrease in periostin immunostaining (*red colour* in **b**). Conversely, infection with caBMPR-1B virus showed significant increase periostin expression in cushion mesencyme (*red colour* in **a**). High magnification view of virus-infected cushion mesenchyme (**c** and **d**). In the caBMPR-1B infected cushion, there was extensive overlapping of viral marker expression (*green*) and periostin expression (*red*) (*yellow arrows* in **c**). Although periostin staining (*red*) was detected in the dnBMPR-1B infected cushion, there was little overlapping of periostin expression (*red*) and viral marker expression (*green arrows* in **d**)

effectively intensified by BMP-2 or caBMPR-1B. Based on these findings, it can be said that the effects of dnBMPR-1B treatment on CMC migration and phosphor-Smad1/5/8 expression are as profound as the effects of noggin treatment. Moreover, caBMPR-1B treatment enhances CMC migration as much as BMP2 treatment does (Fig. 17.5). These facts strongly suggest that BMP signaling plays an important role in mesenchymal cell migration after EMT. We also showed that expressions of Id and Twist-1, which are important transcriptional factors involved in cell migration, are enhanced by BMP signaling [4].

17.3.3 BMP2 Induces Expression of ECM Proteins in the Post-EMT Cushion

We showed the expression patterns of three major ECM components, periostin, versican, and hyaluronic acid (HA), during AV cushion expansion and maturation. We also elucidated the role of BMP2 in the production of periostin, versican, and HA. Using a well-defined 3D CMC aggregate by hanging-drop culture on hydrated collagen gels, we found that BMP2 induces production of periostin, versican, and HA and that these ECM components contribute to BMP2-supported CMC migration during post-EMT cushion expansion and maturation [7].

Taken together, BMP signaling plays a critical role for AV cushion formation and AV valve maturation.

17.4   The Role of BMP2 for Cardiomyocytes Formation

It is well known that the BMP signaling pathway plays a central role in cardiomyogenesis [8–10]. In the past, it was found that administration of BMP-2 counteracts apoptosis of neonatal cardiomyocytes in culture, induced by serum starvation [9]. Moreover, BMP10 expression is restricted to cardiomyocytes in the developing and postnatal heart. Ligands of BMP10 have specificity for ALK1, ALK6, and BMPR2 receptors. Global disruption of BMP10 is embryonic lethal with severely impaired cardiomyocyte proliferative capacity [11]. BMP2 and BMP10 appear to control cardiomyocyte differentiation through activation of transcription factors NKX2.5, MEF2C, and TBX20 [12].

Chakraborty et al. reported that cardiomyocyte-specific Tbx20 overexpression beginning in the fetal period is sufficient for promoting cardiomyocyte proliferation [12].

Therefore, BMP signaling plays special role in linking AV cushion and cardiomyocyte formation.

17.5 Future Direction

At present, many aspects of the molecular mechanisms underlying RV–TV dysplastic syndrome remain to be elucidated. To fully describe the morphogenetic mechanisms underlying TV–RV dysplastic syndrome, we need to determine the molecular mechanisms underlying the interaction between TV and RV myocardium through BMP signaling. Moreover, we need to outline the steps involved in cardiac valve and myocardium morphogenesis.

What is the next target molecule? Here, I would like to discuss about the potential of the "Notch" signaling pathway. The Notch signaling pathway plays multiple roles in cardiac morphogenesis, including regulation of valve formation, outflow tract development, and cardiac chamber maturation [13]. Notch activation is upstream of ephrin- and neuregulin-based modulation of trabeculation and BMP-2 and BMP-10 modulation of cardiomyocyte proliferation. Consistent with the involvement of Notch signaling in multiple aspects of cardiac development, components of the Notch pathway show dynamic spatial and temporal expression patterns in the developing vertebrate heart, and both endocardial and myocardial expression have been observed. Furthermore, the cross reaction between NOTCH and BMP signaling is a well-known fact [14]. Therefore, understanding the complicity of these signaling pathways is imperative for elucidating the relationship between cardiac valve and myocardium formation.

Acknowledgments Special thanks to Dr. Yukiko Sugi and Prof. Roger R. Markwald for their generous support in the experiments and excellent mentorship.

References

1. Inai K. Encyclopedia of molecular mechanisms of disease. Ebstein's anomaly. Berlin: Springer; 2009. p. 222–3.
2. Uhl HSM. A previously undescribed cardiac malformation of the heart: almost total absence of the myocardium of the right ventricle. Bull Johns Hopkins Hosp. 1952;91:197–209.
3. Kanjuh VI, Stevenson JE, Amplatz K, Edwards JE. Congenitally unguarded tricuspid orifice with coexistent pulmonary atresia. Circulation. 1964;30:911–7.
4. Inai K, Norris RA, Hoffman F, Markwald RR, Sugi Y. BMP-2 induces cell migration and periostin expression during atrioventricular valvulogenesis. Dev Biol. 2008;315:383–96.

5. Sugi Y, Yamamura H, Okagawa H, Markwald RR. Bone morphogenetic protein-2 can mediate myocardial regulation of atrioventricular cushion mesenchymal cell formation in mice. Dev Biol. 2004;269:505–18.
6. Okagawa H, Markwald RR, Sugi Y. Functional BMP receptor in endocardial cells is required in atrioventricular cushion mesenchymal cell formation in chick. Dev Biol. 2007;306 (1):179–92.
7. Inai K, Burnside JL, Hoffman S, Toole BP, Sugi Y. BMP-2 Induces versican and hyaluronan that contribute to Post-EMT AV cushion cell migration. PLos ONE 2012;8(10):e77593. doi:10.1371.
8. Zhang H, Bradley A. Mice deficient for BMP-2 are nonviable and have defects in amnion/chorion and cardiac development. Development. 1996;122:2977–86.
9. Izumi M, Fujio Y, Kunisada K, Negoro S, Tone E, Funamoto M, et al. Bone morphogenetic protein-2 inhibits serum deprivation-induced apoptosis of neonatal cardiac myocytes through activation of the Smad1 pathway. J Biol Chem. 2001;276:31133–41.
10. Ghosh-Choudhury N, Abboud SL, Chandrasekar B, Ghosh CG. BMP-2 regulates cardiomyocyte contractility in a phosphatidylinositol 3 kinase-dependent manner. FEBS Lett. 2003;544:181–4.
11. Huang J, Elicker J, Bowens N, Liu X, Chen L, Cappola TP, et al. Myocardin regulates BMP10 expression and is required for heart development. J Clin Invest. 2012;122:3678–91.
12. Chakraborty S, Sengupta A, Yutzey KE. Tbx20 promotes cardiomyocyte proliferation and persistence of fetal characteristics in adult mouse hearts. J Mol Cell Cardiol. 2013;62:203–13.
13. Samsa LA, Yang B, Liu J. Embryonic cardiac chamber maturation: trabeculation, conduction, and cardiomyocyte proliferation. Am J Med Genet Part C Semin Med Genet. 2013;163C:157–68.
14. MacGrogan D, Luxán G, de la Pompa JL. Genetic and functional genomics approaches targeting the Notch pathway in cardiac development and congenital heart disease. Brief Funct Genom. 2014;13(1):15–27.

Molecular Mechanisms of Heart Valve Development and Disease

18

M. Victoria Gomez Stallons, Elaine E. Wirrig-Schwendeman, Ming Fang, Jonathan D. Cheek, Christina M. Alfieri, Robert B. Hinton, and Katherine E. Yutzey

Abstract

The mature heart valves consist of stratified extracellular matrix (ECM) layers, and heart valve disease is characterized by ECM dysregulation and mineralization. There is increasing evidence that regulatory pathways that control heart valve development also are active in disease. In human diseased valves and mouse models, the expression of valve progenitor markers, including Twist1, Msx1/2 and Snail1/2, is induced. Additional markers of osteogenesis, including Runx2, osteocalcin and bone sialoprotein, also are expressed in calcific aortic valve disease (CAVD) in humans and mice. New mouse models have been developed for studies of valve disease mechanisms. Klotho-null mice are a model for premature aging and exhibit calcified nodules in aortic valves with osteogenic gene induction. Osteogenesis Imperfecta mice, bearing a collagen1a2 mutation, develop features of myxomatous valve disease, including thickening, increased proteoglycan deposition and chondrogenic gene induction. Together, these findings demonstrate specific molecular indicators of valve disease progression, including the identification of early disease markers, which represent potential targets for therapeutic intervention.

Keywords

Heart valve • Embryo • Mouse model • Aortic valve disease

M.V. Gomez Stallons • E.E. Wirrig-Schwendeman • M. Fang • J.D. Cheek • C.M. Alfieri •
R.B. Hinton • K.E. Yutzey (✉)
Division of Molecular Cardiovascular Biology, The Heart Institute, Cincinnati Children's Hospital
Medical Center, 240 Albert Sabin Way, ML 7020, Cincinnati, OH 45229, USA
e-mail: Katherine.Yutzey@cchmc.org

© The Author(s) 2016
T. Nakanishi et al. (eds.), *Etiology and Morphogenesis of Congenital Heart Disease*,
DOI 10.1007/978-4-431-54628-3_18

18.1   Introduction

The semilunar and atrioventricular (AV) valves of the heart are made up of highly organized extracellular matrix (ECM) layers populated by quiescent valve interstitial cells (VICs) [1]. In healthy valves, the ECM is compartmentalized into layers composed of collagens, proteoglycans, and elastin, which are maintained by the VICs for proper valve function throughout life (Reviewed in [2]). In diseased valves, the leaflets are thickened as a result of ECM dysregulation and VIC activation. Calcific aortic valve disease (CAVD) includes calcification of the cusps [3], whereas mitral valve prolapse (MVP) is accompanied by increased proteoglycans and myxomatous changes in the leaflets [4]. Currently, the standard treatment for severe heart valve disease is surgical replacement, and new therapies based on molecular mechanisms are needed.

   Molecular mechanisms associated with heart valve disease include activation of signaling pathways involved in progenitor specification, cell proliferation, and differentiation of heart valve and bone precursors [5, 6]. We have reported that pediatric and adult diseased valves are characterized by expression of markers of valve mesenchymal and chondrogenic progenitor cells, while adult diseased aortic valves express markers of osteogenic calcification [7]. We also have identified novel mouse models of calcific and myxomatous valve disease [8] that will be useful in determination of the underlying mechanisms driving disease and in development of pharmacologic-based therapies.

18.2   Heart Valve Development

Heart valve development in vertebrate embryos begins with the formation of endocardial cushions in the AV canal and outflow tract of the primitive heart tube [5]. The mesenchymal cells of the endocardial cushions originate from the endocardium after an endothelial-to-mesenchymal transition (EMT). Valve progenitors are highly proliferative and migratory and express transcription factors Twist1, Tbx20, Sox9, Msx1/2, and Snai1. The endocardial cushion cells diversify into lineages that express distinct ECM profiles regulated by BMP and FGF signaling [9]. Wnt/β-catenin signaling also is active during endocardial cushion maturation, but the specific role for this pathway in valve lineage differentiation is yet to be determined [10]. Valve development continues with the remodeling of the endocardial cushions into thin elongated leaflets composed of stratified ECM, which occurs soon after birth in mice and humans [2]. The ECM layers consist of the collagen-rich fibrosa, proteoglycan-rich spongiosa, and elastin-rich atrialis/ventricularis [2]. These layers are oriented in the semilunar and AV valves with the elastin layer adjacent to blood flow. While it is likely that hemodynamics has a role in leaflet stratification, the regulatory pathways that control ECM organization and compartmentalization during valve maturation are largely unknown.

18.3 Heart Valve Disease

Heart valve disease can result from congenital malformation or gene mutations, or it may be acquired later in life [1]. The prevalence of heart valve disease increases significantly with age, such that ~10 % of people >75 years old have moderate aortic or mitral valve disease [11]. However, the pathogenic mechanisms that drive the development of heart valve disease and that could serve as potential therapeutic targets are not well understood. There is increasing evidence that regulatory pathways that control heart valve and bone development also are active in disease. However, the roles of these pathways in valve pathogenesis and/or repair are not well defined.

18.3.1 Calcific Aortic Valve Disease (CAVD)

CAVD is a progressive disease, initially presenting with aortic valve (AoV) thickening (sclerosis) and resulting in valve stenosis and insufficiency later in life [12]. End-stage disease is characterized by the presence of calcific nodules at the hinge region of the AoV, underlying the pathology of CAVD [3]. In an effort to draw parallels between the progression of disease and the underlying molecular mechanisms, pediatric and adult diseased AoV were analyzed for markers of valve development and endochondral bone formation [7]. Activated VICs in both pediatric and adult valves have increased expression of valvulogenic markers Twist1, Msx2, and Sox9. Strikingly, the formation of calcific nodules was found to be an exclusive feature of adult calcified AoV. Furthermore, phosphorylation of Smads1/5/8, indicative of active BMP signaling, in addition to expression of osteogenic genes, such as Runx2, was observed only in adult calcified AoV. These findings demonstrate that both pediatric and adult diseased AoV express valvulogenic markers, while adult calcified AoV also express markers of osteogenic calcification. Differential expression of these markers suggests that an osteogenic regulatory mechanism contributes specifically to CAVD.

The incidence of human CAVD strongly correlates with aging, which is an independent risk factor for AoV disease. We have recently identified *Klotho*[-/-] mice, a model of premature aging, as a novel mouse model of CAVD [8]. Notably, *Klotho*[-/-] mice develop calcific nodules at the hinge region of the fibrosa side of the AoV (Fig. 18.1a, b), similar to human CAVD. In these mice, calcification occurs independent of inflammation and cusp thickening, providing initial evidence for a valve-intrinsic molecular mechanism for age-related calcification common in elderly patients. *Klotho*[-/-] AoV have increased expression of osteogenic factors *Runx2* and *Osteopontin*, in addition to increased expression of chondrogenic factors *Sox9* and *Col10a1*, consistent with an osteochondrogenic-like mechanism of disease (Fig. 18.1e). Increased activation of pSmad1/5/8 also precedes calcification in the *Klotho*[-/-] mice, and inhibition of BMP signaling represents an attractive new therapeutic approach for CAVD.

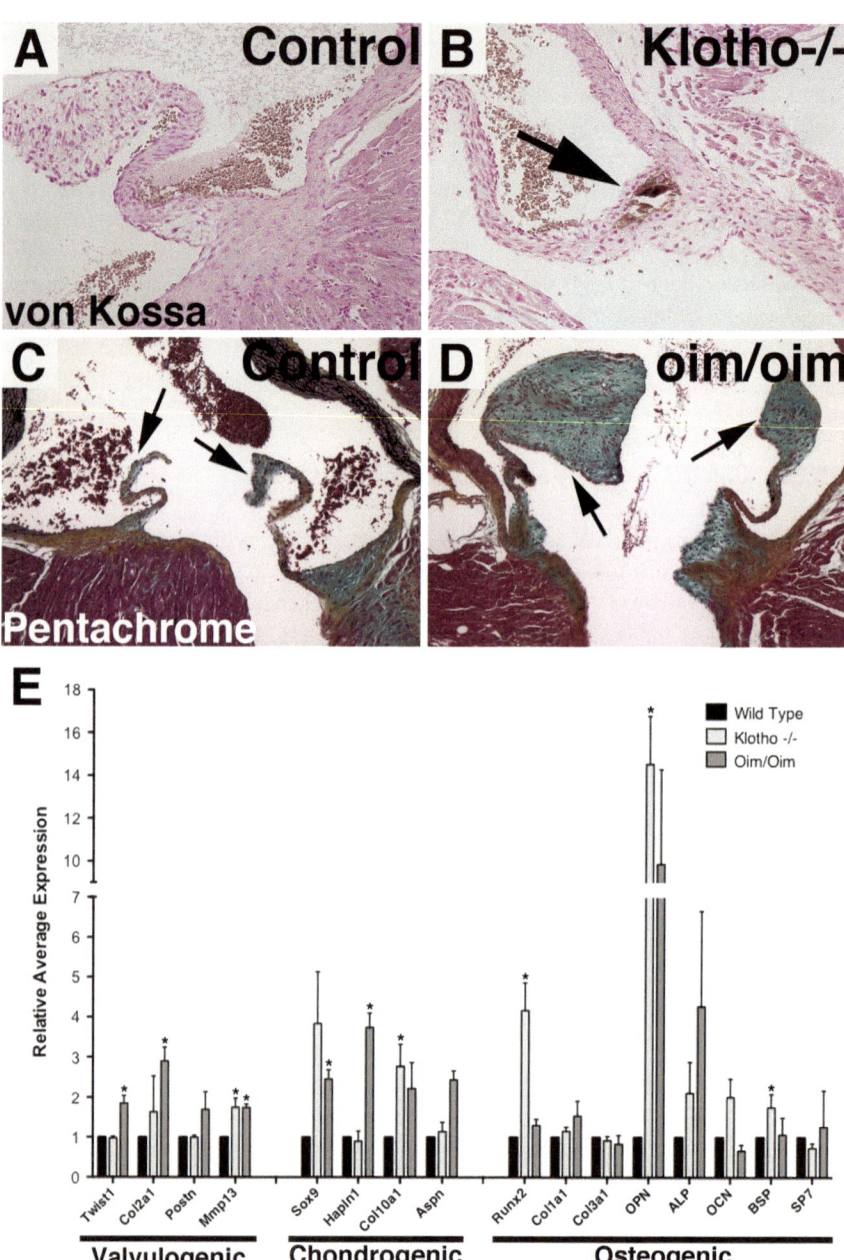

Fig. 18.1 Valvulogenic, chondrogenic, and osteogenic programs are induced in mouse models of calcific (*Klotho⁻/⁻*) and myxomatous (*Oim/Oim*) valve disease. *Klotho⁻/⁻* mice (**b**) exhibit AoV nodular calcification (*arrows*), as compared to wild-type littermate controls (**a**) at 6 weeks of age, as detected by von Kossa staining. *Oim/Oim* AoV cusps (**d**) exhibit distal thickening and increased proteoglycan deposition (aqua, *arrows*), compared to WT littermates (**c**) at 9 months of age as observed by Movat's pentachrome staining. Expression of genes involved in valvulogenesis, chondrogenesis, and osteogenesis was examined by qRT-PCR of RNA isolated from *Klotho⁻/⁻*

18.3.2 Myxomatous Valve Disease

The most common cause of MVP is myxomatous valve disease, which is defined by pathological thickening of the valve leaflets, primarily due to accumulation of proteoglycans [4]. This is accompanied by alterations in the distribution of ECM components, such as disrupted collagen fiber organization and elastic fiber fragmentation. The pathogenesis of MVP is not well understood; however, MVP is often linked to connective tissue disorders or specific mutations in ECM genes, supporting the concept that defects originating during valve development could underlie adult disease.

While myxomatous disease most commonly affects the mitral valves, myxoid AoV have been reported. *Osteogenesis imperfecta murine* (*Oim*) have a spontaneous mutation in the *Col1a2* gene and display bone fragility characteristic of human osteogenesis imperfecta (OI) [13]. Interestingly, humans with OI or *Col1a2* mutations have a predisposition to AoV disease [14]. Likewise, the AoV of *Oim/Oim* mice exhibit distal cusp thickening and increased proteoglycan accumulation, characteristic of myxomatous valve disease (Fig. 18.1c, d) [8]. Furthermore, the *Oim/Oim* mice have increased expression of valve progenitor markers *Twist1*, *Col2a1*, *Mmp13*, *Sox9*, and *Hapln1*, in addition to increased *Col10a1* and *Asporin* expression (Fig. 18.1e). These changes in gene expression are consistent with increased proteoglycan accumulation and cartilage gene induction, which are key features of myxomatous disease.

18.4   Signaling Pathways in Heart Valve Development and Disease

Similar to heart valve and bone development, studies of human explanted valves implicate BMP, Notch, and Wnt signaling pathways in the progression of CAVD. Thus, heart valve disease shares signaling networks with valve and bone developmental pathways. Together, these studies demonstrate that activation of both BMP and Wnt signaling correlates with progression of CAVD [15, 16]. On the other hand, loss-of-function mutations in *NOTCH1* are associated with bicuspid aortic valve (BAV) and CAVD, in humans and mice, suggesting an inhibitory function for the Notch pathway in valve calcification [17]. Human genetic conditions including Marfan syndrome and Loeys-Dietz syndrome lead to myxomatous mitral valve disease and are associated with increased TGF-β signaling [18, 19]. However, the specific mechanisms by which these different pathways contribute to the development and progression of heart valve disease remain unknown.

←──

Fig. 18.1 (continued) and *Oim/Oim* mice aortic valves relative to wild-type littermate controls (**e**). Normalized values are shown as average fold changes compared to wild-type group set at 1.0. * is *p*-value ≤ 0.05 calculated by paired student's *t*-test

18.5   Future Directions and Clinical Implications

Klotho$^{-/-}$ and *Oim/Oim* mice are novel mouse models of CAVD and myxomatous valve disease that will be useful for determination of the underlying pathogenic mechanisms driving valve disease. Understanding how signaling networks contribute to disease will likely have a significant impact on clinical outcomes, since knowledge gained from these studies will allow for the development and design of new drugs/treatments for patients with valve disease.

Acknowledgments This work was supported by funding from NIH HL082716, HL094319, HL114682 (K.E.Y), F32HL110390 (E.E.W.), and the American Heart Association-Great Rivers Affiliate (M.V.G. and M.F.).

References

1. Lincoln J, Yutzey KE. Molecular and developmental mechanisms of congenital heart valve disease. Birth Defects Res A Clin Mol Teratol. 2011;91:526–34.
2. Hinton RB, Yutzey KE. Heart valve structure and function in development and disease. Annu Rev Physiol. 2011;73:29–46.
3. Otto CM, Kuusisto J, Reichenbach DD, Gown AM, O'Brien KD. Characterization of the early lesion of 'degenerative' valvular aortic stenosis. Histological and immunohistochemical studies. Circulation. 1994;90:844–53.
4. Guy TS, Hill AC. Mitral valve prolapse. Annu Rev Med. 2012;63:277–92.
5. Combs MD, Yutzey KE. Heart valve development: regulatory networks in development and disease. Circ Res. 2009;105:408–21.
6. Rajamannan NM, Subramaniam M, Rickard DJ, Stock SR, Donovan J, Springett M, Orszulak T, Fullerton DA, Tajik AJ, Bonow RO, et al. Human aortic valve calcification is associated with an osteoblast phenotype. Circulation. 2003;107:2181–4.
7. Wirrig EE, Hinton RB, Yutzey KE. Differential expression of cartilage and bone-related proteins in pediatric and adult diseased aortic valves. J Mol Cell Cardiol. 2011;50:561–9.
8. Cheek JD, Wirrig EE, Alfieri CM, James JF, Yutzey KE. Differential activation of valvulogenic, chondrogenic, and osteogenic pathways in mouse models of myxomatous and calcific aortic valve disease. J Mol Cell Cardiol. 2012;52:689–700.
9. Lincoln J, Alfieri CM, Yutzey KE. BMP and FGF regulatory pathways control cell lineage diversification of heart valve precursor cells. Dev Biol. 2006;292:290–302.
10. Alfieri CM, Cheek J, Chakraborty S, Yutzey KE. Wnt signaling in heart valve development and osteogenic gene induction. Dev Biol. 2010;338:127–35.
11. Nkomo VT, Gardin JM, Skelton TN, Gottdiener JS, Scott CG, Enriquez-Sarano M. Burden of valvular heart diseases: a population-based study. Lancet. 2006;368:1005–11.

12. Otto CM. Valvular aortic stenosis: disease severity and timing of intervention. J Am Coll Cardiol. 2006;47:2141–51.
13. Chipman SD, Sweet HO, McBride DJ, Davisson MT, Marks SC, Shuldiner AR, Wenstrup RJ, Rowe DW, Shapiro JR. Defective proa2(I) collagen synthesis in a recessive mutation in mice: a model of human osteogenesis imperfecta. Proc Natl Acad Sci U S A. 1993;90:1701–5.
14. Bonita RE, Cohen IS, Berko BA. Valvular heart disease in osteogenesis imperfecta. Echocardiography. 2010;27:69–73.
15. Ankeny RF, Thourani VH, Weiss D, Vega JD, Taylor WR, Nerem RM, Jo H. Preferential activation of SMAD1/5/8 on the fibrosa endothelium in calcified human aortic valves – association with low BMP antagonists and SMAD6. PLoS ONE. 2011;6:e20969.
16. Caira FC, Stock SR, Gleason TG, McGee EC, Huang J, Bonow RO, Spelsberg TC, McCarthy PM, Rahimtoola SH, Rajamannan NM. Human degenerative valve disease is associated with up-regulation of low-density lipoprotein-related protein 5 receptor-mediated bone formation. J Am Coll Cardiol. 2006;47:1707–12.
17. Garg V, Muth AN, Ransom JF, Schluterman MK, Barnes R, King IN, Grossfeld PD, Srivastava D. Mutations in NOTCH1 cause aortic valve disease. Nature. 2005;437:270–4.
18. Dietz HC, Cutting GR, Pyeritz RE, Maslen CL, Sakai LY, Corson GM, Puffenberger EG, Hamosh A, Nanthakumar EJ, Curristin SM, et al. Marfan syndrome caused by a recurrent de novo missense mutation in the fibrillin gene. Nature. 1991;352:337–9.
19. Loeys BL, Chen J, Neptune ER, Judge DP, Podowski M, Holm T, Meyers J, Leitch CC, Katsanis N, Sharifi N, et al. A syndrome of altered cardiovascular, craniofacial, neurocognitive and skeletal development caused by mutations in TGFBR1 or TGFBR2. Nat Genet. 2005;37:275–81.

A Novel Role for Endocardium in Perinatal Valve Development: Lessons Learned from Tissue-Specific Gene Deletion of the Tie1 Receptor Tyrosine Kinase

19

Xianghu Qu and H. Scott Baldwin

Abstract

The mechanisms regulating late-gestational and early postnatal semilunar valve remodeling and maturation are poorly understood. Tie1 is a receptor tyrosine kinase with broad expression in embryonic endothelium. During semilunar valve development, Tie1 expression becomes restricted to the turbulent, arterial surfaces of the valves in the perinatal period. Previous studies in our laboratory have demonstrated that Tie1 can regulate cellular responses to blood flow and shear stress. We hypothesized that Tie1 signaling would regulate the flow-dependent remodeling of the semilunar valves associated with the conversion from maternal/placental to independent neonatal circulation. To circumvent the embryonic lethality of the Tie1 null mutation, we developed a floxed Tie1 allele and crossed it with an $Nfactc1^{en}Cre$ line that mediates gene excision exclusively in the endocardial cushion endothelium. Excision of Tie1 resulted in aortic valve leaflets displaying hypertrophy with perturbed matrix deposition. The valves demonstrated insufficiency and stenosis by ultrasound, and atomic force microscopy documented decreased stiffness in the mutant aortic valve consistent with an increased glycosaminoglycan to collagen ratio. These data suggest that active endocardial to mesenchymal signaling, at least partially mediated by Tie1, is uniquely required for normal remodeling of the aortic but not pulmonary valve in the late gestation and postnatal animal.

X. Qu (✉)
Department of Pediatrics (Cardiology), Vanderbilt University, 2213 Garland Ave, Nashville, TN 37232-0493, USA
e-mail: xianghu.qu@vanderbilt.edu

H.S. Baldwin
Department of Pediatrics (Cardiology), Vanderbilt University, 2213 Garland Ave, Nashville, TN 37232-0493, USA

Department of Cell and Developmental Biology, Vanderbilt University, 2213 Garland Ave, Nashville, TN 37232-0493, USA

© The Author(s) 2016
T. Nakanishi et al. (eds.), *Etiology and Morphogenesis of Congenital Heart Disease*,
DOI 10.1007/978-4-431-54628-3_19

153

Keywords
Semilunar valves • Endocardium • Extracellular matrix • Tie1

19.1   Introduction

Cardiovascular defects are the most common congenital abnormality in the human population, affecting approximately 1 out of every 100 live births worldwide [1]. Defects in valve development account for 25–30 % of these malformations. Therefore, there is significant interest in understanding the mechanisms that underlie the complex process of heart valve development. During heart valve formation, a subset of endothelial cells overlying the future valve site are specified to delaminate, differentiate, and migrate into the cardiac jelly, a process referred to as endothelial-mesenchymal transformation or transdifferentiation (EMT). Locally expanded swellings of cardiac jelly and mesenchymal cells are referred to as cardiac cushions. In a poorly understood process, these cardiac cushions undergo extensive remodeling from bulbous swellings to eventual thinly tapered heart valve leaflets [2]. Numerous studies have focused on identifying the major regulators of valve development using murine and avian embryos and have particularly focused on the early stages of valve development; however, the pathways regulating the final events of remodeling and maturation have not been well defined.

The orphan receptor tyrosine kinase Tie1 is primarily expressed in endothelial cells and is closely related to Tie2, the receptor for the angiopoietins. Both Tie1 and Tie2 are essential for developmental vascularization where they appear to have roles in promoting microvessel maturation and stability. Targeted disruption of Tie1 gene in mice results in a lethal phenotype between E14.5 and P0 characterized by extensive edema, hemorrhage, and defective microvessel integrity [3–5]. Expression of Tie1 is first detected in the endothelium of mice at E8.0, and by E12.5, Tie1 is robustly expressed in the valvular endothelium, vasculature, heart, and lungs [6–9]. Tie1 expression subsides in the postnatal period, although low levels of Tie1 persist into adulthood on the arterial side of the valve leaflet as well as branch points of the descending aorta [10]. Thus, Tie1 is expressed in the vasculature, valves, and other endothelial cell populations in the developing heart during critical periods of valve morphogenesis. However, its potential roles in heart valve development are totally unknown. Here, using a conditional knockout mouse model, we show that lack of Tie1 in endocardial cells leads to hypertrophic semilunar valves in the postnatal and adult heart.

19.2   Model for Valvar Endocardial-Specific Gene Deletion

To specifically determine the potential role of Tie1 in valve morphogenesis, we have developed a novel valve endocardial cell-specific Cre mouse line (*Nfatc1^{en}Cre*). We previously identified a transcriptional enhancer that regulates the sustained expression of *Nfatc1* exclusively in the endocardium overlying the endocardial cushions [11]. We then utilized this enhancer to develop a transgenic mouse line, $Nfactc1^{en}Cre$, which has recently been described by our lab [12]. To determine the utility and efficiency of this model for deletion at latter stages of cardiac development, we have analyzed Cre expression in the developing valves by $lacZ$ staining and qRT-PCR of $Nfactc1^{en}Cre$ transgenic mice bred with R26R reporter mice (Fig. 19.1a–d). $lacZ$ staining confirmed that Cre activation is first detected at E9.5 in the developing AVC region and in OFT region by E10.5 (data not shown). Cre activation is localized to endocardial cushion endocardium of the outflow track at E12.5 and at E14.5 (Fig. 19.1a, b) and remains restricted to the endocardium of overlying the developing valves throughout gestation into the perinatal period (Fig. 19.1c) and in the adult (data not shown). To determine the utility of $Nfactc1^{en}Cre$ for tissue-specific deletion, we crossed these mice with our mice harboring a floxed Tie1 allele previously described [5]. When Tie1 expression as determined by QT-PCR was normalized to expression detected in the adult animal, we observed the normal attenuation of Tie1 previously described by our group [5]. In addition we observed significant attenuation of Tie1 expression as a result of Cre-mediated deletion in the heart that was not detected in other vascular beds (data not shown). Thus our Cre transgenic line $Nfactc1^{en}Cre$ mediates $loxP$ excision exclusively in the endocardium overlying the endocardial cushions and not in mesenchyme derived from EMT or other endothelial populations.

19.3   Tie1 Is Required for Late-Gestational and Early Postnatal Aortic Valve Remodeling

To investigate the potential role of Tie1 in valve morphogenesis, we generated endocardial-specific Tie1 knockout ($Tie1^{fl/lz}$;$Nfactc1^{en}Cre$) mice by breeding mice homozygous for a floxed Tie1 allele [5] with mice heterozygous for a null mutation in $Tie1^{fl/lz}$ ($Tie1^{+/lz}$ is a "knock-in"/knockout" insertion of lacZ into the Tie 1 locus [3]) as well as the $Nfactc1^{en}Cre$ transgene. Analysis of timed matings revealed that no embryos died in utero during early gestation. However, within a few days after birth, $Tie1^{fl/lz}$;$Nfactc1^{en}Cre$ pups could be distinguished from $Tie1^{fl/fl}$ or $Tie1^{fl/lz}$ littermates as the mutants were often growth retarded (Fig. 19.2) and lethargic, both of which are characteristics of human patients with heart defects. Most of mutants did not survive to adulthood. Histological analysis revealed no significant differences in valve formation between mutants and littermates until E16.5 (Fig. 19.3a, b), indicating early events of valvulogenesis occur normally in Tie1 mutant mice. However, an increase in semilunar valve size was noted in $Tie1^{fl/lz}$; $Nfactc1^{en}Cre$ mice from E18.5 (Fig. 19.3c, d) and early perinatal periods with the discrepancy in valve size continuing into the adult (Fig. 19.3e, f). Interestingly, the valve abnormalities were not only limited to the late gestation and postnatal period, but they were also primarily only detected in the aortic valves. The aortic valve areas of adult $Tie1^{fl/lz}$;$Nfactc1^{en}Cre$ mice were nearly twice as large as $Tie1^{fl/fl}$ aortic valves (increased by 88.3 %). The pulmonary valves of $Tie1^{fl/lz}$;$Nfactc1^{en}Cre$ mice were somewhat thicker than the controls, but the difference was not significant. There were no differences in cell proliferation or apoptosis between mutant mice

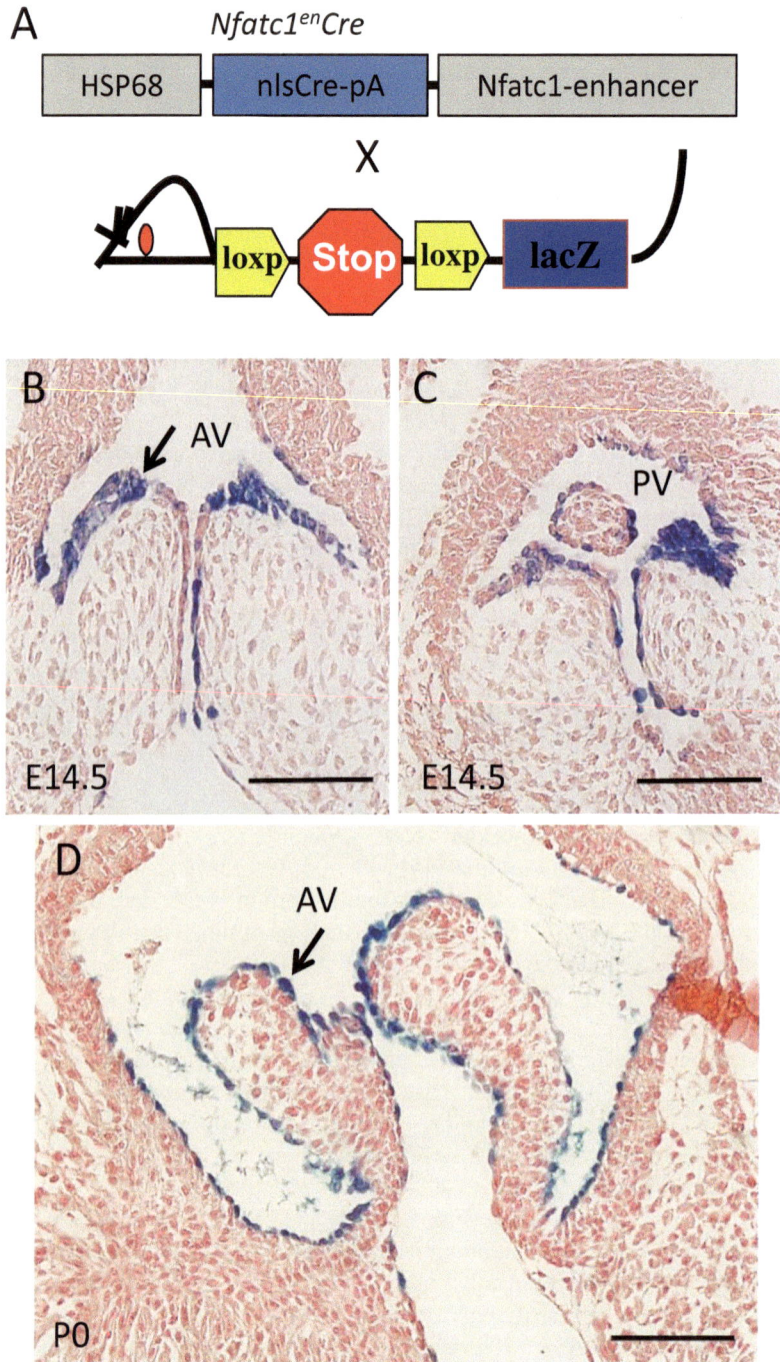

Fig. 19.1 *Nfactc1^{en}Cre* mediates loxP excision exclusively in the valvular endocardium. (**a**) Schematic of breeding of *Nfatc1enCre* and *R26fslz* mice. X-gal staining of *R26^{fslz};Nfactc1^{en}Cre* heart sections shows that expression of the Cre is restricted to the endothelium of the developing

Fig. 19.2 *Tie1$^{fl/lz}$; Nfactc1^{en}Cre* animals are smaller than the controls

and littermates suggesting a difference in extracellular matrix (ECM) composition or production. Consistent with this, Movat's pentachrome stain revealed an increase in proteoglycan/glycosaminoglycan (GAGs) production and/or degradation in the valve interstitium of postnatal *Tie1$^{fl/lz}$;Nfactc1^{en}Cre* mice. As compared to *Tie1$^{fl/fl}$*, GAGs content (Fig. 19.3e, f) was increased and total collagen content (data not shown) was decreased in aortic valve leaflets of *Tie1$^{fl/lz}$;Nfactc1^{en}Cre* animals. Thus, attenuation of Tie1 results in severe abnormalities in ECM remodeling that are characteristic of critical events in late gestation and postnatal valve development in chicken, mouse, and humans [13, 14]. This work suggests that there is a non-cell autonomous defect that results from endocardial-specific Tie1 deletion as the ECM is produced primarily from the valvular interstitial cells (VICS), not the endocardium. This work also provides in vivo support for the in vitro observations that valvular endocardial cells regulate the phenotype of VICS. In addition, Tie1 mutant aortic valves demonstrated a decrease in rigidity as measured by atomic force microscopy and valvular insufficiency, as determined by high-resolution ultrasound (data not shown).

19.4 Future Directions

This study describes a unique dosage-dependant role for Tie1 during later stages of valve remodeling. Tie1 is essential for valve remodeling, and abnormalities within late-gestational valve remodeling lead to flaccid valve leaflets, which in turn do not function correctly in the animal. It is likely that the mortality observed in *Tie1$^{fl/lz}$; Nfactc1^{en}Cre* mice can be attributed to aortic insufficiency and ultimate heart failure. However, the mechanisms regulating late-gestational and early postnatal semilunar valve remodeling and maturation are still poorly understood.

Fig. 19.1 (continued) valves (*arrowhead*) at E14.5 (**b**, **c**) and at P0 (**d**), and weak *lacZ* expression remains in the majority of the valvular endothelial cells through adulthood (**d**). HSP68 indicates heat shock protein 68; *AV* aortic valve, *PV* pulmonary and aortic valve. Scale bars: 50 μm

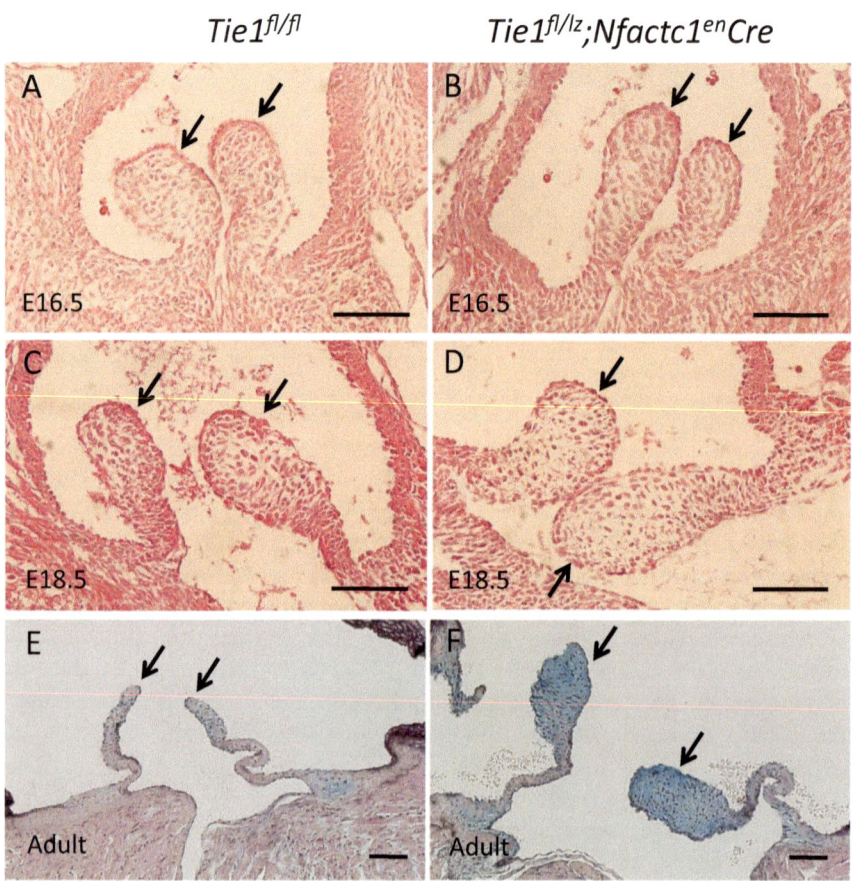

Fig. 19.3 Tie1 attenuation leads to enlarged aortic valves and abnormal ECM deposition and organization. H&E-stained sections of valves of *Tie1$^{fl/fl}$* and *Tie1$^{fl/lz}$;Nfactc1^{en}Cre* mice at E16.5 (**a**, **b**) and at E18.5 (**c**, **d**). Compared to *Tie1$^{fl/fl}$* mice, *Tie1$^{fl/lz}$;Nfactc1^{en}Cre* mice have enlarged aortic valves. Movat's pentachrome stain shows enlarged aortic valves with increased GAGs deposition (*blue*) in adult (9 weeks) *Tie1$^{fl/lz}$;Nfactc1^{en}Cre* mice (**f**) as compared to the controls (**e**). *Arrows* indicate valve leaflets. Scale bars: 50 μm

Tie1, although a close sequence homologue of Tie2, does not interact directly with the angiopoietins, and its in vivo ligands are yet to be identified. Nevertheless, growing reports based on in vitro studies suggest that a primary function of Tie1 might be to modulate Tie2 signaling and function [15–17]. Furthermore, both Tie1 and Tie2 are co-expressed in endocardial cells from very early endocardial cushion stage to the mature valve stage. So, we hypothesize that Tie1 signals independent of Tie2 and also acts an inhibitory co-receptor for Tie2 activation. Further investigation on this project using double (Tie1 and Tie2) conditional knockouts is currently on the way. Additional animal models are being developed to characterize the mechanism of Tie1-Tie2 interactions in modulating critical events in cardiac ontogeny in vivo, and expression profiling is being implemented to delineate downstream signaling cascades that are activated by Tie1 and Tie1-Tie2 interactions. These studies will be essential for understanding the role of the endocardium in modulation valve matrix deposition and remodeling in an effort to unravel basic mechanisms of congenital heart disease.

Acknowledgements This work was supported by grant from NHLB/NIH: RL1HL0952551 (H.S.B.)

References

1. Hoffman JI, Kaplan S. The incidence of congenital heart disease. J Am Coll Cardiol. 2002;39:1890–900.
2. DeLaughter DM, Saint-Jean L, Baldwin HS, et al. What chick and mouse models have taught us about the role of the endocardium in congenital heart disease. Birth Defects Res Part A Clin Mol Teratol. 2011;91:511–25.
3. Puri MC, Rossant J, Alitalo K, Bernstein A, Partanen J. The receptor tyrosine kinase tie is required for integrity and survival of vascular endothelial cells. EMBO J. 1995;14:5884–91.
4. Sato TN, Tozawa Y, Deutsch U, Wolburg-Buchholz K, Fujiwara Y, Gendron-Maguire M, Gridley T, Wolburg H, Risau W, Qin Y. Distinct roles of the receptor tyrosine kinases tie-1 and tie-2 in blood vessel formation. Nature. 1995;376:70–4.
5. Qu X, Tompkins K, Batts LE, Puri M, Baldwin S. Abnormal embryonic lymphatic vessel development in tie1 hypomorphic mice. Development. 2010;137:1285–95.
6. Partanen J, Armstrong E, Makela TP, Korhonen J, Sandberg M, Renkonen R, Knuutila S, Huebner K, Alitalo K. A novel endothelial cell surface receptor tyrosine kinase with extracellular epidermal growth factor homology domains. Mol Cell Biol. 1992;12:1698–707.
7. Sato TN, Qin Y, Kozak CA, Audus KL. Tie-1 and tie-2 define another class of putative receptor tyrosine kinase genes expressed in early embryonic vascular system. Proc Natl Acad Sci U S A. 1993;90:9355–8.
8. Dumont DJ, Fong GH, Puri MC, Gradwohl G, Alitalo K, Breitman ML. Vascularization of the mouse embryo: a study of flk-1, tek, tie, and vascular endothelial growth factor expression during development. Dev Dyn. 1995;203:80–92.
9. Taichman DB, Schachtner SK, Li Y, Puri MC, Bernstein A, Scott Baldwin H. A unique pattern of tie1 expression in the developing murine lung. Exp Lung Res. 2003;29:113–22.
10. Woo KV, Qu X, Babaev VR, Linton MF, Guzman RJ, Fazio S, Baldwin HS. Tie1 attenuation reduces murine atherosclerosis in a dose-dependent and shear stress-specific manner. J Clin Invest. 2011;121:1624–35.
11. Zhou B, Wu B, Tompkins KL, Boyer KL, Grindley JC, Baldwin HS. Characterization of nfatc1 regulation identifies an enhancer required for gene expression that is specific to pro-valve endocardial cells in the developing heart. Development. 2005;132:1137–46.

12. Wu B, Wang Y, Lui W, Langworthy M, Tompkins KL, Hatzopoulos AK, Baldwin HS, Zhou B. Nfatc1 coordinates valve endocardial cell lineage development required for heart valve formation. Circ Res. 2011;109:183–92.
13. Hinton Jr RB, Lincoln J, Deutsch GH, Osinska H, Manning PB, Benson DW, Yutzey KE. Extracellular matrix remodeling and organization in developing and diseased aortic valves. Circ Res. 2006;98:1431–8.
14. Aikawa E, Whittaker P, Farber M, Mendelson K, Padera RF, Aikawa M, Schoen FJ. Human semilunar cardiac valve remodeling by activated cells from fetus to adult: implications for postnatal adaptation, pathology, and tissue engineering. Circulation. 2006;113:1344–52.
15. Saharinen P, Kerkela K, Ekman N, Marron M, Brindle N, Lee GM, Augustin H, Koh GY, Alitalo K. Multiple angiopoietin recombinant proteins activate the tie1 receptor tyrosine kinase and promote its interaction with tie2. J Cell Biol. 2005;169:239–43.
16. Marron MB, Hughes DP, Edge MD, Forder CL, Brindle NP. Evidence for heterotypic interaction between the receptor tyrosine kinases tie-1 and tie-2. J Biol Chem. 2000;275:39741–6.
17. Seegar TC, Eller B, Tzvetkova-Robev D, Kolev MV, Henderson SC, Nikolov DB, Barton WA. Tie1-tie2 interactions mediate functional differences between angiopoietin ligands. Mol Cell. 2010;37:643–55.

The Role of the Epicardium in the Formation of the Cardiac Valves in the Mouse

20

Marie M. Lockhart, Maurice van den Hoff, and Andy Wessels

Abstract

In recent years, insights into the role of the epicardium in cardiac development have significantly changed. An important contribution to this increasing knowledge comes from the availability of mouse models that facilitate the study of the fate of the epicardial cell lineage and that allow epicardial-specific manipulation of expression of genes involved in regulation of epicardial cell behavior. In this contribution we will discuss our growing understanding of the role of the epicardium and epicardially derived cells in the formation of the atrioventricular valve leaflets. We will illustrate how epicardially derived cells specifically contribute to the development of the leaflets that derive from the lateral atrioventricular cushions, and we will discuss the role of Bmp signaling, through the Bmp receptor BmpR1A/Alk3, in the regulation of the preferentially migration of EPDCs into the parietal AV valve leaflets.

Keywords

Epicardium • Valves • Bmp2 • Bmp4 • Alk3

M.M. Lockhart, Ph.D. • A. Wessels, Ph.D. (✉)
Department of Regenerative Medicine and Cell Biology, Medical University of South Carolina, 173 Ashley Avenue, Room BSB-648B, P.O. Box 250508, Charleston, SC 29425, USA
e-mail: wesselsa@musc.edu

M. van den Hoff, Ph.D.
Department of Anatomy, Embryology and Physiology, Academic Medical Center, Heart Failure Research Center, Meibergdreef 15, 1105AZ Amsterdam, The Netherlands

© The Author(s) 2016
T. Nakanishi et al. (eds.), *Etiology and Morphogenesis of Congenital Heart Disease*,
DOI 10.1007/978-4-431-54628-3_20

161

20.1 Introduction

20.1.1 The AV Valves and Their Leaflets

The atrioventricular (AV) valves are important cardiac components that, when properly formed, prevent the retrograde flow of blood through the AV junction during ventricular systole. The precursor tissues of the AV valves are the endocardial AV cushions, formed by (1) the accumulation of cardiac jelly in the AV canal followed by (2) the population of these cushions by endocardially derived cells (ENDCs) resulting from an endocardial epithelial-to-mesenchymal transformation (or endoEMT) [1]. Not all AV cushions develop at the same time. The two "major" AV cushions (the inferior and superior cushion) form first, while the two "lateral" cushions form later [2, 3]. The major AV cushions will eventually fuse together. They will also fuse with the mesenchymal cap of the primary atrial septum and the dorsal mesenchymal protrusion (DMP) to form the AV mesenchymal complex [4]. This process is essential in the partition of the left and right atrial and ventricular components. The AV mesenchymal complex is also involved in the formation of the septal leaflet of the right AV valve and the aortic (or anterior) leaflet of the left AV valve [4, 5]. The lateral AV cushions are significantly smaller than the major AV cushions and develop at the lateral myocardial AV junctions. Just like the major AV cushions, the lateral AV cushions become populated with ENDCs. The right lateral cushion eventually forms the parietal leaflet of the right AV valve, while the left lateral cushion forms the parietal (or mural/posterior) leaflet of the left AV valve (Fig. 20.1).

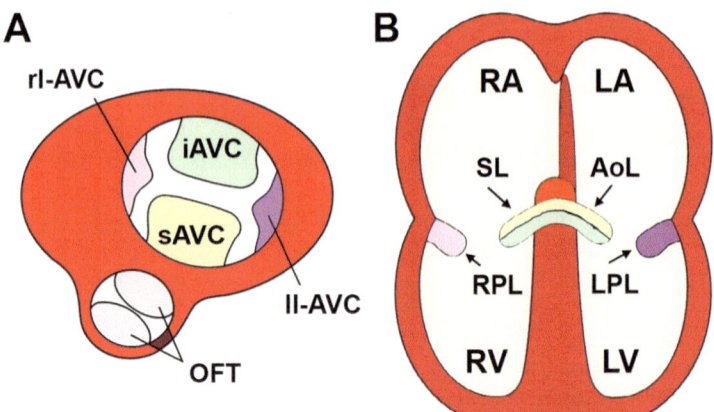

Fig. 20.1 This cartoon shows the fate of the individual AV cushions. The superior and inferior AV cushion (sAVC and iAVC) fuse at the midline and give rise to the septal leaflet of the right AV valve (SL) and the aortic leaflet (AoL) of the left AV valve. The right lateral AV cushion (rlAVC) forms the right parietal leaflet of the right AV valve (RPL), whereas the left lateral AV cushion forms the left parietal leaflet of the left AV valve (LPL)

20.1.2 The Epicardium and Epicardially Derived Cells (EPDCs)

In the mouse, epicardial development starts around embryonic day 9.5 (ED9.5) when the proepicardium, a heterogeneous cluster of cells, can be seen as a "cauliflower-like" mesothelial structure at the inferior margin of the cardiac sinus venosus at the venous pole of the heart. Around ED10, proepicardial cells reach and attach to the myocardium. Subsequently, they migrate out over the surface of the developing heart, ultimately covering the heart entirely. As a result of an epicardial epithelial-to-mesenchymal transition (epiEMT), epicardially derived cells (EPDC) are generated that migrate into the sub-epicardial space and into the underlying myocardium. Over the last 15 years or so, a series of cell fate studies in avian and murine models have been published reporting that EPDCs can, or have the potential to, differentiate in multiple cell types including coronary smooth muscle cells, interstitial fibroblasts, coronary endothelium, and potentially cardiomyocytes [6–9]. It is important to note that the contribution of EPDCs to the coronary endothelium and myocardium remains controversial.

20.1.3 The Contribution of EPDCs to the Developing AV Valves

In recent papers we have discussed the cascade of epicardial-related events involved in the formation and development of the AV junction [7, 10]. This process starts with the establishment of the AV-epicardium. After the formation of the AV-epicardium, subsequent AV-epiEMT leads to the generation of AV-EPDCs, a process regulated by a variety of factors. These AV-EPDCs eventually form the AV sulcus, which can be seen as a mesenchymal wedge between the expanding atrial and ventricular myocardial walls. As AV-EPDCs from the sulcus start to penetrate the AV myocardial junction, the formation of the annulus fibrosus is initiated. The annulus fibrosus is a fibrous tissue plane that physically and electrically separates the atrial from the ventricular working myocardium. From the annulus fibrosus, a subset of AV-EPDCs continues to migrate into the parietal AV valve leaflets, where they eventually comprise a large percentage of the mesenchymal cells found of the leaflet (Fig. 20.2). Just like the ENDCs found in the developing leaflets, the AV-EPDCs in the valves eventually become valve interstitial cells (VICs). It is important to note that very few AV-EPDCs are found in the leaflets that are developing from the fused major AV cushions [7].

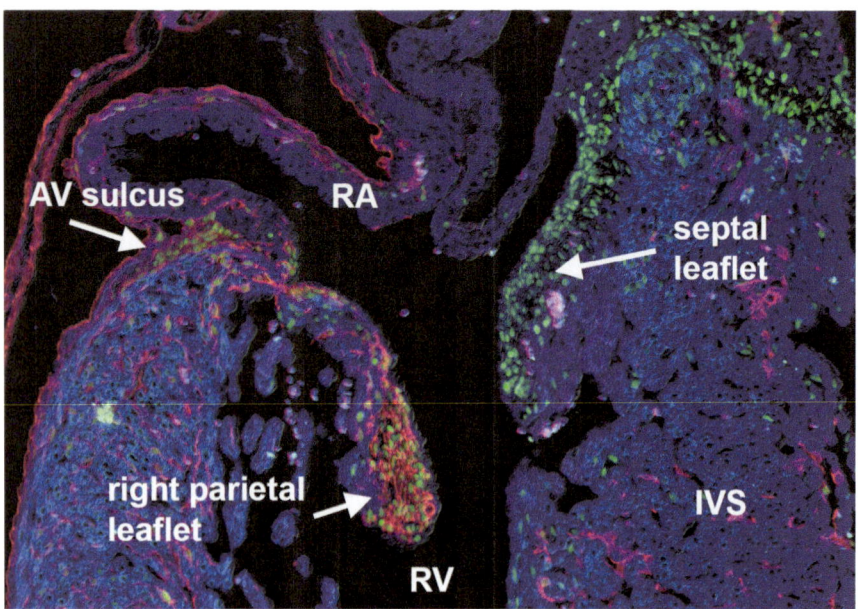

Fig. 20.2 A section of a 17ED Wt1-cre mouse crossed with a ROSA26$^{mT/mG}$ reporter mouse was stained for the presence of epicardial and epicardially derived cells (*red*), the transcription factor Sox9 (*green*), and cardiac myosin heavy chain (*blue*). The section shows the abundant presence of EPDCs in the right parietal leaflets and the virtual absence of EPDCs in the septal leaflet of the right AV valve

20.2 The Role of Bmp Signaling in Regulating the Contribution of EPDC to the AV Valves

20.2.1 Epicardial-Specific Deletion of the Bmp Receptor BmpR1A/ Alk3 Leads to Disruption of AV Junction Development

To elucidate the mechanisms that regulate the events that are responsible for the contribution of AV-EPDCs to the tissues of the AV junction, we investigated the importance of bone morphogenetic protein (Bmp) signaling in the epicardium. Bmp isoforms are known to be involved in a variety of developmental steps in heart development. In particular Bmp2, found at high levels in the AV myocardium, has been shown to be important for the development of the AV cushions by promoting endoEMT [11]. In our study, we determined that Bmp4 is highly expressed in the AV-epicardium and in the AV-EPDCs that form the AV sulcus. Furthermore, strong phospho-Smad-1/5/8 labeling of the AV-EPDCs indicates that canonical Bmp signaling is active in these cells, strongly suggesting that this signaling pathway is important in regulating the events associated with the cascade of epicardial-related events at the AV junction [12]. To test the hypothesis that Bmp

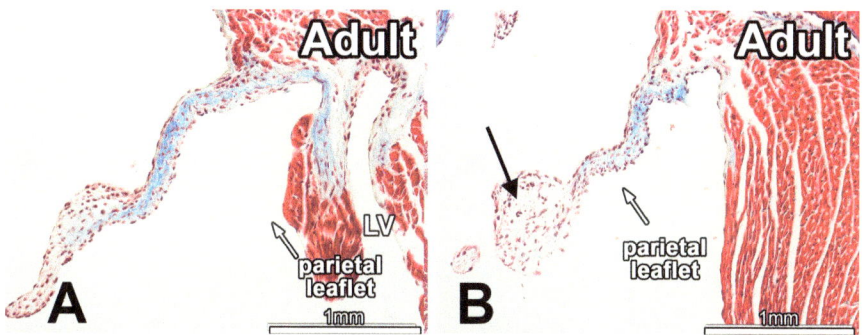

Fig. 20.3 Masson's trichrome staining of adult wild-type (**a**) and Wt1-cre;Alk3 mouse hearts. Panel (**b**) demonstrates the myxomatous phenotype of the left parietal leaflet of the epicardial Alk3 knockout mouse (*arrow* in **b**)

signaling is important for the development of the structures that rely on the contribution of AV-EPDCs, we deleted the Bmp receptor Alk3 from the epicardial and epicardially derived cells using an epicardial-specific cre-mouse (the Wt1-cre mouse [7, 12]). This approach resulted in a decrease in the size of the AV sulcus, an inhibition of the formation of the annulus fibrosus, and a reduction of the number of EPDCs that migrate into the parietal AV valve leaflets. Remarkably, electrophysiological analysis of postnatal Wt1-cre;Alk3 mice did not show, despite the defect in the formation of the annulus fibrosus, ventricular pre-excitation [12]. In addition, and also quite unexpectedly, postnatal Wt1-Cre;Alk3 showed a myxomatous mitral valve phenotype, particularly of the left parietal leaflet (Fig. 20.3).

20.2.2 Discussion

Our studies convincingly show that Bmp signaling plays a critical role in the establishment of the mesenchymal and fibrous tissues at the AV junction. We propose that during normal development, AV-epiEMT is a crucial process as it is responsible for generating the critical amount of AV-EPDCs needed for the subsequent formation of the annulus fibrosus and the associated migration of AV-EPDCs into the parietal leaflets. We believe that perturbation of Bmp signaling by conditionally deleting Alk3 from the epicardial cell population inhibits AV-epiEMT resulting in a domino effect leading to (1) inhibition of AV sulcus formation, (2) failure of the annulus fibrosus to form, and (3) reduction of the number of EPDCs that migrate into the AV valve leaflets (Fig. 20.4).

20.2.3 Future Direction and Clinical Implications

Myxomatous valve disease is generally considered an acquired disease, often observed in patients with mitral valve prolapse. The observed myxomatous changes

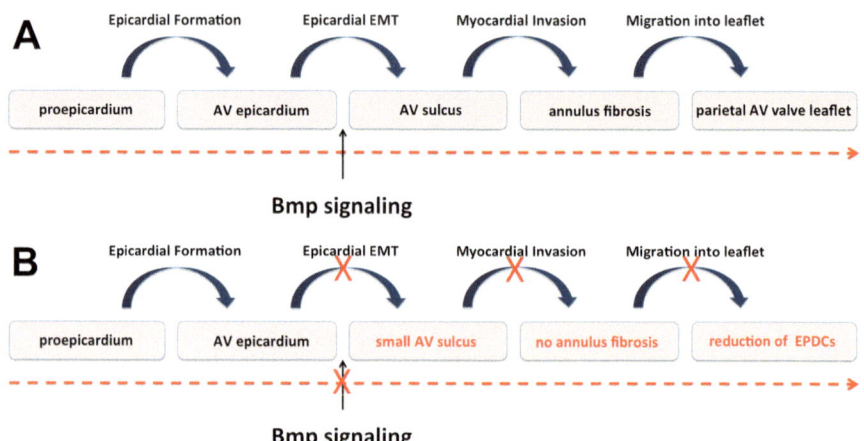

Fig. 20.4 Simplified hypothetical model showing the cascade of epicardial-related events at the AV junction. (**a**) In normal development, Bmp signaling positively regulates epiEMT resulting in the formation of the AV sulcus and subsequent development of annulus fibrosus, followed by the migration of EPDCs into the parietal leaflets. (**b**) When Bmp signaling is perturbed, AV sulcus formation and all AV-epicardial events downstream of this are inhibited

in the leaflets of the postnatal Wt1-cre;Alk3 mice, however, are associated with a reduced influx of EPDCs within those leaflets during embryonic development. We therefore propose that perturbation in the normal development of EPDCs in the AV junction might play the role in the etiology of myxomatous valve degeneration in human heart disease.

Acknowledgments The authors would like to acknowledge the financial support by NIH grants P30 GM103342 (A.W.), UL1RR029882 and UL1TR000062 (A.W.), R01HL033756 (A.W.) American Heart Association Grant-in-Aid 13GRNT16220004 (A.W.), American Heart Association Predoctoral grant 12PRE11340000 (M.M.L.), European Community's Sixth Framework Program Grant LSHM-CT-2005-018630 (MJBvdH), and Netherlands Heart Foundation Grant 1996M002 (MJBvdH).

References

1. Markwald RR, Fitzharris TP, Manasek FJ. Structural development of endocardial cushions. Am J Anat. 1977;148:85–119.
2. Wessels A, Markman MW, Vermeulen JL, Anderson RH, Moorman AF, Lamers WH. The development of the atrioventricular junction in the human heart. Circ Res. 1996;78:110–7.
3. Wessels A, Sedmera D. Developmental anatomy of the heart: a tale of mice and man. Physiol Genomics. 2003;15:165–76.
4. Snarr BS, Wirrig EE, Phelps AL, Trusk TC, Wessels A. A spatiotemporal evaluation of the contribution of the dorsal mesenchymal protrusion to cardiac development. Dev Dyn Off Publ Am Assoc Anat. 2007;236:1287–94.
5. Lamers WH, Viragh S, Wessels A, Moorman AF, Anderson RH. Formation of the tricuspid valve in the human heart. Circulation. 1995;91:111–21.
6. Cai CL, Martin JC, Sun Y, Cui L, Wang L, Ouyang K, Yang L, Bu L, Liang X, Zhang X, et al. A myocardial lineage derives from tbx18 epicardial cells. Nature. 2008;454:104–8.
7. Wessels A, van den Hoff MJ, Adamo RF, Phelps AL, Lockhart MM, Sauls K, Briggs LE, Norris RA, van Wijk B, Perez-Pomares JM, et al. Epicardially derived fibroblasts preferentially contribute to the parietal leaflets of the atrioventricular valves in the murine heart. Dev Biol. 2012;366:111–24.
8. Zhou B, Ma Q, Rajagopal S, Wu SM, Domian I, Rivera-Feliciano J, Jiang D, von Gise A, Ikeda S, Chien KR, et al. Epicardial progenitors contribute to the cardiomyocyte lineage in the developing heart. Nature. 2008;454:109–13.
9. Zhou B, Pu WT. More than a cover: epicardium as a novel source of cardiac progenitor cells. Regen Med. 2008;3:633–5.
10. Lockhart MM, Phelps AL, van den Hoff MJ, Wessels A. The epicardium and the development of the atrioventricular junction in the murine heart. J Dev Biol. 2014;2:1–17.
11. Ma L, Lu MF, Schwartz RJ, Martin JF. Bmp2 is essential for cardiac cushion epithelial-mesenchymal transition and myocardial patterning. Development. 2005;132:5601–11.
12. Lockhart MM, Boukens BJ, Phelps AL, Brown CL, Toomer KA, Burns TA, Mukherjee RD, Norris RA, Trusk TC, van den Hoff MJ et al. Alk3 mediated bmp signaling controls the contribution of epicardially derived cells to the tissues of the atrioventricular junction. Dev Biol. 2014;396:8–18.

21

TMEM100: A Novel Intracellular Transmembrane Protein Essential for Vascular Development and Cardiac Morphogenesis

Ken Mizuta, Masahide Sakabe, Satoshi Somekawa, Yoshihiko Saito, and Osamu Nakagawa

Keywords
TMEM100 • BMP • ALK1 • Cardiovascular development

Among members of the TGFβ superfamily, bone morphogenetic protein (BMP) 9 and BMP10 regulate vascular endothelium differentiation and morphogenesis by activating the specific receptor complex, which consists of ALK1 (or ACVRL1), BMPR2, and endoglin. Mutations in *ACVRL1*, *BMPR2*, or *ENG* are associated with hereditary hemorrhagic telangiectasia and pulmonary arterial hypertension in humans [1, 2]. We previously identified *TMEM100* as a downstream target gene of BMP9/BMP10-ALK1 signaling pathway [3]. TMEM100 is a novel intracellular protein with two putative transmembrane domains, and its amino acid sequence is highly conserved from fish to humans.

To clarify the physiological significance of TMEM100, we generated *Tmem100*-deficient mice and found that all mutant embryos died in utero around embryonic day 10.5 (E10.5). LacZ reporter driven by the *Tmem100* locus was predominantly expressed in endothelial cells of developing arteries and endocardium. *Tmem100* null embryos showed severe vascular dysmorphogenesis and cardiac enlargement at E9.5 and massive pericardial effusion and growth retardation at E10.5 (Fig. 21.1). These phenotypic abnormalities were virtually identical to those observed in *Alk1/Acvrl1*-deficient mice, suggesting that *Tmem100* is an important downstream gene of BMP9/BMP10-ALK1 signaling during cardiovascular

---

Note: This Chapter is also related to Part VI.

K. Mizuta • M. Sakabe • O. Nakagawa (✉)
Laboratory for Cardiovascular System Research, Nara Medical University Advanced Medical Research Center, 840 Shijo-cho, Kashihara-city, Nara 634-8521, Japan
e-mail: osamu.nakagawa@ncvc.go.jp

S. Somekawa • Y. Saito
First Department of Internal Medicine, Nara Medical University, 840 Shijo-cho, Kashihara, Nara 634-8522, Japan

T. Nakanishi et al. (eds.), *Etiology and Morphogenesis of Congenital Heart Disease*, DOI 10.1007/978-4-431-54628-3_21

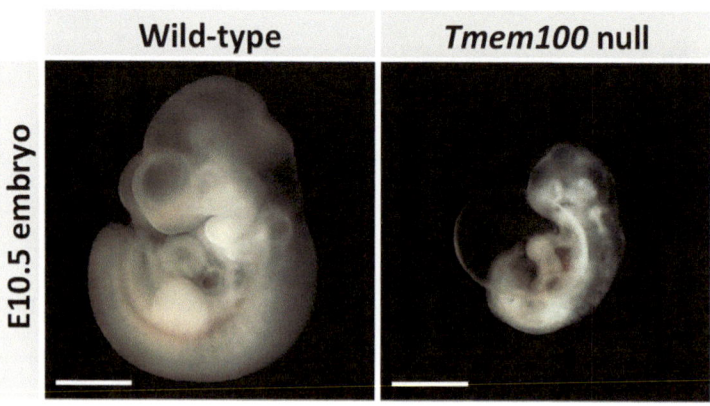

Fig. 21.1 *Tmem100* null embryos at E10.5 show severe cardiovascular dysmorphogenesis, massive pericardial effusion, and growth retardation (*scale bar*, 1 mm)

development. We also demonstrated that *Tmem100* null embryos showed atrioventricular canal cushion formation defect, indicating Tmem100 works also as an important factor for valve and septum morphogenesis.

Taken together, our studies indicate that TMEM100 is a novel endothelial-specific protein for cardiovascular morphogenesis downstream of BMP9/BMP10-ALK1 signaling. Clarifying the function of TMEM100 will lead to a better understanding of the mechanisms of cardiovascular morphogenesis and the etiologies of human congenital diseases.

References

1. Govani FS, Shovlin CL. Hereditary haemorrhagic telangiectasia: a clinical and scientific review. Eur J Hum Genet. 2009;17:860–71.
2. Lowery JW, de Caestecker MP. BMP signaling in vascular development and disease. Cytokine Growth Factor Rev. 2010;21:287–98.
3. Somekawa S, et al. Tmem100, an ALK1 receptor signaling-dependent gene essential for arterial endothelium differentiation and vascular morphogenesis. Proc Natl Acad Sci U S A. 2012;109:12064–9.

The Role of Cell Autonomous Signaling by BMP in Endocardial Cushion Cells in AV Valvuloseptal Morphogenesis

22

Yukiko Sugi, Bin Zhou, Kei Inai, Yuji Mishina, and Jessica L. Burnside

Keywords
Atrioventricular (AV) canal • Endocardial cushion • BMP-2 • BMP receptors •
Valvuloseptal morphogenesis

Distal outgrowth and fusion of the mesenchymalized AV endocardial cushions are essential morphogenetic events in AV valvuloseptal morphogenesis. BMP-2 myocardial conditional knockout (cKO) mice die by embryonic day (ED) 10.5 [1] at the initial stage for the formation of endocardial cushions, hampering investigation of the role of BMP-2 in AV valvuloseptal morphogenesis at the later stages. In our previous study, we localized BMP-2 and type I BMP receptors, *BMPR1A* and *Alk2*, in AV endocardial cushions [2, 3]. Based on their expression patterns, we hypothesize that autocrine signaling by BMP-2 within mesenchymalized AV cushions plays a critical role during AV valvuloseptal morphogenesis. To test this hypothesis, we employed recently generated endocardial/endocardial cushion-specific cre-driver line $Nfact1^{Cre}$. Unlike a previously generated $Nfatc1^{enCre}$ line whose cre-mediated recombination is restricted to AV

Y. Sugi (✉) • J.L. Burnside
Department of Regenerative Medicine and Cell Biology, Medical University of South Carolina, 171 Ashley Avenue, Charleston, SC 2925, USA
e-mail: sugiy@musc.edu

B. Zhou
Department of Genetics, Pediatrics, and Medicine (Cardiology), Albert Einstein College of Medicine of Yeshiva University, Bronx, NY, USA

K. Inai
Pediatric Cardiology, Heart Institute, Tokyo Women's Medical University, Tokyo, Japan

Y. Mishina
Department of Biological Materials Science, School of Dentistry, University of Michigan, Ann Arbor, MI, USA

© The Author(s) 2016   171
T. Nakanishi et al. (eds.), *Etiology and Morphogenesis of Congenital Heart Disease*,
DOI 10.1007/978-4-431-54628-3_22

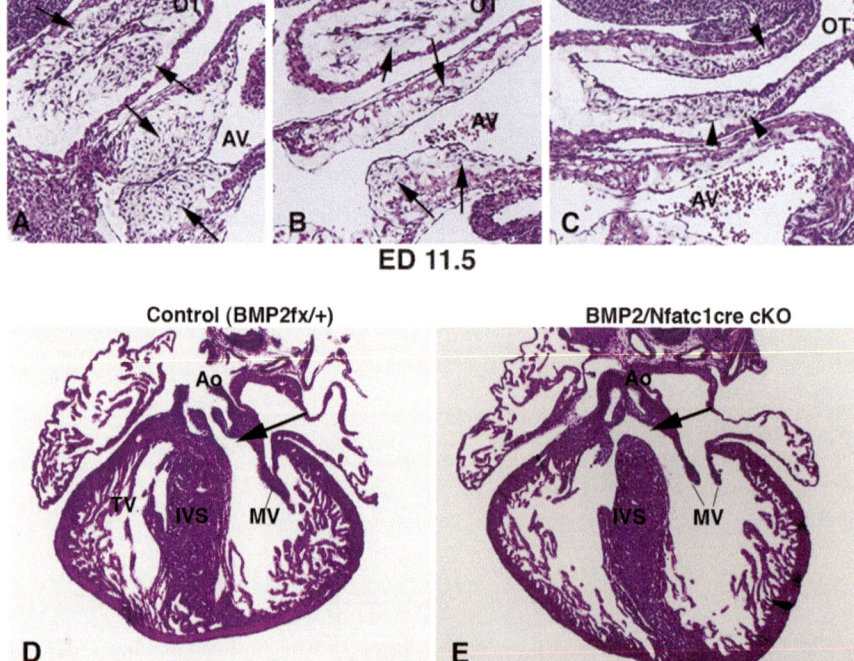

Fig. 22.1 (**a–c**) *BMPR1A cKO^Endo* mouse embryos exhibit failure of cellularization in AV cushions. *Arrows* indicate abundant mesenchymal cells in AV and outflow tract (OT) cushions in a control heart (**A**). *Arrows* show a few mesenchymal cells in AV and OT in a mutant heart (**B**). *Arrowheads* indicate abundant cells at the distal part of the mutant OT, which appear to be neural crest-derived and not reduced by *BMPR1A* endocardial/cushion-specific inactivation (**C**). (**d, e**) Endocardial/cushion-specific deletion of BMP-2 results in perimembranous septal defects (*arrow* in **E**), whereas control mouse heart shows well-formed ventricular septum (*arrow* in **D**). *Ao* aorta, *IVS* interventricular septum, *TV* tricuspid valves, *MV* mitral valves

and OT endocardium, this *Nfatc1^Cre* line confers cre-mediated recombination within the endocardial cells as well as their mesenchymal progeny. Using the *Nfactc1^Cre* driver line, we disrupted BMPR1A (Alk3) and BMP-2 specifically from AV endocardium and endocardial cushions. *BMPR1A* endocardial cushion cKO (*cKO^Endo*) mouse embryos died by ED 12.5 and exhibited failure of cellularization of AV cushions (Fig. 22.1a–c) and disruption of extracellular matrix (ECM) protein deposition in the cushion mesenchyme. On the other hand, AV cushion formation occurred in the *BMP-2 cKO^Endo* mice that survived beyond the AV cushion formation stage because BMP-2 expression remained intact in the AV myocardium during AV cushion formation. *BMP-2 cKO^Endo* mice exhibited perimembranous ventricular septal defects (VSDs) (Fig. 22.1d, e), defective deposition of ECMs in the membranous septum, and AV mitral valve dysplasia, suggesting the cell autonomous requirement of BMP-2 in AV endocardial cushions.

BMP-2 cKO^{Endo} did not exhibit muscular VSDs. These data strongly support our hypothesis that cell autonomous signaling by BMP-2 in the endocardial lineage plays a critical role in mesenchymalized AV cushions during AV valvuloseptal morphogenesis.

References

1. Ma L, Lu MF, Schwartz RJ, Martin JF. Bmp2 is essential for cardiac cushion epithelial-mesenchymal transition and myocardial patterning. Development. 2005;132:5601–11.
2. Sugi Y, Yamamura H, Okagawa H, Markwald RR. Bone morphogenetic protein-2 can mediate myocardial regulation of atrioventricular cushion mesenchymal cell formation in mice. Dev Biol. 2004;269:505–18.
3. Inai K, Norris RA, Hoffman S, Markwald RR, Sugi Y. BMP-2 induces cell migration and periostin expression during atrioventricular valvulogenesis. Dev Biol. 2008;315:383–96.

Perspective

Seema Mital

Outflow tract defects account for around 30 % of all congenital heart disease and are associated with a significant health burden and ongoing late mortality despite complete repair. Although outflow tract defects manifest as a range of anatomically distinct and apparently heterogeneous lesions – tetralogy of Fallot, double-outlet right ventricle, interrupted aortic arch, transposition of great arteries, truncus arteriosus, and other related defects – embryologically they appear to have a common origin related to abnormal development of the embryonic conotruncus during the septation of the arterial outflow tract. In the vast majority of cases (80 %), the genetic etiology is not known. The most common defect causing outflow tract defects is the 22q11del that only accounts for 10–15 % of cases overall.

In the past decade, major advances using mouse models have created a paradigm shift in our understanding of cardiac development through the discovery of the second heart field. While the primary heart field gives rise to the left ventricle and most other cardiac structures, the right ventricle and the outflow tract arise from the embryologically distinct second heart field. Proliferation and differentiation of the second heart field are controlled by a complex transcriptional network. Pioneering work in the understanding of the second heart field including lineage-tracing studies has shown that perturbations in genes of the second heart field result in abnormal formation and septation of the arterial outflow tract resulting in outflow tract defects. Computer-assisted 3D reconstruction analysis to assess spatial and

S. Mital (✉)
Division of Cardiology, Department of Pediatrics, Hospital for Sick Children, University of Toronto, 555 University Avenue, Toronto, Ontario M5G 1X8, Canada
e-mail: seema.mital@sickkids.ca

developmental gene expression patterns, morphogenesis, and proliferation in situ has further helped understand outflow tract morphogenesis and effect of dysregulation of genes regulating its development.

The foregoing chapters discuss the history of the second heart field, the central role of multipotent second heart field progenitors in cardiac development and their contribution to the definitive arterial pole, the major signaling pathways that control second heart field progenitors including the role of genes like *Mef2c*, *Cadm4*, and *Foxc2*, as well as the environmental factors in outflow tract defect development. Eventually, knowledge gained from these model organisms can be used to guide genetic screening of human subjects to identify causal genes associated with outflow tract defects in humans.

Properties of Cardiac Progenitor Cells in the Second Heart Field

Alexandre Francou and Robert G. Kelly

Abstract

Heart tube elongation occurs by progressive addition of cells from pharyngeal mcsodcrm to the poles of the heart. These progenitor cells, termed the second heart field, contribute to right ventricular and outflow tract myocardium at the arterial pole of the heart and to atrial myocardium at the venous pole. Perturbation of this process results in congenital heart defects. Since the discovery of this progenitor cell population, much has been learned about the signaling pathways and transcription factors regulating second heart field deployment. However, fundamental questions about the molecular and cellular mechanisms underlying heart tube elongation remain. Here we briefly review a selection of recent findings in the area of second heart field biology and discuss the clinical implications of these new studies for our understanding of the etiology of congenital heart defects.

Keywords

Cardiac progenitor cells • Congenital heart defect • Second heart field

23.1   Introduction

In 2001 a cardiac progenitor cell population situated in pharyngeal mesoderm was found to give rise to myocardium of the right ventricle and outflow tract. In the intervening period, it has become apparent that this progenitor cell population (1) is part of a larger population of cardiac progenitor cells termed the second heart field (SHF) that also contributes to atrial myocardium, (2) corresponds to a genetic

A. Francou • R.G. Kelly (✉)

University of Aix-Marseilles, Developmental Biology Institute of Marseilles, CNRS UMR 7288, Campus de Luminy Case 907, 13288 Marseilles Cedex 9, France

e-mail: Robert.Kelly@univ-amu.fr

T. Nakanishi et al. (eds.), *Etiology and Morphogenesis of Congenital Heart Disease*, DOI 10.1007/978-4-431-54628-3_23

lineage distinct from the first heart field (FHF) that gives rise to the cardiac crescent and left ventricle, and (3) is progressively deployed in the pharyngeal region during embryonic heart development through the activity of multiple signaling pathways and transcription factors. Of particular relevance is the discovery that perturbation of SHF development in animal models and human patients results in a spectrum of congenital heart defects (CHD) affecting the poles of the heart, including conotruncal and atrial septal defects. Details of these features of SHF development have been documented in a series of recent reviews [1, 2]. Here we will discuss a selection of recently published studies that impact on our understanding of second heart field biology, with focus on mechanistic insights into the etiology of CHD.

23.2   Demarcating the First and Second Heart Fields and Their Contributions to the Heart

The distinction between the FHF and SHF has been controversial, despite evidence from clonal analysis and genetic lineage studies that these progenitor populations correspond to separate lineages [1, 2]. Further support for a two-lineage model of heart development has been provided by a study involving Boolean modeling of gene regulatory networks in heart development [3]. This work identified two stable states corresponding to the FHF and SHF and suggests that the differences between these states are hardwired into the signaling and transcription factor interactions operative in the early embryo. This modeling approach highlights the temporal distinction between the FHF and SHF and can be used to predict gene function, providing an important step toward integrated understanding of regulatory networks during early heart development.

The discovery that *Hcn4*, encoding a nucleotide-gated channel protein, is expressed in the FHF, in a complementary pattern to the SHF gene *Isl1*, has reinforced the concept that the vertebrate heart is built from distinct progenitor cell populations [4, 5]. An inducible Cre allele of *Hcn4* has allowed evaluation of the contribution of the FHF to the definitive heart. Unlike the SHF, which contains multipotent cardiovascular progenitor cells, FHF derivatives appear to be restricted to myocardium [4]. Interestingly, the two lineages contribute differently and in a complementary manner to different components of the cardiac conduction system [5]. In particular, the right bundle branch and majority of the right Purkinje fiber system have a SHF origin. Similar observations have recently been made based on retrospective clonal analysis and regionalized transgene expression data, supporting dual contributions of the FHF and SHF to the conduction system [6]. These findings are of relevance in understanding the origins of arrhythmias resulting from perturbed development of particular segments of the conduction axis.

23.3   New Insights into the Role and Regulation of Noncanonical Wnt Signaling in the Second Heart Field and the Origins of Conotruncal CHD

Continued proliferation and delayed differentiation are defining properties of the SHF; indeed, separation of the sites of proliferation and differentiation provides a mechanism allowing rapid growth of the embryonic heart. The role and regulation of the noncanonical Wnt ligands Wnt5 and Wnt11 in these processes have been the focus of a number of recent studies. While both ligands were known to be individually required for outflow tract morphogenesis, they have now been shown to be co-required for SHF development [7], by downregulating the canonical Wnt pathway during myocardial differentiation and activating noncanonical Wnt planar cell polarity (PCP) signaling. *Wnt5* is expressed before *Wnt11* in SHF cells in the posterior dorsal pericardial wall and has been shown to be directly activated by the DiGeorge syndrome candidate gene *Tbx1*, a key regulator of proliferation and differentiation in the SHF [8]. The severity of conotruncal defects in *Wnt5a* null mice is increased in the presence of a *Tbx1* null allele, while double mutant embryos lack the right ventricle and outflow tract and die at midgestation. TBX1 directly regulates *Wnt5a* through interaction with BAF60a, a progenitor cell-specific component of the BAF chromatin remodeling complex, as well as the histone methyltransferase SETD7, interactions shown to be necessary for activation of a number of TBX1 transcriptional targets in the SHF [8]. *Tbx1* itself is regulated by the activity of a histone acetyltransferase, MOZ, loss of function of which partially phenocopies DiGeorge syndrome [9]. The intersection of *Tbx1* with chromatin regulators is highly significant given the recent finding that de novo mutations in genes encoding such molecules are overrepresented in human CHD patients [10].

In support of a role for Tbx1 upstream of noncanonical Wnt signaling, *wnt11r* is downregulated in *tbx1* mutant zebrafish; heart looping and differentiation defects in the absence of *tbx1* can be partially rescued by ectopic wnt11r or a wnt11r target gene encoding a cell adhesion molecule, alcama [11]. Altered cell shape in *tbx1* mutant fish hearts suggests that noncanonical Wnt signaling regulates cardiomyocyte cell polarity downstream of tbx1 [11]. Whether Tbx1 and noncanonical Wnt signaling also regulate cell polarity in the SHF remains to be seen. SHF and conotruncal development is impaired in mice lacking Dvl1 and Dvl2, regulators of both canonical Wnt signaling and the noncanonical Wnt PCP pathway [12]. In this study, the PCP signaling function of *Dvl* genes was shown to be specifically required in the SHF lineage and the cardiac phenotype to resemble that of embryos lacking the core PCP gene *Vangl2*; furthermore, *Wnt5a* and *Vangl2* interact to increase the severity of outflow tract defects [12]. Loss of PCP gene function results in disorganization of progenitor cells in the posterior dorsal pericardial wall, potentially resulting in impaired SHF deployment toward the outflow tract and conotruncal anomalies.

Among the last myocardial derivatives of the SHF to be added to the elongating heart tube is future subpulmonary myocardium, a cell population specifically affected in *Tbx1* mutant embryos [13]. Asymmetric addition of SHF progenitor cells giving rise to subpulmonary myocardium continues on the left side of the outflow tract up until embryonic day 12.5 in the mouse; furthermore, this late contribution drives rotation of the outflow tract, positioning subpulmonary myocardium on the ventral side of the heart and aligning the ascending aorta with the left ventricle [14]. Failure of this process, termed the "pulmonary push," results in outflow tract alignment defects such as double outlet right ventricle. Underdevelopment of subpulmonary myocardium has been implicated in the etiology of tetralogy of Fallot [15], and further analysis of the regulation of future subpulmonary myocardium is an important step toward deciphering the etiology of CHD.

23.4 Involvement of the Second Heart Field in Atrial and Atrioventricular Septal Defects

At the venous pole of the heart, SHF cells contribute to the dorsal mesenchymal protrusion that bridges the atrioventricular cushions and primary atrial septum. Failure of proliferation or precocious differentiation of these cells results in primum atrial septal defects when canonical Wnt or hedgehog signaling, respectively, is compromised [1, 2]. BMP signaling has now been shown to promote the proliferation of DMP progenitor cells, in contrast to the pro-differentiation role of this signaling pathway during SHF addition at the arterial pole [16]. Isl1, Wnt2, and the transcription factor Gli1 are expressed in this region of the SHF and have been shown to identify a multipotent cardiopulmonary progenitor cell population that contributes not only to the venous pole of the heart but also to diverse pulmonary lineages, including smooth muscle and endothelium [17]. The development of these cells that ensure the vascular connection between the heart and lung is coordinated by hedgehog signaling from future pulmonary endoderm. The Holt-Oram syndrome gene *Tbx5* also operates in this posterior component of the SHF and is required for normal atrial septation through regulating proliferation and hedgehog signal reception by direct activation of cell cycle progression genes such as *Cdk6* and the hedgehog signaling component Gas1 [18]. Genetic and retrospective lineage studies have revealed that venous pole and future subpulmonary myocardium are clonally related, suggesting that a population of common SHF progenitor cells segregates to the arterial and venous pole of the heart [19]. Where such common progenitor cells are located and how their segregation to the poles is regulated remains unknown.

23.5 Future Directions and Clinical Implications

The importance of perturbed SHF development in the etiology of common forms of CHD affecting both poles of the heart is now clear. The studies discussed here provide new insights into the underlying mechanisms, although much remains to be learnt about the regulatory interactions between signaling pathways and transcription factors controlling cellular properties of the SHF and how different regions of the heart are prepatterned and segregate within the progenitor cell population. Future research will address these questions in the context of dynamic heart tube elongation. The human SHF, as defined by ISl1 expression, coincides with that observed in avian and mouse embryos [20], and thus findings from these models are directly relevant to a better understanding of the origins of CHD in man. Furthermore, insights into how the differentiation of cardiac progenitor cells is controlled are essential for future regenerative therapies.

Acknowledgments Supported by the Fondation pour la Recherche Médicale, Agence National pour la Recherche and Association Française contre les Myopathies.

References

1. Vincent SD, Buckingham ME. How to make a heart: the origin and regulation of cardiac progenitor cells. Curr Top Dev Biol. 2010;90:1–41.
2. Kelly RG. The second heart field. Curr Top Dev Biol. 2012;100:33–65.
3. Herrmann F, Gross A, Zhou D, et al. A boolean model of the cardiac gene regulatory network determining first and second heart field identity. PLoS One. 2012;7:e46798.
4. Spater D, Abramczuk MK, Buac K, et al. A HCN4+ cardiomyogenic progenitor derived from the first heart field and human pluripotent stem cells. Nat Cell Biol. 2013;15:1098–106.
5. Liang X, Wang G, Lin L, et al. HCN4 dynamically marks the first heart field and conduction system precursors. Circ Res. 2013;113:399–407.
6. Miquerol L, Bellon A, Moreno N, et al. Resolving cell lineage contributions to the ventricular conduction system with a Cx40-GFP allele: a dual contribution of the first and second heart fields. Dev Dyn. 2013;242:665–77.
7. Cohen ED, Miller MF, Wang Z, et al. Wnt5a and Wnt11 are essential for second heart field progenitor development. Development. 2012;139:1931–40.
8. Chen L, Fulcoli FG, Ferrentino R, et al. Transcriptional control in cardiac progenitors: Tbx1 interacts with the BAF chromatin remodeling complex and regulates Wnt5a. PLoS Genet. 2012;8:e1002571.
9. Voss AK, Vanyai HK, Collin C, et al. MOZ regulates the Tbx1 locus, and Moz mutation partially phenocopies DiGeorge syndrome. Dev Cell. 2012;23:652–63.
10. Zaidi S, Choi M, Wakimoto H, et al. De novo mutations in histone-modifying genes in congenital heart disease. Nature. 2013;498:220–3.
11. Choudhry P, Trede NS. DiGeorge syndrome gene tbx1 functions through wnt11r to regulate heart looping and differentiation. PLoS One. 2013;8:e58145.
12. Sinha T, Wang B, Evans S, et al. Disheveled mediated planar cell polarity signaling is required in the second heart field lineage for outflow tract morphogenesis. Dev Biol. 2012;370:135–44.

13. Theveniau-Ruissy M, Dandonneau M, Mesbah K, et al. The del22q11.2 candidate gene Tbx1 controls regional outflow tract identity and coronary artery patterning. Circ Res. 2008;103:142–8.
14. Scherptong RW, Jongbloed MR, Wisse LJ, et al. Morphogenesis of outflow tract rotation during cardiac development: the pulmonary push concept. Dev Dyn. 2012;241:1413–22.
15. Van Praagh R. The first Stella van Praagh memorial lecture: the history and anatomy of tetralogy of Fallot. Semin Thorac Cardiovasc Surg Pediatr Card Surg Ann. 2009:19–38.
16. Briggs LE, Phelps AL, Brown E, et al. Expression of the BMP receptor Alk3 in the second heart field is essential for development of the dorsal mesenchymal protrusion and atrioventricular septation. Circ Res. 2013;112:1420–32.
17. Peng T, Tian Y, Boogerd CJ, et al. Coordination of heart and lung co-development by a multipotent cardiopulmonary progenitor. Nature. 2013;500:589–92.
18. Xie L, Hoffmann AD, Burnicka-Turek O, et al. Tbx5-hedgehog molecular networks are essential in the second heart field for atrial septation. Dev Cell. 2012;23:280–91.
19. Lescroart F, Mohun T, Meilhac SM, et al. Lineage tree for the venous pole of the heart: clonal analysis clarifies controversial genealogy based on genetic tracing. Circ Res. 2012;111:1313–22.
20. Sizarov A, Ya J, de Boer BA, et al. Formation of the building plan of the human heart: morphogenesis, growth, and differentiation. Circulation. 2011;123:1125–35.

Nodal Signaling and Congenital Heart Defects

24

Ralston M. Barnes and Brian L. Black

Abstract

Nodal is a TGF-β family member ligand that is critical for early embryonic patterning in vertebrates. Nodal signaling functions through core TGF-β receptors to activate a Smad transcription factor signaling cascade. However, unlike other TGF-β ligands, Nodal signaling requires an additional co-receptor of the EGF-CFC family to activate intracellular signaling. Nodal signaling is also subject to extensive negative regulation by Lefty and other factors. Work in numerous model organisms, including mouse, chicken, and zebrafish, established that Nodal signaling plays an essential role during germ layer formation, anterior-posterior axis patterning, and left-right axis determination. Incomplete or delayed loss of Nodal signaling results in defective organogenesis and birth defects, including congenital heart defects, and clinical studies have linked aberrant Nodal signaling in humans with many common congenital malformations, including congenital heart defects. Congenital heart defects associated with disrupted Nodal signaling in mammals include those that arise due to global defects in left-right patterning of the embryo, such as heterotaxy. Other Nodal-associated heart defects appear to occur as more subtle isolated malformations of the great arteries and atrioventricular septum, which may not be related to overall perturbations in laterality. A more detailed understanding of the Nodal signaling pathway and its targets in the heart is required to more fully understand the etiology of Nodal signaling pathway-associated congenital heart defects.

Keywords

Nodal signaling • *Tdgf1* • Cripto • Congenital heart defects • TGF-β • Laterality defects

Note: This chapter is also related to Part II.

R.M. Barnes • B.L. Black (✉)
Cardiovascular Research Institute, University of California, San Francisco, CA, USA
e-mail: brian.black@ucsf.edu

T. Nakanishi et al. (eds.), *Etiology and Morphogenesis of Congenital Heart Disease*,
DOI 10.1007/978-4-431-54628-3_24

183

24.1   Introduction

The Nodal signaling pathway is critical for early embryonic patterning in all vertebrates. Nodal is a secreted signaling molecule that belongs to the transforming growth factor-β (TGF-β) family and functions primarily to activate downstream signaling through a receptor-mediated response [1–3]. The Nodal pathway plays a critical role in mesoderm specification and anterior-posterior patterning during gastrulation and is also essential for establishing the left-right axis in the developing embryo [4–7]. Changes in Nodal expression or dosage can disrupt left-right patterning and result in a range of congenital defects that affect development of the forebrain, the craniofacial skeleton, and several other organs, including the heart.

24.2   The Nodal Signaling Pathway

The Nodal signaling pathway shares many similarities with other TGF-β signaling pathways in that it utilizes core Smad-dependent signaling components (Fig. 24.1). Nodal is secreted extracellularly as a proprotein homodimer like other TGF-β ligands [8]. Once cleaved into the mature ligand, the Nodal homodimer binds tightly to a TGF-β receptor heterodimer consisting of both Type I and Type II receptors [9]. Nodal ligand binding causes the constitutively active Type II receptor serine/threonine kinases to associate with the inactive Type I receptor kinases and leads to phosphorylation and the subsequent dissociation of R-Smad from the TGF-β receptor [10]. Following formation of a trimeric complex composed of two R-Smads and the common partner Smad, Smad4, the Smad oligomer translocates to the nucleus where it regulates gene expression through direct and indirect DNA binding (Fig. 24.1) [1, 11, 12].

The Nodal signaling pathway has important distinctions from other TGF-β-mediated signaling pathways. Nodal can only activate TGF-β receptor signaling in the presence of an EGF-CFC protein (Fig. 24.1). There are two EGF-CFC proteins, Cripto and Cryptic, which function as co-receptors for Nodal [13]. Cripto and Cryptic are extracellular proteins that contain an epidermal growth factor-like motif and a novel cystine-rich domain named the CFC [13]. EGF-CFC proteins function primarily by binding to the Type II TGF-β receptor through the CFC domain and by binding to Nodal through the EGF domain [14].

In response to Nodal signaling, the Smad complex cooperates specifically with FoxH1, a winged helix transcription factor, or Mixer, a member of the Mix subclass of homeodomain proteins [15, 16]. These cofactors are critical for Nodal-dependent downstream gene activation and act to stabilize Smad-DNA interactions, since Smads have relatively weak DNA-binding affinity [17, 18]. FoxH1 and Mixer recruit the Smad complex to promoter and enhancer elements and help to establish temporal and spatial regulation of Nodal-dependent target genes.

In addition, the Nodal signaling pathway is subject to specific negative regulation not found for other TGF-β family members. Proteins of the Lefty family, specifically Lefty1 and Lefty2, inhibit Nodal-dependent activation of TGF-β

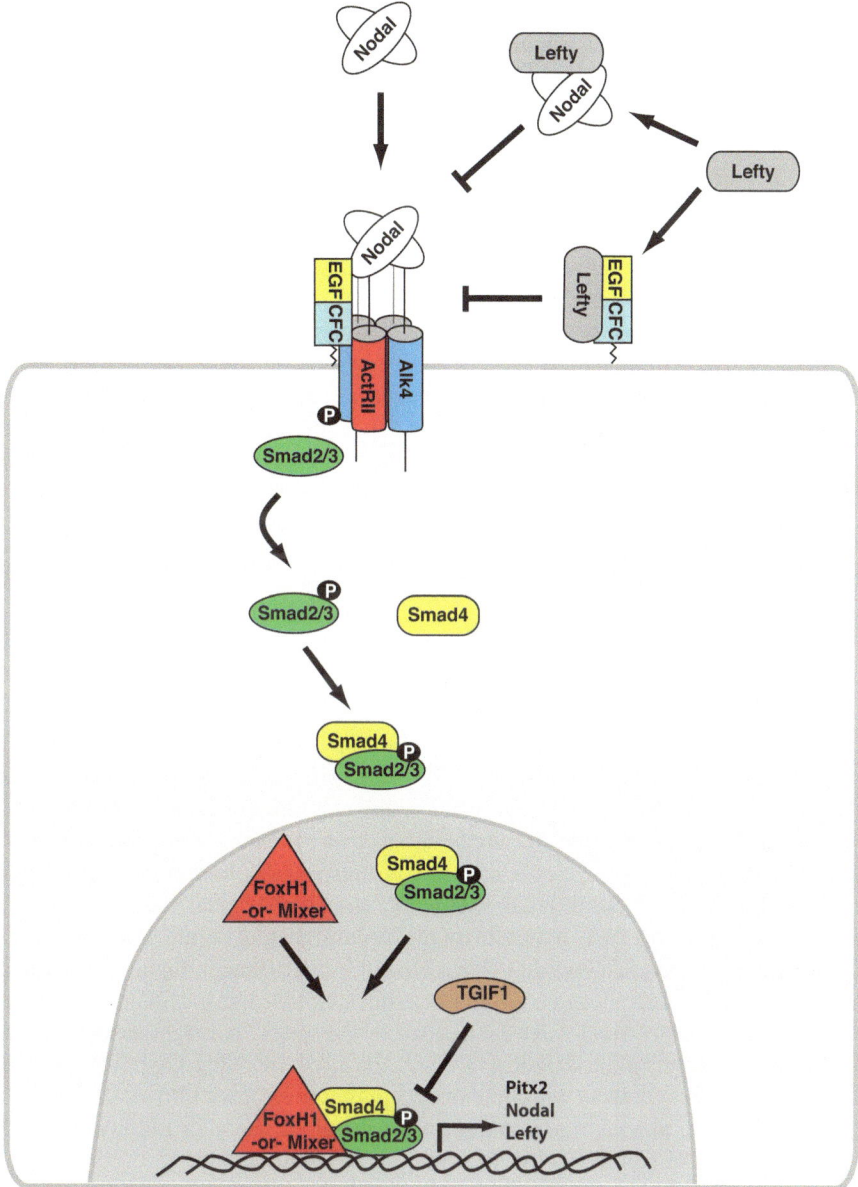

Fig. 24.1   The Nodal signaling pathway. The Nodal ligand binds to a dimer of the TGF-β Type I and Type II receptors. In association with an EGF-CFC co-receptor (Cripto or Cryptic), Nodal activates the receptor complex causing the phosphorylation of either Smad2 or Smad3 followed by oligomerization with the common partner Smad4. The Smad2/Smad3-Smad4 complex then translocates to the nucleus where it binds to DNA with the transcriptional cofactor FoxH1 or Mixer, leading to transcription of downstream target genes. The Nodal signaling pathway is negatively regulated by proteins such as Lefty, which can bind to either EGF-CFC or the Nodal dimer to prevent activation of the receptor complex, or TGIF1, which recruits histone deacetylases to Smad2/Smad3 and represses transcriptional activation

receptors by interacting directly with the Nodal ligand and preventing binding to the TGF-β receptor heterodimer and by binding to the EGF-CFC Nodal co-receptors Cripto and Cryptic and preventing their association with the TGF-β receptor complex (Fig. 24.1) [19–21].

24.3   Requirement for Nodal in Development

Nodal signaling is critical for the patterning of the developing embryo [4, 22]. Loss-of-function studies in vertebrate model organisms indicate that Nodal signaling is first required during gastrulation. Germline loss of *Nodal* in mice results in severe patterning and differentiation defects and embryonic lethality due to a failure to induce the primitive streak from the ectoderm and to disrupted specification of mesoderm and endoderm from the epiblast [5]. Additionally, loss of *Nodal* in mice results in impaired anterior-posterior axis formation due to the lack of anterior visceral endoderm formation [22]. The requirement of Nodal signaling for early embryonic pattern formation was further highlighted by loss-of-function studies of the EGF-CFC Nodal co-receptor Cripto. Inactivation of *Tdgf1*, the gene encoding Cripto, in mice results in a phenocopy of early Nodal defects, including lethality shortly after gastrulation. *Tdgf1* mutants lack a primitive streak, fail to form embryonic mesoderm, and exhibit anterior-posterior axis defects [6]. The *one-eyed pinhead* (*oep*) mutation in zebrafish results in a complete loss of function of the fish ortholog of the Nodal co-receptor Cripto [23]. These mutants exhibit a phenocopy of the early Nodal defects seen in mice, including an absence of mesoderm and anterior-posterior axis abnormalities [23]. Together, both animal models establish that EGF-CFC proteins are required for Nodal signaling and support an early requirement for the Nodal pathway in embryo morphogenesis.

Following gastrulation, Nodal signaling is indispensable for the establishment of left-right asymmetry. Conditional deletion of *Nodal* in the lateral plate mesoderm in mice circumvents the early requirement for Nodal signaling during gastrulation and results in heterotaxy, a condition characterized by left-right ambiguity of thoracic and abdominal visceral organs [24]. These mice exhibit transposition of the great arteries of the heart, right-sided isomerism of the lungs, and right-sided stomach [25]. Similarly, germline deletion of *Cfc1*, the gene encoding Cryptic, results in left-right laterality defects with mutants exhibiting heterotaxy [26].

In addition to playing a role in the establishment of laterality following gastrulation, Nodal signaling is also required for midline patterning of the ventral forebrain [27]. Zebrafish mutants for the Nodal ligands and for the ortholog of Cripto result in holoprosencephaly, a condition where bifurcation of the ventral forebrain fails to occur and results in fusion of the two brain hemispheres [28]. Similarly, mice heterozygous for germline knockout alleles of both *Nodal* and *Smad2* have cyclopia, a rare and severe form of holoprosencephaly [29], further supporting the relationship between Nodal signaling and forebrain development. Mechanistically, this phenotype is thought to occur due to the patterning of *Sonic Hedgehog* (*Shh*) expression in the forebrain by Nodal [30, 31].

24.4 Congenital Heart Defects Associated with Perturbations in Nodal Signaling

It is perhaps not surprising to find that the Nodal signaling pathway is associated with pathogenesis in humans, given its critical role in patterning during embryonic development. Nodal signaling was first linked to human congenital defects through the identification of mutations associated with left-right laterality defects. These include mutations in genes encoding Lefty family members, the EGF-CFC co-receptor Cryptic, and the Type II TGF-β receptors [32–34]. In addition, mutations in genes encoding transcriptional inhibitors of Smad2, such as TGIF1, are associated with holoprosencephaly, a defect strongly associated with disrupted Nodal signaling [30, 35]. Interestingly, these human pathologies are similar to the defects observed in animal models with defective (but not incomplete) Nodal signaling.

Congenital heart defects have also been linked to aberrant Nodal signaling (Table 24.1) [36]. Loss-of-function mutations in genes encoding numerous Nodal signaling components, including Nodal, Cripto, Cryptic, and FoxH1, have been identified in patients with heart defects [37, 38]. The spectrum of heart defects in these patients can be roughly grouped into two broadly defined classes: (1) those that occur as a result of overall isomerism or heterotaxy and (2) those that occur as isolated congenital heart defects. The isomerisms of the heart can be classified as *situs inversus totalis*, a complete mirror image of the visceral organs of the body including the heart, or *situs inversus ambiguous*, where the abdominal and visceral organs are distributed abnormally and randomly in a condition more commonly called heterotaxy [39]. Both of these conditions can be linked to aberrant Nodal signaling [40]. Heterotaxy often results in complex congenital heart defects [41, 42]. These defects include levo-transposition of the great arteries (l-TGA) and atrial isomerism [36, 39]. The feature characteristics of l-TGA are improper positioning of the aorta and the pulmonary artery such that the arteries are switched in conjunction with the ventricles such that they still have the normal relationship between the ventricles and the arteries [43].

Isolated congenital heart defects associated with Nodal mutations can result in structural and functional abnormalities that appear to be independent of overall left-right ambiguity. These isolated defects include dextro-transposition of the great arteries (d-TGA), double outlet right ventricle (DORV), tetralogy of Fallot, and isolated ventricular septal defects [37, 38]. Unlike l-TGA, which results in a proper alignment of the arteries with respect to the ventricles, d-TGA results in a switching of aorta and the pulmonary artery such that the pulmonary artery connects to the left ventricle and the aorta emanates from the right ventricle [43]. DORV, as the name implies, occurs when the aorta and the pulmonary artery connect to the right ventricle. Tetralogy of Fallot is essentially a milder form of DORV in which the aorta overrides the ventricular septum and empties blood from both ventricles [44]. There is also pulmonary artery stenosis and a hypertrophic right ventricle secondary to pulmonary artery blockage associated with tetralogy of Fallot [44]. These defects appear to be independent of overall laterality, although the

Table 24.1 Some of the congenital defects associated with altered Nodal signaling in mammals

Component	Species	CV defects	Other defects	References
Nodal	Human	Dextrocardia, d-TGA, DORV, VSD, ASD, l-TGA, DILV, PA, TOF	Asplenia, bilateral trilobed lungs, hydronephrosis, HPE, intestinal malrotation	[36, 38]
	Mouse	l-TGA, VSD	Heterotaxy, asplenia, isomerisms, HPE, cyclopia, disrupted endoderm and mesoderm specification, defective A-P axis	[5, 22, 25, 27]
Cryptic	Human	Dextrocardia, TGA, d-TGA, PA, VSD, ASD, TOF, DORV	Heterotaxy, isomerisms, polysplenia, asplenia	[32, 37, 49]
	Mouse	l-TGA, ASD	Heterotaxy, r-isomerism of the lung, hyposplenia	[26]
Cripto	Human	TOF, VSD, ASD	HPE	[37, 50, 51]
	Mouse		HPE, disrupted mesoderm specification, defective A-P axis	[6, 52]
Smad2	Human	Dextrocardia, d-TGA, DORV, ASD	Heterotaxy, asplenia, HPE	[37, 53]
TGF-βR2	Mouse	TGA, DORV, dextrocardia, levocardia, VSD, ASD, arch artery defects	R-isomerism of the lung, axial skeleton abnormalities	[54]
FoxH1	Human	VSD, TGA, TOF	HPE	[37, 55]
	Mouse		Defective elongation of the primitive streak, defective A-P axis	[16]

mechanistic basis for why these defects occur when Nodal signaling is disrupted is unclear. Interestingly, *Nodal* expression occurs well prior to the patterning and final alignment of the aorta and pulmonary artery [40]. Furthermore, the temporal expression pattern of *Nodal* observed during development is tightly regulated within a narrow developmental window due, at least in part, to the extensive negative feedback by Lefty proteins and other factors. The diverse nature of heterotaxy in humans suggests that some isolated congenital heart defects associated with perturbed Nodal signaling may still be secondary to overall laterality defects.

A role for Nodal signaling in isolated congenital heart defects is also supported by studies in mice in which subtle perturbations of Nodal signaling result in less severe defects (Table 24.1), suggesting that the Nodal ligand functions in a dosage-dependent manner. For example, deletion of an intronic enhancer of *Nodal* resulted in decreased expression of *Nodal* in the lateral plate mesoderm. These mouse embryos developed laterality defects that were less severe than the heterotaxy observed in *Nodal* conditional knockouts where *Nodal* expression in lateral plate mesoderm was completely abolished [25, 45].

Consistent with Nodal function in left-right patterning, Nodal-dependent target genes are also critical for left-right patterning and heart development [21]. *Pitx2* is perhaps the best-described target gene of Nodal signaling and is expressed asymmetrically on the left side after gastrulation [46]. Misexpression of Pitx2 on the embryonic right side in mice results in heterotaxia with conditions such as aberrant heart looping, cardiac isomerism, and visceral organ laterality defects [46]. In the mouse, germline loss of function of *Pitx2* results in left-right asymmetry defects in specific organs, such as the lung [47]. Interestingly, *Pitx2*-null and isoform-specific *Pitx2c*-null embryos undergo normal heart looping but have a subset of congenital cardiovascular anomalies such as DORV and ventricular and atrial septal defects [47, 48]. These observations suggest that Pitx2 functions in heart development after left-right determination and that other Nodal-dependent target genes may be required for cardiac laterality. Together, these observations suggest that other unappreciated Nodal-dependent target genes are involved in the establishment of left-right identity and cardiac development. A more detailed elucidation of this fundamental pathway, including target genes in the cardiac mesoderm, is required to more fully understand the role of Nodal signaling in heart development and in congenital heart defects.

References

1. Massague J, Seoane J, Wotton D. Smad transcription factors. Genes Dev. 2005;19:2783–810.
2. Schier AF. Nodal morphogens. Cold Spring Harb Perspect Biol. 2009;1:a003459.
3. Shen MM. Nodal signaling: developmental roles and regulation. Development. 2007;134:1023–34.
4. Brennan J, Norris DP, Robertson EJ. Nodal activity in the node governs left-right asymmetry. Genes Dev. 2002;16:2339–44.
5. Conlon FL, Lyons KM, Takaesu N, Barth KS, Kispert A, Herrmann B, et al. A primary requirement for nodal in the formation and maintenance of the primitive streak in the mouse. Development. 1994;120:1919–28.
6. Ding J, Yang L, Yan YT, Chen A, Desai N, Wynshaw-Boris A, et al. Cripto is required for correct orientation of the anterior-posterior axis in the mouse embryo. Nature. 1998;395:702–7.

7. Perea-Gomez A, Vella FD, Shawlot W, Oulad-Abdelghani M, Chazaud C, Meno C, et al. Nodal antagonists in the anterior visceral endoderm prevent the formation of multiple primitive streaks. Dev Cell. 2002;3:745–56.
8. Walton KL, Makanji Y, Harrison CA. New insights into the mechanisms of activin action and inhibition. Mol Cell Endocrinol. 2012;359:2–12.
9. Reissmann E, Jornvall H, Blokzijl A, Andersson O, Chang C, Minchiotti G, et al. The orphan receptor ALK7 and the activin receptor ALK4 mediate signaling by nodal proteins during vertebrate development. Genes Dev. 2001;15:2010–22.
10. Moustakas A, Heldin CH. The regulation of TGFbeta signal transduction. Development. 2009;136:3699–714.
11. Ali MH, Imperiali B. Protein oligomerization: how and why. Bioorg Med Chem. 2005;13:5013–20.
12. Moustakas A. Smad signalling network. J Cell Sci. 2002;115:3355–6.
13. Shen MM, Schier AF. The EGF-CFC gene family in vertebrate development. Trends Genet. 2002;16:303–9.
14. Yeo C, Whitman M. Nodal signals to Smads through Cripto-dependent and Cripto-independent mechanisms. Mol Cell. 2001;7:949–57.
15. Kunwar PS, Zimmerman S, Bennett JT, Chen Y, Whitman M, Schier AF. Mixer/Bon and FoxH1/Sur have overlapping and divergent roles in Nodal signaling and mesendoderm induction. Development. 2003;130:5589–99.
16. Yamamoto M, Meno C, Sakai Y, Shiratori H, Mochida K, Ikawa Y, et al. The transcription factor FoxH1 (FAST) mediates Nodal signaling during anterior-posterior patterning and node formation in the mouse. Genes Dev. 2001;15:1242–56.
17. Germain S, Howell M, Esslemont GM, Hill CS. Homeodomain and winged-helix transcription factors recruit activated Smads to distinct promoter elements via a common Smad interaction motif. Genes Dev. 2000;14:435–51.
18. Osada SI, Saijoh Y, Frisch A, Yeo CY, Adachi H, Watanabe M, et al. Activin/nodal responsiveness and asymmetric expression of a Xenopus nodal-related gene converge on a FAST-regulated module in intron 1. Development. 2000;127:2503–14.
19. Chen C, Shen MM. Two modes by which Lefty proteins inhibit nodal signaling. Curr Biol. 2004;14:618–24.
20. Cheng SK, Olale F, Brivanlou AH, Schier AF. Lefty blocks a subset of TGFbeta signals by antagonizing EGF-CFC coreceptors. PLoS Biol. 2004;2:E30.
21. Shiratori H, Hamada H. The left-right axis in the mouse: from origin to morphology. Development. 2006;133:2095–104.
22. Brennan J, Lu CC, Norris DP, Rodriguez TA, Beddington RS, Robertson EJ. Nodal signalling in the epiblast patterns the early mouse embryo. Nature. 2001;411:965–9.
23. Gritsman K, Zhang J, Cheng S, Heckscher E, Talbot WS, Schier AF. The EGF-CFC protein one-eyed pinhead is essential for nodal signaling. Cell. 1999;97:121–32.
24. Lin AE, Ticho BS, Houde K, Westgate MN, Holmes LB. Heterotaxy: associated conditions and hospital-based prevalence in newborns. Genet Med. 2000;2:157–72.
25. Kumar A, Lualdi M, Lewandoski M, Kuehn MR. Broad mesodermal and endodermal deletion of Nodal at postgastrulation stages results solely in left/right axial defects. Dev Dyn. 2008;237:3591–601.
26. Gaio U, Schweickert A, Fischer A, Garratt AN, Muller T, Ozcelik C, et al. A role of the cryptic gene in the correct establishment of the left-right axis. Curr Biol. 1999;9:1339–42.
27. Lowe LA, Yamada S, Kuehn MR. Genetic dissection of nodal function in patterning the mouse embryo. Development. 2001;128:1831–43.
28. Hayhurst M, McConnell SK. Mouse models of holoprosencephaly. Curr Opin Neurol. 2003;16:135–41.
29. Nomura M, Li E. Smad2 role in mesoderm formation, left-right patterning and craniofacial development. Nature. 1998;393:786–90.

30. Gripp KW, Wotton D, Edwards MC, Roessler E, Ades L, Meinecke P, et al. Mutations in TGIF cause holoprosencephaly and link NODAL signalling to human neural axis determination. Nat Genet. 2000;25:205–8.
31. Rohr KB, Barth KA, Varga ZM, Wilson SW. The nodal pathway acts upstream of hedgehog signaling to specify ventral telencephalic identity. Neuron. 2001;29:341–51.
32. Bamford RN, Roessler E, Burdine RD, Saplakoglu U, dela Cruz J, Splitt M, et al. Loss-of-function mutations in the EGF-CFC gene CFC1 are associated with human left-right laterality defects. Nat Genet. 2000;26:365–9.
33. Kosaki K, Bassi MT, Kosaki R, Lewin M, Belmont J, Schauer G, et al. Characterization and mutation analysis of human LEFTY A and LEFTY B, homologues of murine genes implicated in left-right axis development. Am J Hum Genet. 1999;64:712–21.
34. Kosaki R, Gebbia M, Kosaki K, Lewin M, Bowers P, Towbin JA, et al. Left-right axis malformations associated with mutations in ACVR2B, the gene for human activin receptor type IIB. Am J Med Genet. 1999;82:70–6.
35. Massague J, Blain SW, Lo RS. TGFbeta signaling in growth control, cancer, and heritable disorders. Cell. 2000;103:295–309.
36. Mohapatra B, Casey B, Li H, Ho-Dawson T, Smith L, Fernbach SD, et al. Identification and functional characterization of NODAL rare variants in heterotaxy and isolated cardiovascular malformations. Hum Mol Genet. 2009;18:861–71.
37. Roessler E, Ouspenskaia MV, Karkera JD, Velez JI, Kantipong A, Lacbawan F, et al. Reduced NODAL signaling strength via mutation of several pathway members including FOXH1 is linked to human heart defects and holoprosencephaly. Am J Hum Genet. 2008;83:18–29.
38. Roessler E, Pei W, Ouspenskaia MV, Karkera JD, Velez JI, Banerjee-Basu S, et al. Cumulative ligand activity of NODAL mutations and modifiers are linked to human heart defects and holoprosencephaly. Mol Genet Metab. 2009;98:225–34.
39. Bisgrove BW, Morelli SH, Yost HJ. Genetics of human laterality disorders: insights from vertebrate model systems. Annu Rev Genomics Hum Genet. 2003;4:1–32.
40. Lowe LA, Supp DM, Sampath K, Yokoyama T, Wright CV, Potter SS, et al. Conserved left-right asymmetry of nodal expression and alterations in murine situs inversus. Nature. 1996;381:158–61.
41. Shiraishi I, Ichikawa H. Human heterotaxy syndrome - from molecular genetics to clinical features, management, and prognosis. Circ J. 2012;76:2066–75.
42. Bowers PN, Brueckner M, Yost HJ. The genetics of left-right development and heterotaxia. Semin Perinatol. 1996;20:577–88.
43. Warnes CA. Transposition of the great arteries. Circulation. 2006;114:2699–709.
44. Fox D, Devendra GP, Hart SA, Krasuski RA. When 'blue babies' grow up: what you need to know about tetralogy of Fallot. Cleve Clin J Med. 2010;77:821–8.
45. Norris DP, Brennan J, Bikoff EK, Robertson EJ. The Foxh1-dependent autoregulatory enhancer controls the level of Nodal signals in the mouse embryo. Development. 2002;129:3455–68.
46. Logan M, Pagan-Westphal SM, Smith DM, Paganessi L, Tabin CJ. The transcription factor Pitx2 mediates situs-specific morphogenesis in response to left-right asymmetric signals. Cell. 1998;94:307–17.
47. Kitamura K, Miura H, Miyagawa-Tomita S, Yanazawa M, Katoh-Fukui Y, Suzuki R, et al. Mouse Pitx2 deficiency leads to anomalies of the ventral body wall, heart, extra- and periocular mesoderm and right pulmonary isomerism. Development. 1999;126:5749–58.
48. Liu C, Liu W, Palie J, Lu MF, Brown NA, Martin JF. Pitx2c patterns anterior myocardium and aortic arch vessels and is required for local cell movement into atrioventricular cushions. Development. 2002;129:5081–91.
49. Goldmuntz E, Bamford R, Karkera JD, de la Cruz J, Roessler E, Muenke M. CFC1 mutations in patients with transposition of the great arteries and double-outlet right ventricle. Am J Hum Genet. 2002;70:776–80.

50. de la Cruz JM, Bamford RN, Burdine RD, Roessler E, Barkovich AJ, Donnai D, et al. A loss-of-function mutation in the CFC domain of TDGF1 is associated with human forebrain defects. Hum Genet. 2002;110:422–8.

51. Wang B, Yan J, Peng Z, Wang J, Liu S, Xie X, et al. Teratocarcinoma-derived growth factor 1 (TDGF1) sequence variants in patients with congenital heart defect. Int J Cardiol. 2011;146:225–7.

52. Chu J, Ding J, Jeays-Ward K, Price SM, Placzek M, Shen MM. Non-cell-autonomous role for Cripto in axial midline formation during vertebrate embryogenesis. Development. 2005;132:5539–51.

53. Zaidi S, Choi M, Wakimoto H, Ma L, Jiang J, Overton JD, et al. De novo mutations in histone-modifying genes in congenital heart disease. Nature. 2013;498:220–3.

54. Oh SP, Li E. The signaling pathway mediated by the type IIB activin receptor controls axial patterning and lateral asymmetry in the mouse. Genes Dev. 1997;11:1812–26.

55. Wang B, Yan J, Mi R, Zhou S, Xie X, Wang J, et al. Forkhead box H1 (FOXH1) sequence variants in ventricular septal defect. Int J Cardiol. 2010;145:83–5.

Utilizing Zebrafish to Understand Second Heart Field Development

25

H.G. Knight and Deborah Yelon

Abstract

Heart formation relies on two sources of cardiomyocytes: the first heart field (FHF), which gives rise to the linear heart tube, and the second heart field (SHF), which gives rise to the right ventricle, the outflow tract, parts of the atria, and the inflow tract. The development of the SHF is of particular importance due to its relevance to common congenital heart defects. However, it remains unclear how the SHF is maintained in a progenitor state while the FHF differentiates. Likewise, the factors that trigger SHF differentiation into specific cardiac cell types are poorly understood. Investigation of SHF development can benefit from the utilization of multiple model organisms. Here, we review the experiments that have identified the SHF in zebrafish and investigated its contribution to the poles of the zebrafish heart. Already, zebrafish research has illuminated novel positive and negative regulators of SHF development, cementing the utility of zebrafish in this context.

Keywords

Second heart field • Zebrafish • Outflow tract • Inflow tract

25.1 Introduction

The embryonic origins of the heart have been a topic of intense interest due to the prevalence of congenital heart defects [1]. Cardiac progenitors (CPs) from the first heart field (FHF) form the initial heart tube, and CPs from the second heart field (SHF) contribute to most of the structures of the mature heart including the outflow tract, right ventricle, and much of the atria [2]. The SHF is generally defined as a

H.G. Knight • D. Yelon, Ph.D. (✉)
Division of Biological Sciences, University of California, San Diego, La Jolla, CA 92093, USA
e-mail: dyelon@ucsd.edu

© The Author(s) 2016   193
T. Nakanishi et al. (eds.), *Etiology and Morphogenesis of Congenital Heart Disease*,
DOI 10.1007/978-4-431-54628-3_25

population of CPs that originates adjacent to the FHF, differentiates after the initial heart tube has formed, and is responsible for cardiomyocyte accretion at both poles of the heart tube [2]. The SHF is particularly significant to congenital heart disease; many common cardiac abnormalities are caused by defects in SHF-derived tissues, including ventricular and atrial septal defects, transposition of the great arteries, and double outlet right ventricle [3]. Despite the importance of the SHF, the mechanisms that distinguish FHF and SHF development remain unclear. What signals or factors prevent the SHF from differentiating while the FHF is deployed, and what eventual change triggers SHF differentiation? Recent advances in zebrafish research offer new approaches that can complement work in mice to deepen our comprehension of SHF regulation.

Several lines of evidence indicate the presence of a population of late-differentiating CPs in zebrafish that is likely to be analogous to the mammalian SHF. The conservation of the SHF provides exciting opportunities to advance our understanding using the distinct advantages of zebrafish embryos [4]. Zebrafish embryos develop rapidly and have small hearts that are particularly tractable for cellular resolution of cardiogenesis. Furthermore, the transparency of the zebrafish embryo facilitates exceptional opportunities for time-lapse imaging of heart formation and tracking of cardiac cell fates. Finally, zebrafish are particularly well suited for conducting both genetic and chemical screens, which have the potential to identify novel regulators of heart development. Here, we review the studies that support the existence of a zebrafish SHF and demonstrate the utility of the zebrafish for opening new avenues in SHF research.

25.2   Late-Differentiating Cardiomyocytes Originate from the SHF in Zebrafish

Two types of assays have demonstrated that late-differentiating cardiomyocytes are recruited to the poles of the zebrafish heart tube. First, a developmental timing assay that relies on the different kinetics of GFP and DsRed fluorescence was used to visualize the dynamics of cardiomyocyte differentiation. Analysis of Tg(myl7: GFP); Tg(myl7:DsRed) embryos showed that newly differentiated cardiomyocytes populate the cardiac poles at 48 h postfertilization (hpf), whereas cardiomyocytes in the middle of the heart differentiate at an earlier stage (Fig. 25.1a; [5]). Second, photoconversion assays have consistently revealed late-differentiating cardiomyocytes in the outflow tract. UV exposure of Tg(myl7:kaede) or Tg(myl7: KikGR) embryos after the heart tube has formed, followed by imaging at 48 hpf, showed addition of cardiomyocytes to the outflow tract after the time of photoconversion (Fig. 25.1b, [5, 6]). Together, these experiments revealed the existence of late-differentiating cardiomyocytes at the arterial pole of the zebrafish heart that seem to be analogous to SHF-derived cardiomyocytes in mammals.

Fate mapping in zebrafish has shown that early SHF precursors seem to neighbor the FHF. Prior to gastrulation, arterial pole progenitors are found adjacent to ventricular progenitors at the embryonic margin (Fig. 25.1c; [7]). After

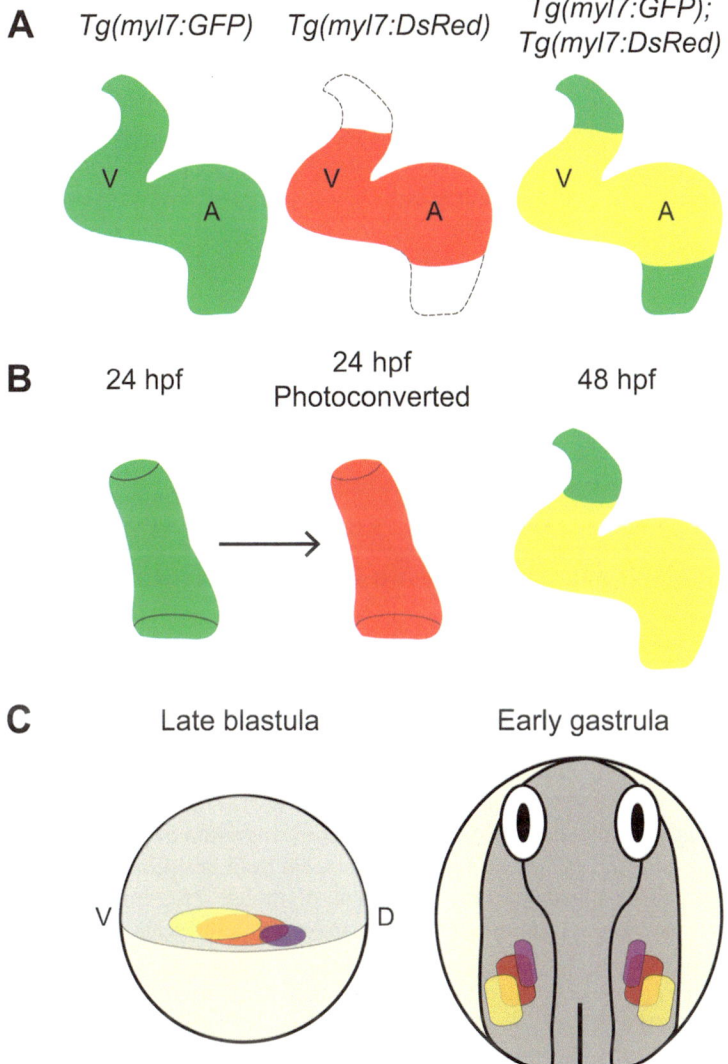

Fig. 25.1 Late-differentiating cardiomyocytes originate from the zebrafish SHF. (**a**) A developmental timing assay reveals late-differentiating cardiomyocytes displaying GFP, but not DsRed [5]. (**b**) Green-to-red conversion of photoconvertible proteins expressed in differentiated cardiomyocytes at 24 hpf, followed by imaging at 48 hpf, reveals newly added green cardiomyocytes in the outflow tract [6]. (**c**) Fate mapping in the late blastula shows that outflow tract progenitors (*purple*) are located close to the margin, adjacent to ventricular progenitors (*red*), and separate from atrial progenitors (*yellow*) [7]. In the early gastrula, outflow tract progenitors are located in a medial cranial portion of the ALPM [7]

gastrulation, arterial pole progenitors map to a medial cranial region next to the FHF in the anterior lateral plate mesoderm (ALPM) (Fig. 25.1c; [7]). Finally, DiI labeling has shown that the SHF resides adjacent to the heart tube in older embryos: pericardial cells just outside the outflow tract at 24 hpf move into the arterial pole at later stages [7]. The SHF has also been identified using Cre-mediated lineage tracing. This technique has shown that arterial pole progenitors express both *gata4* and *nkx2.5* during somitogenesis, confirming that SHF progenitors originate in the ALPM [8]. Furthermore, Cre-mediated lineage tracing has confirmed that cells from the pericardial mesenchyme adjacent to the heart tube migrate into the outflow tract [9]. Taken together, these analyses show that the late-differentiating cardiomyocytes at the zebrafish arterial pole meet the criteria that define the SHF. Outflow tract cells remain undifferentiated until after the linear heart tube has formed, are recruited to the arterial pole from outside the heart, and map to an area adjacent to the FHF. These data, combined with conserved molecular mechanisms regulating mouse and zebrafish arterial pole development, suggest that the SHF is a conserved vertebrate feature.

25.3 Mechanisms Regulating Outflow Tract Development in Zebrafish

Studies of the regulation of outflow tract formation have demonstrated conservation of the transcription factors utilized in zebrafish and mice. Zebrafish embryos deficient in *mef2cb* lack late-differentiating cells that form the outflow tract [6], which is strikingly similar to the phenotype of *Mef2c* mutant mice that lack the SHF-derived outflow tract and right ventricle [10]. Zebrafish *tbx1* mutants have several outflow tract defects, including reduced migration of cells into the heart [7] and reduced proliferation of cells at the arterial pole, resulting in a small outflow tract [11]. This phenotype is reminiscent of mouse *Tbx1* mutants, which also display outflow tract abnormalities due to severely reduced proliferation in the SHF [12].

Signaling pathways also seem to have conserved roles in the mouse and zebrafish SHF. Hedgehog signaling is important for zebrafish SHF development; migration of cells into the heart is impaired in *smoothened* mutants, resulting in a small outflow tract [7]. Similarly, hedgehog signaling is crucial for mammalian SHF survival and outflow tract septation [13]. In zebrafish, reduced FGF signaling eliminates accretion of cardiomyocytes at the arterial pole [5] and blocks *mef2cb* expression in the SHF [6]. This requirement for FGF signaling mimics mouse *Fgf8* mutants, which have a severely hypoplastic outflow tract and right ventricle [13]. These findings underscore the conserved mechanisms regulating outflow tract development and suggest that new discoveries in the zebrafish SHF are likely to be relevant to mammals.

Importantly, novel insights into outflow tract development have emerged through studies in zebrafish. The role of Ltbp3, a secreted protein that regulates TGF-β ligand availability, has been of particular interest. *ltbp3* is expressed in the zebrafish SHF, and Cre-mediated lineage tracing has shown that *ltbp3*-expressing cells give rise to outflow tract cardiomyocytes [9]. Ltbp3-deficient embryos lack an outflow tract due to reduced SHF proliferation, a consequence of reduced TGF-β signaling [9]. This work not only illuminated Ltbp3 as a new SHF regulator but also uncovered a novel role for TGF-β signaling in SHF development. Additional studies have revealed that Nkx2.5 promotes maintenance of *ltbp3* expression [8]. This is exciting, as it elucidates a new pathway downstream of Nkx2.5: Nkx2.5 facilitates the activation of TGF-β signaling through regulation of *ltbp3* and thereby drives SHF proliferation. Since Nkx2.5 is highly relevant to congenital heart disease, factors downstream of Nkx2.5 are excellent candidates for translational research. Thus, investigations in zebrafish can lead to the discovery of novel regulators of SHF development and provide new insight into connections between important factors.

25.4 Mechanisms Regulating Inflow Tract Development in Zebrafish

In mice, the SHF has been shown to contribute to the venous pole in addition to the arterial pole [2]. The mammalian SHF is thought to be subdivided into the anterior SHF, which gives rise to the right ventricle and outflow tract, and the posterior SHF, which gives rise to the atria and the inflow tract [2]. The zebrafish heart has a distinct population of inflow tract cells that express the canonical SHF marker Isl1 [14]. In addition, developmental timing assays have shown that the zebrafish inflow tract contains a population of late-differentiating cardiomyocytes (Fig. 25.1a; [5]). However, the degree of overlap between these two populations has not been examined, and the precise timing of when inflow tract cells are added to the heart is unclear. Furthermore, it is not known where zebrafish inflow tract cells originate in the early embryo and if inflow and outflow tract progenitors share a common lineage. Future experiments will be valuable to elucidate the zebrafish equivalent of the mammalian posterior SHF.

Studies of inflow tract development in zebrafish have revolved around the role of Isl1. Zebrafish Isl1 mutants lack late-differentiating cardiomyocytes at the venous pole [5]. This phenotype is similar to that of Isl1 null mouse embryos, which lack SHF-derived atrial cardiomyocytes [15]. Interestingly, studies in zebrafish have identified a novel requirement for the LIM domain protein Ajuba, which directly interacts with Isl1 [14]. Ajuba-deficient embryos have large hearts with an excess of Isl1-expressing cells and an expansion of SHF markers in the ALPM. Conversely, Ajuba overexpression eliminates Isl1 in the inflow tract [14]. Ajuba is one of the first factors that has been shown to limit SHF development, and the presence of Ajuba may determine whether Isl1 activity promotes or limits cardiomyocyte formation. The identification of Ajuba as a negative regulator of inflow tract formation further illustrates the utility of zebrafish for the discovery of novel factors involved in SHF development.

25.5 Future Directions and Clinical Implications

Altogether, the studies summarized here support the value of the zebrafish for the investigation of SHF development. It will be particularly exciting for future work in zebrafish to probe important open questions in this area. For example, zebrafish studies may be valuable for elucidating the mechanisms that pattern the SHF into its anterior and posterior subdivisions. In addition, it will be interesting to use zebrafish to examine the factors that control differentiation of multipotent SHF cells into myocardial, endocardial, and smooth muscle lineages [9]. Zebrafish will also be valuable for exploring whether multipotent SHF cells are maintained after embryogenesis, perhaps to be deployed after injury. In the long term, use of the zebrafish for analysis of SHF development is likely to illuminate pathways that facilitate our understanding of the etiology of congenital heart disease.

Acknowledgments Research in the Yelon laboratory is supported by grants from the NIH/NHLBI, American Heart Association, and March of Dimes. H.G.A. is supported by the UCSD Training Program in Cell and Molecular Genetics (NIH T32 GM007240).

References

1. Hoffman JIE, Kaplan S. The incidence of congenital heart disease. J Am Coll Cardiol. 2002;39:1890–900.
2. Kelly RG. The second heart field. Curr Top Dev Biol. 2012;100:33–65.
3. Gittenberger-de Groot AC, Bartelings MM, Poelmann RE, et al. Embryology of the heart and its impact on understanding fetal and neonatal heart disease. Semin Fetal Neonatal Med. 2013;18:237–44.
4. Scott IC, Yelon D. Cardiac development in the zebrafish. In: Rosenthal N, Harvy RP, editors. Heart development and regeneration. Oxford: Academic; 2010. p. 103–20.
5. De Pater E, Clijsters L, Marques SR, et al. Distinct phases of cardiomyocyte differentiation regulate growth of the zebrafish heart. Development. 2009;136:1633–41.
6. Lazic S, Scott IC. Mef2cb regulates late myocardial cell addition from a second heart field-like population of progenitors in zebrafish. Dev Biol. 2011;354:123–33.
7. Hami D, Grimes AC, Tsai H-J, et al. Zebrafish cardiac development requires a conserved secondary heart field. Development. 2011;138:2389–98.
8. Guner-Ataman B, Paffett-Lugassy N, Adams MS, et al. Zebrafish second heart field development relies on progenitor specification in anterior lateral plate mesoderm and nkx2.5 function. Development. 2013;140:1353–63.

9. Zhou Y, Cashman TJ, Nevis KR, et al. Latent TGF-β binding protein 3 identifies a second heart field in zebrafish. Nature. 2011;474:645–8.
10. Lin Q, Schwarz J, Bucana C, et al. Control of mouse cardiac morphogenesis and myogenesis by transcription factor MEF2C. Science. 1997;276:1404–7.
11. Nevis K, Obregon P, Walsh C, et al. Tbx1 is required for second heart field proliferation in zebrafish. Dev Dyn. 2013;242:550–9.
12. Greulich F, Rudat C, Kispert A. Mechanisms of T-box gene function in the developing heart. Cardiovasc Res. 2011;91:212–22.
13. Rochais F, Mesbah K, Kelly RG. Signaling pathways controlling second heart field development. Circ Res. 2009;104:933–42.
14. Witzel H, Jungblut B, Choe CP, et al. The LIM protein ajuba restricts the second heart field progenitor pool by regulating Isl1 activity. Dev Cell. 2012;23:58–70.
15. Cai C-L, Liang X, Shi Y, et al. Isl1 identifies a cardiac progenitor population that proliferates prior to differentiation and contributes a majority of cells to the heart. Dev Cell. 2003;5:877–89.

A History and Interaction of Outflow Progenitor Cells Implicated in "Takao Syndrome"

26

Hiroyuki Yamagishi, Kazuki Kodo, Jun Maeda, Keiko Uchida, Takatoshi Tsuchihashi, Akimichi Shibata, Reina Ishizaki, Chihiro Yamagishi, and Deepak Srivastava

Abstract

Progenitor cells, derived from the cardiac neural crest (CNC) and the second heart field (SHF), play key roles in development of the cardiac outflow tract (OFT), and their interaction is essential for establishment of the separate pulmonary and systemic circulation in vertebrates. 22q11.2 deletion syndrome (22q11DS) or Takao syndrome is the most common human chromosomal deletion syndrome that is highly associated with OFT defects. Historically, based on the observations in animal models, OFT defects implicated in the 22q11/Takao syndrome are believed to result primarily from abnormal development of CNC that populate into the conotruncal region of the heart. In the twenty-first century, elegant efforts to model 22q11/Takao syndrome in mice succeeded in the identification of T-box-containing transcription factor, Tbx1, as an etiology of OFT defects in this syndrome. Subsequent investigations of the Tbx1 expression pattern revealed that Tbx1 was surprisingly not detectable in CNC but was expressed in the SHF and provided a new concept of molecular and cellular basis for OFT defects associated with 22q11/Takao syndrome. More recently, it was reported that mutations in the gene encoding the transcription factor GATA6 caused CHD characteristic of OFT defects. Genes encoding the neurovascular guiding molecule semaphorin 3C (SEMA3C) and its receptor plexin A2 (PLXNA2) appear to be regulated directly by GATA6. Elucidation

H. Yamagishi (✉) • K. Kodo • J. Maeda • K. Uchida • T. Tsuchihashi • A. Shibata • R. Ishizaki • C. Yamagishi
Department of Pediatrics, Division of Pediatric Cardiology, Keio University School of Medicine, Tokyo, Japan
e-mail: hyamag@keio.jp

D. Srivastava
Gladstone Institute of Cardiovascular Disease, San Francisco, CA 94158, USA

Department of Pediatrics and Department of Biochemistry and Biophysics, University of California, San Francisco, San Francisco, CA 94158, USA

© The Author(s) 2016
T. Nakanishi et al. (eds.), *Etiology and Morphogenesis of Congenital Heart Disease*, DOI 10.1007/978-4-431-54628-3_26

201

of molecular mechanism involving GATA6, SEMA3C, PLXNA2, and TBX1 in the interaction between the CNC and the SHF would provide new insights into the OFT development.

Keywords
Congenital heart disease • 22q11.2 deletion syndrome • Neural crest • Second heart field • GATA6

26.1 Introduction

Cardiac outflow tract (OFT) defects account for approximately 30 % of congenital heart disease (CHD) and usually require an intervention during the first year of life [1, 2]. A variety of OFT defects results from disturbance of the morphogenetic process for the establishment of separated systemic and pulmonary circulation. Despite their clinical importance, the etiology of most OFT defects remains unknown because of the multifactorial nature of the diseases.

Progenitor cells derived from the cardiac neural crest (CNC) and the second heart field (SHF) play key roles in development of the OFT. The SHF cells give rise to the OFT myocardium along with subpulmonary conus, and CNC cells give rise to the OFT septum during development. Defects of these progenitor cells may lead to a variety of OFT defects, including tetralogy of Fallot (TOF), characterized by malalignment of the major vessels with the ventricular chambers; interrupted aortic arch type B (IAA-B), resulting from maldevelopment of the left fourth pharyngeal arch artery; and persistent truncus arteriosus (PTA), resulting from failure of septation of the OFT into the aorta and pulmonary artery [3, 4].

26.2 The 22q11.2 Deletion Syndrome (Takao Syndrome)

The 22q11.2 deletion syndrome (22q11DS) is the most common genetic cause of a spectrum of OFT defects with an incidence of 1 in 4000–5000 births [5, 6]. Most are sporadic in origin, while 10–20 % of deletions are inherited as an autosomal dominant trait. 22q11DS involves three distinct syndromes, namely, DiGeorge syndrome (DGS; OMIM#188400), velocardiofacial syndrome (VCFS; OMIM#192430), and conotruncal anomaly face syndrome (CAFS; OMIM#217095) which is so-called Takao syndrome. Historically, DGS was originally characterized by CHD, hypopara-thyroidism, and immune deficiency reported in 1965 from the field of immunology [7]; VCFS was associated with cleft palate, CHD, a distinct facial appearance, and learning difficulties reported in 1978 from the field of plastic surgery [8]; and CAFS or Takao syndrome was characterized by conotruncal CHD (OFT defects), a distinct facial appearance and hyper-nasal voice reported in 1976 (in Japanese) from the field of pediatric cardiology [9]. In 1993, clinical genetics revealed that these syndromes shared a common heterozygous deletion of 22q11.2 region and thus had overlapping phenotype [10–12].

Although the acronym "CATCH22 (cardiac defects, abnormal facies, thymic hypoplasia, cleft palate, hypocalcemia, and 22q11 deletions)" was proposed to encompass these syndromes in 1993 [13], clinical use of this term is restricted today, because (1) the term "CATCH22" has a negative meaning which represents a situation where it is impossible for you to do anything, originally from a novel entitled "Catch-22" by Heller [14]; (2) the term "abnormal facies" represented by "A" is difficult to be accepted by patients and their family; and (3) the clinical spectrum associated with 22q11DS is much wider than was previously recognized as "CATCH" [15].

Approximately 75 % of patients with 22q11 DS have CHD. The type of CHD are characterized as OFT defects including TOF, estimated about 30 %; IAA-B, estimated about 15 %; ventricular septal defect (VSD), estimated about 15 %; PTA, estimated about 10 %; and others, estimated about 5 %. Alternatively, 22q11.2 deletion is present in approximately 60 % of patients with IAA-B, 35 % of patients with PTA, and 15 % of patients with TOF. Specifically, it is detected in 55 % of patients with TOF plus pulmonary atresia and major aortopulmonary collateral arteries (MAPCA) [16–18].

Although CHD are the major cause of mortality in 22q11DS, survivors have an exceptionally high incidence of psychiatric illness, including schizophrenia and bipolar disorder, in adolescents and adults, making del22q11 the most frequent genetic cause of such psychiatric disorders [16, 19, 20]. In our experience of 18 adults with 22q11DS, common school and employment were observed in 11 of 18 cases, and 2 females got married; however, difficulties with social interaction and employment were observed in 7 cases. The main reason of difficulties for social interaction and employment was incomplete repair of CHD in four cases, and all of them had TOF with pulmonary atresia and MAPCA. One case was also diagnosed as schizophrenia. Other three cases had repaired VSD and are away from hospital care. Taken together, lifelong comprehensive evaluation and management of patients with 22q11DS, like as shown in Table 26.1, are required for multisystem disorders [6]. The primary care physician, a pediatric cardiologist in most cases, has an important role in the follow-up for the patients and their families and needs to collaborate with many specialists for the associated abnormalities.

26.3   Identification of TBX1

Because of the high incidence and association with OFT defects, 22q11DS has attracted attention as a model for investigating the genetic basis for OFT defects [21, 22]. The structures primarily affected in patients with 22q11DS are derivatives of the embryonic pharyngeal arches, or neural crest cells, suggesting that haploinsufficiency of the gene(s) on the 22q11.2 deleted region is essential for pharyngeal arch and/or CNC development [1, 2, 21, 22]. Extensive gene searches have been successful in identifying more than 30 genes in the deleted segment.

Table 26.1 Management program for patients with 22q11.2 deletion syndrome at Keio University Hospital

	Newborn and infant	Toddler	School age	Puberty
Congenital heart diseases	Palliative operation Regular follow-up	Corrective operation	Regular follow-up	Exercise guidance (if necessary)
Immunodeficiency	Blood exam (immunodeficiency screening) Immune reconstruction (severe cases)	Vaccination	Vaccination	
Velopharyngeal dysfunction	Otolaryngeal exam Plastic surgery (if necessary)	Assessment of velopharyngeal function and ear problem Pharyngoplasty (if necessary) and speech therapy Assessment of speech (preschool)	Continuous speech therapy	
Hypocalcemia	Periodic serum calcium exam Oral administration of active vitamin D (if necessary) → urinary calcium (and renal ultrasound) check Calcium supplement at perioperative periods		Evaluation of latent hypocalcemia	
Developmental delay	Assessment of developmental quality Intervention for developmental delay	Assessment of intelligence quality (preschool)	Intervention for learning disabilities	
Psychiatric disorder				Psychiatrist consultation
Short stature	Regular measurements of body size		Hormonal evaluation	
Others	Screening, renal ultrasound Teeth care (dentist consultation) Pediatric surgery (anal atresia, inguinal hernia, etc.) Ophthalmology (squint, etc.) Orthopedics (scoliosis, talipes equinovarus, etc.)			

Although standard positional cloning has failed to demonstrate a role for any of these genes in the syndrome, elegant efforts by several groups to model 22q11DS in mice by creating orthologous chromosomal deletions were successful in revealing the T-box-containing transcription factor, Tbx1, as the etiology of OFT defects associated with 22q11DS [23–26]. Heterozygosity of Tbx1 in mice alone also caused aortic arch defects, while homozygous mutation of Tbx1 in mice resulted in most main clinical presentations of 22q11DS, including OFT defects, abnormal facial features, cleft palate, and hypoplasia of the thymus and parathyroid glands.

26.4 Expression of TBX1

The delineation of the expression pattern of Tbx1 provided a new concept on the molecular and cellular basis of normal and abnormal development of the OFT. We and other group found that Tbx1 was expressed in the SHF but not in the CNC [27–29]. This finding was surprising because CHD associated with 22q11DS had been believed to result primarily from abnormal development of CNC as mentioned above. Interestingly, in mouse and chick embryos, Tbx1 is preferentially expressed in the pharyngeal arches, in the ventral half of the otic vesicle, and in the head (Fig. 26.1) [27, 28]. Within the pharyngeal arch region, Tbx1 is expressed in the pharyngeal mesoderm, including the SHF, the pharyngeal endoderm, and the head mesenchyme. These results suggest that defects of neural crest-derived tissues in 22q11DS may occur in a non-cell autonomous fashion. Our cre-mediated murine

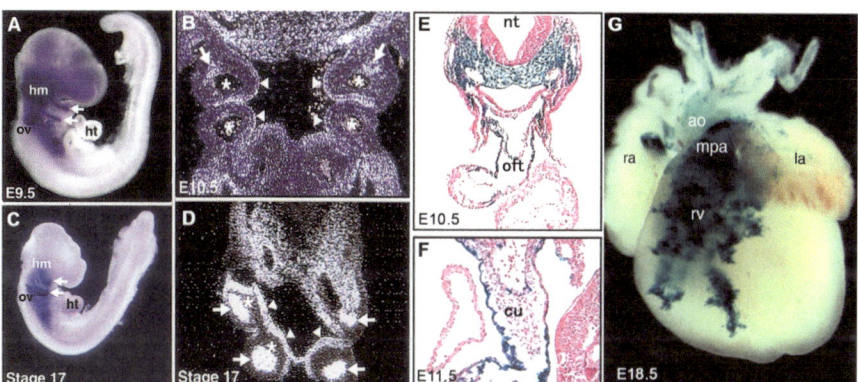

Fig. 26.1 Expression patterns of Tbx1. (**a–d**) RNA in situ hybridizations for whole mount (**a–c**) and section (**b, d**) in mouse (**a, b**) and chick (**c, d**) embryos demonstrate Tbx1 expression (*purple or white signals*) in the mesodermal core (*arrows*) and endodermal epithelium (*arrowheads*) of pharyngeal arches, head mesenchyme (*hm*), and otic vesicle (*ov*). Asterisks indicate pharyngeal arch arteries. (**e, f**). Transverse sections demonstrate Tbx1-lacZ expression (*blue signals*) in the myocardial layer of the cardiac outflow tract (*oft*) but not in the oft cushion (*cu*) which is mainly contributed by cardiac neural crest cells. (**g**) Tbx1-descendant cells marked by Tbx1-Cre/Rosa26R mouse system are localized in the anterior portion (*oft*) of the right ventricle (*rv*) and the main trunk of the pulmonary artery (*mpa*). *ao* Aorta, *h* head, *ht* heart, *la* left atrium, *ra* right atrium

transgenic system revealed that Tbx1-expressing descendants representing a subset of cells derived from the SHF contribute predominantly to the pulmonary infundibulum (Fig. 26.1) [30].

Although precise embryological mechanisms underlying OFT defects remain uncertain, the anatomical defects in TOF are believed to result from malrotation of the OFT that leads to misalignment of the outlet and trabecular septum and consequent overriding of the aorta above the malaligned ventricular septum [2, 3]. Contribution of CNC is thought to be essential for proper rotation and septation of the OFT. Alternatively, hypoplasia and underdevelopment of the pulmonary infundibulum may also be responsible for the infundibular obstruction and malalignment of the outlet septum [2, 3]. Accordingly, our data suggest that developmental defects of the SHF may cause hypoplasia of the pulmonary infundibulum, resulting in TOF [30]. More severe decreased number or absence of this subset of cells may affect development and/or migration of CNC, resulting in PTA. This hypothetical model is supported by the observation that the OFT defects ranging from TOF to TA are highly associated with 22q11DS (Fig. 26.2).

26.5   Mutations of GATA6

Recently, we identified and characterized mutations of GATA6 in our series of Japanese patients with OFT defects [31]. Mutations in GATA6 disrupted its transcriptional activity on downstream target genes involved in the development of the OFT. We also found that the expression of SEMA3C and PLXNA2 was directly regulated during development of the OFT through the consensus GATA binding sites well conserved across species. Mutant GATA6 proteins failed to transactivate SEMA3C and PLXNA2, and mutation of the GATA sites on enhancer elements of Sema3c and Plxna2 abolished their activity, specifically in the OFT/subpulmonary myocardium and CNC derivatives in the OFT region, respectively. These results indicate that mutations of GATA6 are implicated in genetic causes of OFT defects, as a result of the disruption of the direct regulation of semaphorin-plexin signaling (Fig. 26.2).

26.6   Future Direction: Elucidating the Interaction Between CNC and SHF

Recent studies have demonstrated that reciprocal epithelial-mesenchymal signaling is essential for proper development of the pharyngeal arches and that the primary impairment of epithelial endoderm may secondarily affect migration or differentiation of neural crest cells during the pharyngeal arch development [32–34]. As for the development of the OFT, clear roles of CNC- and SHF-derived cells have been established [35]. Future direction in this research field is to reveal how the CNC and SHF interact using complex reciprocal signaling essential for precise morphogenesis of the OFT. Importantly, mutations in genes expressed in either CNC or SHF can

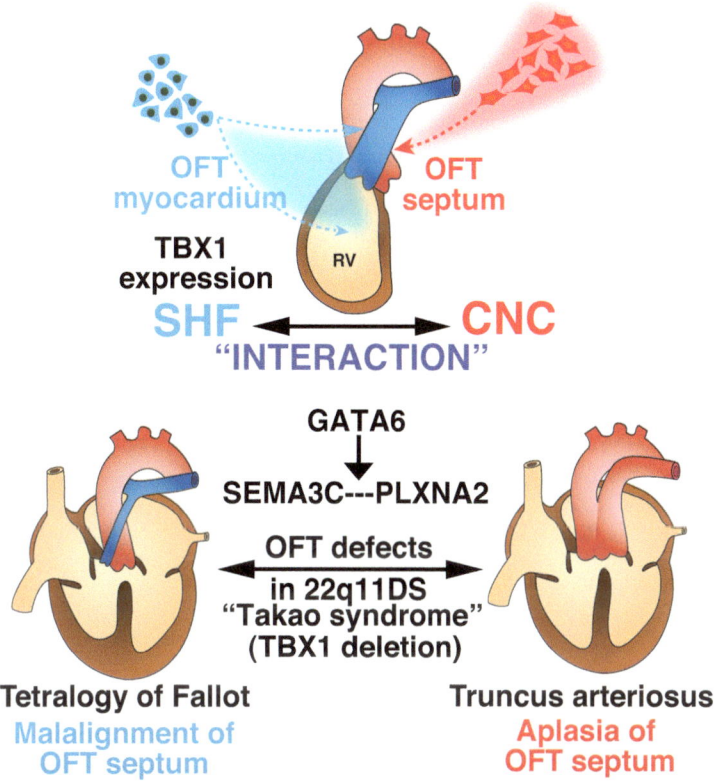

Fig. 26.2 Cellular interaction of the second heart field (*SHF*) and cardiac neural crest (*CNC*) for the outflow tract (*OFT*) development and diseases. Progenitor cells derived from the SHF and the CNC give rise to the *OFT* myocardium and septum, respectively. TBX1 is exclusively expressed in the SHF cells. TBX1 deletion in 22q11DS may affect not only the SHF cells but also the interaction between the SHF cells and CNC, resulting in OFT defects ranging from TOF, which is characterized by malalignment of the OFT septum, to PTA, which results from aplasia of the OFT septum. GATA6-SEMA3C (ligand)-PLXNA2 (receptor) pathway also plays a role in interaction between the SHF and CNC during the OFT development (Modified from [36])

result in similar OFT defects in mice. For example, Pax3 is expressed in the CNC, and Tbx1 is expressed in the SHF, and both Pax3-null mice and Tbx1-null mice show PTA. Studies are, thus, required to focus on the signals that mediate interactions between CNC and SHF in order to uncover the developmental mechanisms underlying various types of the OFT defect. Our result from the research of mutation of GATA6, described above, is an example that revealed such a molecular mechanism. Our recent preliminary data suggest that a molecular cascade involving Gata6, Foxc1/2, Tbx1, Sema3C, and Fgf8 may play roles in reciprocal signaling between SHF and CNC that are essential for the migration of CNC toward the OFT myocardium derived from the SHF (*in revision*). Further study utilizing our model system may provide new insights into the OFT development and embryogenesis of OFT defects.

References

1. Hoffman JI, Kaplan S. The incidence of congenital heart disease. J Am Coll Cardiol. 2002;39:1890–900.
2. Thom T, et al. Heart disease and stroke statistics–2006 update: a report from the American Heart Association Statistics Committee and Stroke Statistics Subcommittee. Circulation. 2006;14:e85–151.
3. Siwik ES, Patel CR, Zahka KG. Tetralogy of fallot. In: Allen HD, Gutgesell HP, Clark EB, Driscoll DJ, editors. Moss and Adams' heart disease in infants, children, and adolescents including the fetus and young adult. Philadelphia: Lippincott: Williams and Wilkins; 2001. p. 880–902.
4. Yamagishi H, Yamagishi C. Embryology. In: Saremi F, editor. Cardiac CT and MR for adult congenital heart disease. New York: Springer; 2014. p. 7–21.
5. Scambler PJ. The 22q11 deletion syndromes. Hum Mol Genet. 2000;9:2421–6.
6. Yamagishi H. The 22q11.2 deletion syndrome. Keio J Med. 2002;51:77–88.
7. DiGeorge AM. Discussion on a new concept of the cellular basis of immunology. J Pediatr. 1965;67:907.
8. Shprintzen RJ, GoldbergR B, Lewin ML, Sidoti EJ, Berkman MD, Argamaso RV, Young D. A new syndrome involving cleft palate, cardiac anomalies, typical facies, and learning disabilities: velo-cardio-facial syndrome. Cleft Palate J. 1978;15:56–62.
9. Kinouchi A, Mori K, Ando M, Takao A. Facial appearance of patients with conotruncal anomalies. Shonika (Pediatr Jpn). 1976;17:84 (in Japanese).
10. Driscoll DA, Budarf ML, Emanuel BS. A genetic etiology for DiGeorge syndrome: consistent deletions and microdeletions of 22q11. Am J Hum Genet. 1992;50:924–33.
11. Scambler PJ, Kelly D, Lindsay E, Williamson R, Goldberg R, Shprintzen R, Wilson DI, Goodship JA, Cross IE, Burn J. Velo-cardio-facial syndrome associated with chromosome 22 deletions encompassing the DiGeorge locus. Lancet. 1992;339:1138–9.
12. Burn J, Takao A, Wilson D, Cross I, Momma K, Wadey R, Scambler P, Goodship J. Conotruncal anomaly face syndrome is associated with a deletion within chromosome 22q11. J Med Genet. 1993;30:822–4.
13. Wilson DI, Burn J, Scambler P, Goodship J. DiGeorge syndrome: part of CATCH 22. J Med Genet. 1993;30:852–6.
14. Hellar J. CATCH 22. London: Jonathan Cape; 1962.
15. Burn J. Closing time for CATCH22. J Med Genet. 1999;36:737–8.
16. Ryan AK, Goodship JA, Wilson DI, Philip N, Levy A, Seidel H, Schuffenhauer S, Oechsler H, Belohradsky B, Prieur M, et al. Spectrum of clinical features associated with interstitial chromosome 22q11 deletions: a European collaborative study. J Med Genet. 1997;34:798–804.

17. Goldmuntz E, Clark BJ, Mitchell LE, Jawad AF, Cuneo BF, Reed L, McDonald-McGinn D, Chien P, Feuer J, Zackai EH, et al. Frequency of 22q11 deletions in patients with conotruncal defects. J Am Coll Cardiol. 1998;32:492–8.

18. Maeda J, Yamagishi H, Matsuoka R, Ishihara J, Tokumura M, Fukushima H, Ueda H, Takahashi E, Yoshiba S, Kojima Y. Frequent association of 22q11.2 deletion with tetralogy of Fallot. Am J Med Genet. 2000;92:269–72.

19. Karayiorgou M, Morris MA, Morrow B, Shprintzen RJ, Goldberg R, Borrow J, Gos A, Nestadt G, Wolyniec PS, Lasseter VK, et al. Schizophrenia susceptibility associated with interstitial deletions of chromosome 22q11. Proc Natl Acad Sci U S A. 1995;92:7612–6.

20. Sugama S, Namihira T, Matsuoka R, Taira N, Eto Y, Maekawa K. Psychiatric inpatients and chromosome deletions within 22q11.2. J Neurol Neurosurg Psychiatry. 1999;67:803–6.

21. Lindsay EA, Baldini A. Congenital heart defects and 22q11 deletions: which genes count? Mol Med Today. 1998;4:350–7.

22. Yamagishi H, Srivastava D. Unraveling the genetic and developmental mysteries of 22q11 deletion syndrome. Trends Mol Med. 2003;9:383–9.

23. Lindsay EA, Vitelli F, Su H, Morishima M, Huynh T, Pramparo T, Jurecic V, Ogunrinu G, Sutherland HF, Scambler PJ, et al. Tbx1 haploinsufficiency in the DiGeorge syndrome region causes aortic arch defects in mice. Nature. 2001;410:97–101.

24. Lindsay EA. Chromosome microdeletions: dissecting del22q11 syndrome. Nat Rev Genet. 2001;2:858–68.

25. Merscher S, Funke B, Epstein JA, Heyer J, Puech A, Lu MM, Xavier RJ, Demay MB, Russell RG, Factor S, et al. TBX1 is responsible for cardiovascular defects in velo-cardio-facial/DiGeorge syndrome. Cell. 2001;104:619–29.

26. Jerome LA, Papaioannou VE. DiGeorge syndrome phenotype in mice mutant for the T-box gene, Tbx1. Nat Genet. 2001;27:286–91.

27. Garg V, Yamagishi C, Hu T, Kathiriya IS, Yamagishi H, Srivastava D. Tbx1, a DiGeorge syndrome candidate gene, is regulated by sonic hedgehog during pharyngeal arch development. Dev Biol. 2001;235:62–73.

28. Yamagishi H, Maeda J, Hu T, McAnally J, Conway SJ, Kume T, Meyers EN, Yamagishi C, Srivastava D. Tbx1 is regulated by tissue-specific forkhead proteins through a common Sonic hedgehog-responsive enhancer. Genes Dev. 2003;17:269–81.

29. Xu H, Morishima M, Wylie JN, Schwartz RJ, Bruneau BG, Lindsay EA, Baldini A. Tbx1 has a dual role in the morphogenesis of the cardiac outflow tract. Development. 2004;131:3217–27.

30. Maeda J, Yamagishi H, McAnally J, Yamagishi C, Srivastava D. Tbx1 is regulated by forkhead proteins in the secondary heart field. Dev Dyn. 2006;235:701–10.

31. Kodo K, Nishizawa T, Furutani M, Arai S, Yamamura E, Joo K, Takahashi T, Matsuoka R, Yamagishi H. GATA6 mutations cause human cardiac outflow tract defects by disrupting semaphoring-plexin signaling. Proc Natl Acad Sci U S A. 2009;106:13933–8.

32. Trumpp A, Depew MJ, Rubenstein JL, Bishop JM, Martin GR. Cre-mediated gene inactivation demonstrates that FGF8 is required for cell survival and patterning of the first branchial arch. Genes Dev. 1999;13:3136–48.

33. Thomas T, Kurihara H, Yamagishi H, Kurihara Y, Yazaki Y, Olson EN, Srivastava D. A signaling cascade involving endothelin-1, dHAND and msx1 regulates development of neural-crest-derived branchial arch mesenchyme. Development. 1998;125:3005–14.

34. Wendling O, Dennefeld C, Chambon P, Mark M. Retinoid signaling is essential for patterning the endoderm of the third and fourth pharyngeal arches. Development. 2000;127:1553–62.

35. Neeb Z, Lajiness JD, Bolanis E, Conway SJ. Cardiac outflow tract anomalies. Dev Biol. 2013;2:499–530.

36. Yamagishi H. Human genetics of truncus arteriosus. In: Rickert-Sperling S, Kelly RG, Driscoll EJ, editors. Congenital heart diseases: the broken heart. Clinical features, human genetics and molecular pathways. Vienna: Springer; 2016. p. 559–67

The Loss of Foxc2 Expression in the Outflow Tract Links the Interrupted Arch in the Conditional Foxc2 Knockout Mouse

Mohammad Khaja Mafij Uddin, Wataru Kimura, Mohammed Badrul Amin, Kasumi Nakamura, Mohammod Johirul Islam, Hiroyuki Yamagishi, and Naoyuki Miura

Keywords
Foxc2 • Interruption of aortic arch • Conditional knockout

Congenital heart disease is the most common birth defects, affecting 1 % live births [1]. The cardiovascular system undergoes a series of morphogenetic events to form a heart and an aorta in fetuses. Formation of the heart and aorta requires migration, differentiation, and precise interactions among multiple cells from several embryonic origins [2]. Forkhead box2 (Foxc2) encodes a transcription factor and is expressed in mesodermal tissues, such as the pharyngeal artery, outflow tract endothelial/surrounding mesenchyme, bone, and kidney [3]. Simple knockout of Foxc2 in mouse causes an interrupted aortic arch, ventricular septal defect, cleft palate, and skeletal malformation [4]. The heart is made from primary and secondary heart field progenitors. The primary heart field gives rise to the left ventricle and atria, while the secondary heart field contributes mainly to the right ventricle and outflow tract [5] (Fig. 27.1).

To explore the tissue-specific roles of Foxc2 in aortic arch remodeling, we generated mice carrying a floxed allele of *Foxc2* (Foxc2^{flox}) and crossed them with several Cre mice, including the primary heart field (Nkx2.5-Cre knock-in)-specific and secondary heart field (Islet1-Cre knock-in and Tbx1-Cre transgenic)-

M.K.M. Uddin • M.B. Amin • K. Nakamura • M.J. Islam • N. Miura (✉)
Department of Biochemistry, Hamamatsu University School of Medicine, Hamamatsu, Japan
e-mail: nmiura@hama-med.ac.jp

W. Kimura
Department of Biochemistry, Hamamatsu University School of Medicine, Hamamatsu, Japan

Division of Cardiology, Department of Internal Medicine, UT Southwestern Medical Center, Dallas, TX, USA

H. Yamagishi
Department of Pediatrics, Keio University School of Medicine, Tokyo, Japan

© The Author(s) 2016 211
T. Nakanishi et al. (eds.), *Etiology and Morphogenesis of Congenital Heart Disease*,
DOI 10.1007/978-4-431-54628-3_27

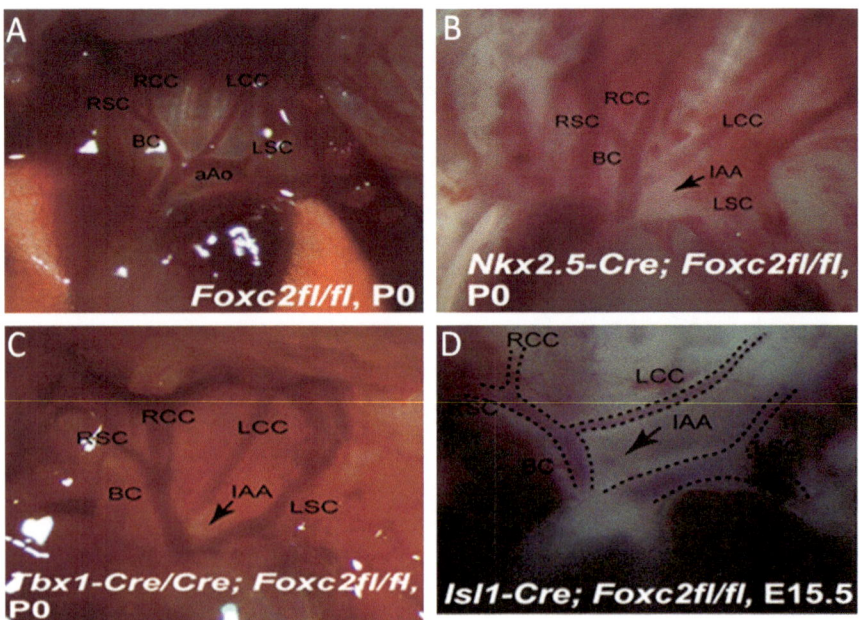

Fig. 27.1 Aortic arch abnormalities in Foxc2 conditional knockout mice. (**a**) Normal aortic arch formation in control mice (Foxc2$^{flox/flox}$). (**b–d**) Conditional knockout mice showed interrupted aortic arch (IAA) type B where part of the aorta between LCC and LSC is missing (*arrow*). *LCC* left common carotid artery, *LSC* left subclavian artery, *aAo* arch of the aorta, *Pt* pulmonary trunk, *RCC* right common carotid artery, *RSC* right subclavian artery, *BC* brachiocephalic artery

specific Cre lines. Surprisingly, conditional knockout (cKO) of *Foxc2* in the primary heart field (Nkx2.5-Cre;Foxc2$^{flox/flox}$) and secondary heart field (Islet1-Cre;Foxc2$^{flox/flox}$ and Tbx1-Cre;Foxc2$^{flox/flox}$) resulted in an interrupted aortic arch and perinatal lethality in mice. X-gal staining and immunostaining with anti-Foxc2 antibody confirmed that Foxc2 expression in the aortic arch was intact but deleted in the outflow tract in these cKO embryos. These results indicate that the Foxc2 expression in the outflow tract, rather than direct role in the aortic arch, is crucial for the aortic arch remodeling. It assumed that Foxc2 in the outflow tract regulates aortic arch remodeling via secreted factors such as Fgf8, Fgf10, and other genes.

References

1. Bruneau BG. The developmental genetics of congenital heart disease. Nature. 2008;451:943–8.
2. Olson EN, Srivastava D. Molecular pathways controlling heart development. Science. 1996;272:671–6.
3. Miura N, Wanaka A, et al. MFH-1, a new member of the fork head domain family, is expressed in developing mesenchyme. FEBS Lett. 1993;326:171–6.
4. Iida K, Miura N, et al. Essential roles of the winged helix transcription factor MFH-1 in aortic arch patterning and skeletogenesis. Development. 1997;124:4627–38.
5. Kelly RG, Buckingham ME. The anterior heart-forming field: voyage to the arterial pole of the heart. Trends Genet. 2002;18:210–6.

Modification of Cardiac Phenotype in Tbx1 Hypomorphic Mice

28

Takatoshi Tsuchihashi, Reina Ishizaki, Jun Maeda, Akimichi Shibata, Keiko Uchida, Deepak Srivastava, and Hiroyuki Yamagishi

---

Keywords

Tbx1 • Truncus arteriosus • Environmental modification

Congenital heart disease is still the leading cause of death within the first year of life. Our lab forces on understanding the morphology of congenital heart disease. Outflow tract anomalies, including abnormal alignment or septation, account for 30 % of all congenital heart disease. To solve the developmental problem of these defects, we are interested in the role of the second heart field (SHF) that gives rise to the outflow tract structure.

TBX1, a member of the T-box family of transcription factors, is a major genetic determinant of 22q11 deletion syndrome (22q11DS) in human. 22q11DS is the most frequent chromosomal microdeletion syndrome in human and characterized by abnormal development of the cardiac outflow tract, such as persistent truncus arteriosus (PTA), tetralogy of Fallot, interrupted aortic arch, and ventricular septal defects.

In the developing murine heart, Tbx1 is expressed in the SHF, but not in the cardiac neural crest cells (NCCs). Our past experiments suggested that sonic hedgehog signal was necessary for maintenance of the *Tbx1* expression in the pharyngeal mesoderm including the SHF [1]. *Tbx1* null (*Tbx1$^{-/-}$*) mice demonstrated PTA reminiscent of the 22q11DS heart phenotype. We generated

---

T. Tsuchihashi • R. Ishizaki • J. Maeda • A. Shibata • K. Uchida • H. Yamagishi (✉)
Department of Pediatrics, Division of Pediatric Cardiology, Keio University School of Medicine, Tokyo, Japan
e-mail: hyamag@keio.jp

D. Srivastava
Gladstone Institute of Cardiovascular Disease, San Francisco, CA 94158, USA

Department of Pediatrics and Department of Biochemistry and Biophysics, University of California, San Francisco, San Francisco, CA 94158, USA

© The Author(s) 2016
T. Nakanishi et al. (eds.), *Etiology and Morphogenesis of Congenital Heart Disease*, DOI 10.1007/978-4-431-54628-3_28

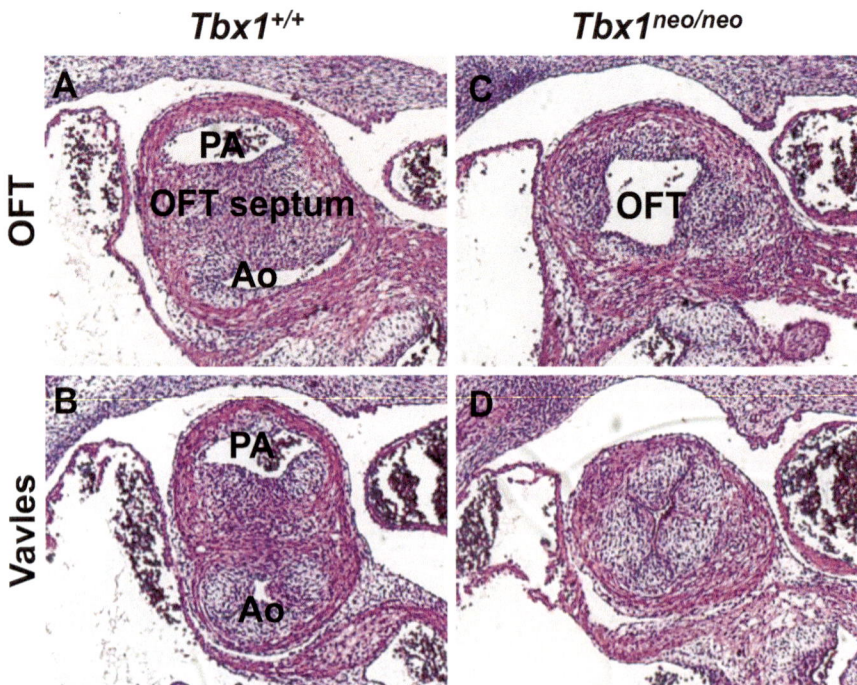

Fig. 28.1 Coronal sections of $Tbx1^{+/+}$ (**a**, **b**) and $Tbx1^{neo/neo}$ (**c**, **d**) embryos at E13.5. $Tbx1^{+/+}$ showed the normal outflow tract (OFT) septation, whereas $Tbx1^{neo/neo}$ demonstrated PTA. *Ao* Aorta, *PA* pulmonary artery

Tbx1 hypomorphic allele ($Tbx1^{neo/+}$) [2] for attempting to recapitulate the human genotype and phenotype correlation. Mice homozygous for this hypomorphic allele expressed around 25 % of Tbx1 mRNA compared to wild-type mice. We demonstrated that Tbx1 is a dosage-dependent gene and believe that the Tbx1 dosage can be affected by genetic and/or environmental modifiers because of highly variable phenotype of 22q11DS instead of the relatively uniform chromosomal microdeletion. We are trying to create the phenotype variability of PTA in this hypomorphic model (Fig. 28.1) by application of environmental modifiers. Through this study, we would better understand the interaction between the gene dosage and environmental factors during the development of outflow tract defects.

References

1. Yamagishi H, Maeda J, Hu T, et al. Tbx1 is regulated by tissue-specific forkhead proteins through a common Sonic hedgehog-responsive enhancer. Gene Dev. 2003;17:269–81.
2. Hu T, Yamagishi H, Maeda J, et al. Tbx1 regulates fibroblast growth factors in the anterior heart field through a reinforcing autoregulatory loop involving forkhead transcription factors. Development. 2005;131:5491–502.

Perspective

H. Scott Baldwin

While the critical pathways that are important for normal cardiac development have focused extensively on transcriptional regulation of myocyte differentiation, critical mediators of vascular development have received much less attention. One reason for this has been the inability in the past to manipulate gene expression in a temporal and tissue-specific manner. There is no doubt that both normal vascular and normal myocardial development are essential for early embryonic survival and the two are inextricably linked; normal vascular development requires normal flow, and maturation of the myocardium requires simultaneous maturation and remodeling of the extracardiac vasculature. Ubiquitous or global gene deletions, resulting in both cardiac and extracardiac mutations, have resulted in numerous "chicken and egg" quandaries: Did the heart fail because of a primary defect in heart development, or were the defects merely secondary to upstream perturbations in extracardiac vascular defects? In this section, investigators used tissue-specific mutagenesis strategies as well as a focus on cell membrane and extracellular matrix regulation to begin to elucidate important aspects of extracardiac vascular development that are particularly relevant to human disease. Sakebe et al. generated an endothelial-specific deletion of Hrt2/Hey2, repressors of Notch signaling, to demonstrate that both Hrt1 and Hrt2 are essential for vascular development independent of their role in myocardial development. Furthermore, they suggest that the

H.S. Baldwin (✉)
Department of Pediatrics (Cardiology), Vanderbilt University, 2213 Garland Ave, Nashville, TN 37232-0493, USA

Department of Cell and Developmental Biology, Vanderbilt University, 2213 Garland Ave, Nashville, TN 37232-0493, USA
e-mail: scott.baldwin@vanderbilt.edu

endothelial or vascular processes mediated by these factors, rather than the defects in myocardial development, might be the primary mechanism for embryonic demise. Exploring the role of calcium signaling in extraembryonic vascular development, Uchida and colleagues were able to document dramatic defects in placentation as early as E9.5 in the mouse as a result of combinatorial deletion of the inositol IP3 receptors. This work clearly establishes a role for calcium handling in cardiovascular viability. Changing the focus to later stages of vascular development, Dr. Imanaka-Yoshida provides a detailed description of the role of the extracellular matrix protein, tenascin-C, in smooth muscle cell recruitment of both the descending aorta and coronary arteries and provides in vitro evidence that tenascin-C promotes SMC precursor expansion and differentiation by augmenting PDG-BB/PDGFR-β signaling. Finally, Yoshikane et al. show the potential importance of delineating the role of tenascin-C in normal and abnormal coronary artery remodeling as they discuss a model of the most common acute systemic vasculitis in children, Kawasaki disease. By studying the inflammatory and abnormal vascular remodeling induced by *Candida albicans*, they demonstrate accentuation of tenascin-C expression associated with aneurysm formation. Furthermore, they document that inhibition of JNK signaling attenuated aneurysm formation potentially providing a mechanistic link between JNK signaling and tenascin-C signaling that could provide a therapeutic target for treatment of Kawasaki disease. In summary, the investigations presented in this section provide an overview of exciting work that expands the focus of cardiovascular development and disease beyond myocyte transcriptional regulation and provides new insights into extracardiac vascular development and remodeling while emphasizing the importance that the extracellular matrix is ontogeny of cardiovascular disease.

Extracellular Matrix Remodeling in Vascular Development and Disease

29

Kyoko Imanaka-Yoshida

Abstract

Blood vessels constantly subjected to mechanical stress have well-developed elastic fiber-rich frameworks, which contribute to the elasticity and distensibility of the vascular wall. Destruction of the fibrous structure due to genetic predisposition as well as acquired disorders such as Kawasaki disease often induces irreversible dilation of blood vessels, e.g., aneurysm formation. In addition to their structural role, extracellular matrix molecules also provide important biological signaling, which influences various cellular functions. Among them, increased attention has been focused on matricellular proteins, a group of non-structural extracellular matrix (ECM) proteins highly upregulated in active tissue remodeling, serving as biological mediators by interacting directly with cells or regulating the activities of growth factors, cytokines, proteases, and other ECM molecules. Tenascin-C (TNC) is a typical matricellular protein expressed during embryonic development and tissue repair/regeneration in a spatiotemporally restricted manner. Various growth factors, pro-inflammatory cytokines, and mechanical stress upregulate its expression. TNC controls cell adhesion, migration, differentiation, and synthesis of ECM molecules. Our recent results suggest that TNC may not only play a significant role in the recruitment of smooth muscle/mural cells during vascular development, but also regulate the inflammatory response during pathological remodeling. TNC may be a key molecule during vascular development, adaptation, and pathological tissue remodeling.

Keywords

Tenascin • Extracellular matrix • Coronary artery • Aorta

K. Imanaka-Yoshida (✉)
Department of Pathology and Matrix Biology, Mie University Graduate School of Medicine, Tsu, Mie 514-8507, Japan

Mie University Research Center for Matrix Biology, Tsu, Mie 514-8507, Japan
e-mail: imanaka@doc.medic.mie-u.ac.jp

T. Nakanishi et al. (eds.), *Etiology and Morphogenesis of Congenital Heart Disease*,
DOI 10.1007/978-4-431-54628-3_29

221

29.1 Introduction

Tissue, including the cardiovascular system, is composed of diverse cells and the extracellular matrix (ECM) synthesized by those cells. Several ECM molecules form a fibrous framework and provide structural support for the tissue. Blood vessels constantly subjected to mechanical stress have a well-developed fibrous framework, which contributes to the elasticity and distensibility of the vascular wall in concert with vascular smooth muscle cells. Highly ordered structures consisting of cells and fibrous elements are formed during development and are remodeled during tissue repair/regeneration after injury. In addition to their physical role, several ECM molecules provide important biological signaling, which influences various cellular functions in physiological and pathological tissue remodeling. In particular, ECM, termed matricellular protein, has attracted increasing attention as a biological mediator. Tenascin-C (TNC) is a prototype matricellular protein expressed during embryonic development and tissue repair after injury. This chapter will focus on the role of TNC in vascular development, especially coronary arteries and the aorta.

29.2 Extracellular Matrix in Vascular Wall

Blood vessels have abundant fibrous matrix tissue: well-developed elastic fibers in the medial layer and rich collagen fibers in adventitia. It is known that several gene mutations related to these fibrous components cause vascular fragility, eventually leading to aneurysm formation or dissection. For example, the collagen gene and fibrillin-1 gene, which is important for microfibril formation, have been identified as the genes responsible for Ehlers-Danlos syndrome (reviewed in [1]) and Marfan's syndrome [2], respectively. In addition to genetic predisposition, inflammation of blood vessels in acquired disease may induce fragmentation and destruction of normal elastic fibers in the vascular wall and causes irreversible dilation of blood vessels. For example, coronary aneurysm formation is sometimes seen in patients with Kawasaki vasculitis, one of the most common acquired heart diseases in children. Evidently, the structural support by fibrous ECM is essential to maintain the proper morphology and function of blood vessels.

Besides these fibrous elements, unique ECM molecules, matricellular protein [3], have attracted considerable attention. The matricellular proteins have common unique properties: (1) do not contribute directly to structures such as fibrils or basement membranes; (2) high levels of expression during embryonic development and in response to injury; and (3) binding to many cell surface receptors, components of ECM, growth factors, cytokines, and proteases [4]. This is a growing family originally including SPARC, tenascin, and thrombospondin [3].

29.3   Tenascin-C in Vascular System

Tenascins are a family of four multimeric extracellular matrix glycoproteins: tenascin-C, X, R, and W [5]. The first member, tenascin-C (TNC), is a typical matricellular protein. It is a huge molecule of about 220–400 kDa as an intact monomer and is assembled with a hexamer. The molecule consists of an N-terminal assembly domain, followed by EGF-like repeats, constant and alternatively spliced fibronectin type III repeats, and a C-terminal fibrinogen-like globular domain. Several receptors including integrins, EGFR, annexin II, syndecan-4, and toll-like receptor 4 (TLR-4) bind to the respective domains of TNC and transmit multiple signals (see [6]). Numerous studies have shown that TNC can control the balance of cell adhesion and de-adhesion, cell motility, proliferation, differentiation, and survival (reviewed in [5–7]). Recently, the role of TNC in the modulation of inflammation is highlighted [8].

Tenascin-C is found in many developing organs, including the cardiovascular system, but is often restricted transiently to specific sites, for example, near migrating cells and at sites of epithelial–mesenchymal/mesenchymal–epithelial transition. In normal adults, tenascin-C expression is sparsely detected; however, marked expression is seen in injury, regeneration, and cancer at sites where the tissue structure is being dynamically remodeled. Various factors, including growth factors and pro-inflammatory cytokines, can activate TNC expression (reviewed in [9]). It is particularly of interest that mechanical stress is an important inducer of TNC. Moreover, it is also noteworthy that TNC itself is an elastic molecule and may contribute to tissue elasticity [10].

As well as in other tissue, the expression of TNC in the normal vascular wall is low and upregulated in pathological conditions. The major source is medial smooth muscle cells [11]. However, TNC in the vascular system appears more complex in contrast to the heart [7]. For example, constitutive expression of TNC is observed in the medial layer of the abdominal aorta of normal adult mice but not in the thoracic aorta [12].

29.3.1 Development of Aorta and Tenascin-C

The origin of vascular smooth muscle cells (VSMC) of the aorta is heterogeneous [13]. The second heart field gives rise to VSMC of the root of the aorta. The cardiac neural crest contributes ascending and arch portions of the aorta. The origin of VSMC of the descending aorta is more complex. Primitive VSMC of the thoracic aorta originate from the lateral plate mesoderm and are replaced by cells derived from the paraxial mesoderm (somites). Moreover, individual somites build up restricted spatial domains of the "segmental" aortic wall. However, no evident segmental expression pattern of TNC is observed during development of the aorta. In E12–13 mouse embryos, very weak expression of TNC is observed in the ascending aorta and pulmonary truncus. Whereas elastic fibers in the medial layer of the aorta become mature around E12–13, the expression of TNC is upregulated after ED14–15 (Fig. 29.1) when the systemic circulatory system is established. This upregulated expression of TNC may reflect the increased hemodynamic stress on the aortic wall.

Tenascin-C in developing aorta

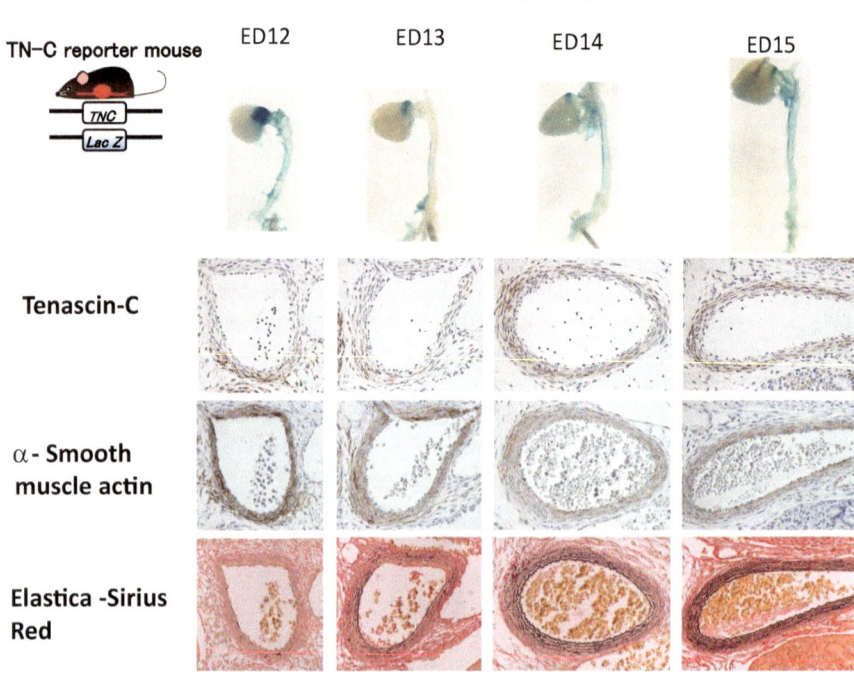

Fig. 29.1 Expression pattern of TNC during development of the aorta. Whole mount lacZ staining and histological sections of descending portion of the thoracic aorta of TNC-reporter mouse embryos at ED12–15. The sections were immunostained with anti-TNC or anti-α-smooth muscle actin or stained with elastica sirius red

29.3.2 Development of Coronary Artery and Tenascin-C

Coronary vessels are formed with the cells originating mostly from extracardiac tissue known as the proepicardial organ (PE) (see [14] for review). During coronary development, strong expression of TNC is observed, closely associated with thickening of the medial layer when the primitive coronary vasculature connects with the aortic sinuses [15], suggesting a significant role of TNC in maturation of the wall of coronary arteries. Indeed, TNC accelerates the differentiation of mesenchymal cells of PE to smooth muscle cells in culture [15]. Maturation of the vascular wall is regulated by various signaling pathways. In particular, the PDGF-BB/PDGFR-β signaling loop is known to be a key regulator of smooth muscle cell recruitment. In vitro, TNC amplifies crosstalk signaling between integrin αvβ3 and PDGF receptor (PDGFR) -β in smooth muscle cells, followed by enhancing cell proliferation and migration [16]. TNC may promote smooth muscle precursor expansion and differentiation in maturation of the vascular wall by enhancing PDGF-BB/PDGFR-β signaling (Fig. 29.2).

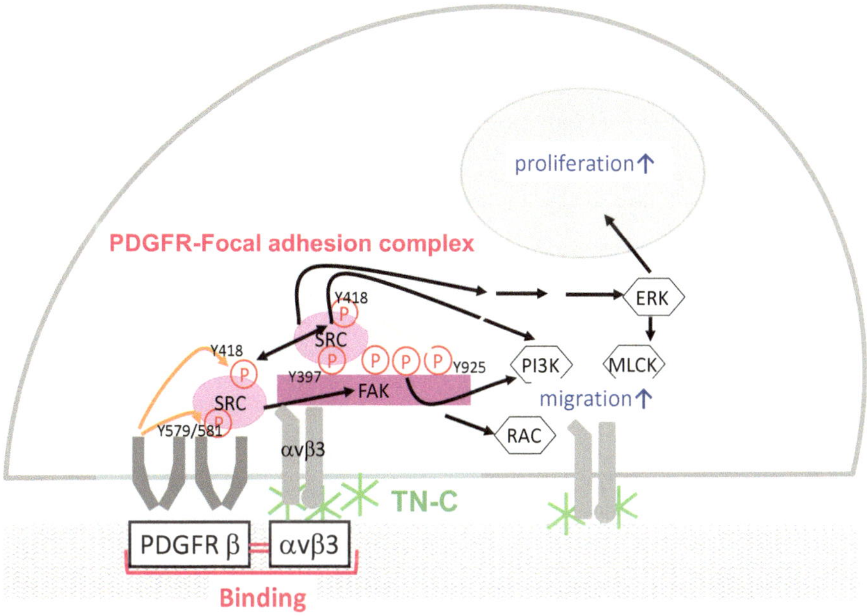

Fig. 29.2 Crosstalk signaling between TNC/integrin αvβ3 and PDGF-BB/PDGFR-β in vascular smooth muscle cells (Adapted from Ref. [7])

29.4   Future Direction and Clinical Implications

The characteristic spatiotemporally restricted expression of TNC has suggested its significant role during embryonic development. Several in vitro functional assays support this possibility. Although the grossly normal phenotype of knockout mice suggests the importance of redundancy and compensatory mechanisms during embryonic development, it is not straightforward to understand its molecular function. Meanwhile, TNC expression is linked to a range of vascular diseases, such as aortic aneurysm, acute aortic dissection, and Kawasaki disease (reviewed in [11, 17], also see Yoshikane et al. in this proceeding). Increasing numbers of studies have reported that TNC is highly upregulated, associated with inflammation and destruction of the vascular wall, suggesting that TNC may be a diagnostic biomarker. Furthermore, we have succeeded in endovascular treatment of a rat aneurysm model with a TNC-coated coil [18]. Although TNC could contribute to both favorable and undesirable effects during pathological processes in a context-dependent manner, it could be a potential therapeutic target for vascular disease.

Acknowledgments The author thanks M. Hara and M. Namikata for providing technical assistance. This work was supported in part by a research grant for intractable diseases from the Ministry of Health, Labor and Welfare of Japan, and a Grant-in-Aid for Scientific Research from the Ministry of Education, Culture, Sports, Science and Technology of Japan.

References

1. Byers PH, Murray ML. Ehlers-Danlos syndrome: a showcase of conditions that lead to understanding matrix biology. Matrix Biol. 2014;33:10–5.
2. Dietz HC, Cutting GR, Pyeritz RE, et al. Marfan syndrome caused by a recurrent de novo missense mutation in the fibrillin gene. Nature. 1991;352:337–9.
3. Sage EH, Bornstein P. Extracellular proteins that modulate cell-matrix interactions. Sparc, tenascin, and thrombospondin. J Biol Chem. 1991;266:14831–4.
4. Bornstein P. Matricellular proteins: an overview. J Cell Commun Signal. 2009;3:163–5.
5. Chiquet-Ehrismann R, Tucker RP. Tenascins and the importance of adhesion modulation. Cold Spring Harb Perspect Biol. 2011;3.
6. Midwood KS, Orend G. The role of tenascin-c in tissue injury and tumorigenesis. J Cell Commun Signal. 2009;3:287–310.
7. Imanaka-Yoshida K. Tenascin-c in cardiovascular tissue remodeling. Circ J. 2012;76:2513–20.
8. Udalova IA, Ruhmann M, Thomson SJ, et al. Expression and immune function of tenascin-c. Crit Rev Immunol. 2011;31:115–45.
9. Tucker RP, Chiquet-Ehrismann R. The regulation of tenascin expression by tissue microenvironments. Biochim Biophys Acta. 2009;1793:888–92.
10. Marin JL, Muniz J, Huerta M, et al. Folding-unfolding of fn-iii domains in tenascin: an elastically coupled two-state system. J Biomech. 2003;36:1733–7.
11. Imanaka-Yoshida K, Yoshida T, Miyagawa-Tomita S. Tenascin-c in development and disease of blood vessels. Anat Rec. 2014;297:1747–57.
12. Kimura T, Furusho A, Ito S et al. Tenascin c protects aorta from acute dissection in mice. Sci Rep. 2014;4:4051.
13. Majesky MW. Developmental basis of vascular smooth muscle diversity. Arterioscler Thromb Vasc Biol. 2007;27:1248–58.
14. Nakajima Y, Imanaka-Yoshida K. New insights into the developmental mechanisms of coronary vessels and epicardium. Int Rev Cell Mol Biol. 2013;303:263–317.
15. Ando K, Takahashi M, Yamagishi T, et al. Tenascin c may regulate the recruitment of smooth muscle cells during coronary artery development. Differentiation. 2011;81:299–306.
16. Ishigaki T, Imanaka-Yoshida K, Shimojo N, et al. Tenascin-c enhances crosstalk signaling of integrin alphavbeta3/pdgfr-beta complex by src recruitment promoting pdgf-induced proliferation and migration in smooth muscle cells. J Cell Physiol. 2011;226:2617–24.
17. Gaengel K, Genove G, Armulik A, et al. Endothelial-mural cell signaling in vascular development and angiogenesis. Arterioscler Thromb Vasc Biol. 2009;29:630–8.
18. Toma N, Imanaka-Yoshida K, Takeuchi T, et al. Tenascin-c-coated platinum coils for acceleration of organization of cavities and reduction of lumen size in a rat aneurysm model. J Neurosurg. 2005;103:681–6.

Sachiko Miyagawa-Tomita, Yuichiro Arima, and Hiroki Kurihara

Abstract

Neural crest cells (NCCs) are a unique stem cell population, which originate from the border between the neural plate and surface ectoderm and migrate throughout the body to give rise to multiple cell lineages during vertebrate embryonic development. The NCCs that contribute to heart development, referred to as the cardiac NCCs, have been assigned to the neural crest at the level of the postotic hindbrain. Recently, we found that the NCCs from the preotic region migrate into the heart and partially differentiate into coronary artery smooth muscle cells. This finding indicates that the origin of the cardiac

S. Miyagawa-Tomita, D.V.M., Ph.D. (✉)
Department of Pediatric Cardiology, Tokyo Women's Medical University, 8-1 Kawada-cho, Shinjuku-ku, Tokyo 162-8666, Japan

Department of Veterinary Technology, Yamazaki Gakuen University, 4-7-2 Minami-osawa, Hachiouji, Tokyo 192-0364, Japan
e-mail: ptomita@hij.twmu.ac.jp

Y. Arima, M.D., Ph.D.
Department of Physiological Chemistry and Metabolism, Graduate School of Medicine, The University of Tokyo, 7-3-1 Hongo, Bunkyo-ku, Tokyo 113-0033, Japan

Department of Cardiovascular Medicine, Faculty of Life Sciences, Kumamoto University, 2-2-1 Honjo, Kumamoto, Kumamoto 860-0811, Japan

H. Kurihara, M.D., Ph.D. (✉)
Department of Physiological Chemistry and Metabolism, Graduate School of Medicine, The University of Tokyo, 7-3-1 Hongo, Bunkyo-ku, Tokyo 113-0033, Japan

Core Research for Evolutional Science and Technology (CREST), Japan Science and Technology Agency (JST), Chiyoda-ku, Tokyo 102-0076, Japan

Institute for Biology and Mathematics of Dynamical Cell Processes (iBMath), The University of Tokyo, 3-8-1 Komaba, Tokyo 153-8914, Japan
e-mail: kuri-tky@umin.net

© The Author(s) 2016
T. Nakanishi et al. (eds.), *Etiology and Morphogenesis of Congenital Heart Disease*,
DOI 10.1007/978-4-431-54628-3_30

227

NCCs appears more widely extended to the anterior direction than Kirby et al. first designated.

Keywords
Neural crest • Preotic • Postotic • Coronary artery • Endothelin

30.1   Introduction

The neural crest (NC) was first identified by Wilhelm His as "Zwischenstrang," the intermediate cord, in 1868 [1], the year of Meiji Ishin, the westernizing revolution of Japan. It is located at the border between the developing neural plate and surface ectoderm and serves as a source of migratory cells spreading throughout the body. The NC research was greatly accelerated by the establishment of quail-chick chimera technique accomplished by Nicole Le Douarin [2]. This technique enabled tracing the origin and fate of the NC during embryonic development and revealed that NC cells (NCCs) differentiate into a wide variety of cell types including neurons, glia, pigment cells, and craniofacial bones and cartilages in different developmental contexts [2]. Thus, NCCs are nowadays regarded as a multipotent stem cell population with unique differentiation capacities.

30.2   Cardiac Neural Crest Arising from the Postotic Region

Since Margaret Kirby discovered that NCCs at the level of occipital somites 1–3 migrate to the region of the aorticopulmonary septum [3], the concept "cardiac neural crest" has prevailed to cover NCCs contributing to the formation of the heart and great vessels. NCCs arising from the postotic hindbrain posterior to the mid-otic vesicle, corresponding to rhombomeres (r) 6–8, migrate into the third, fourth, and sixth pharyngeal arches and contribute to the formation of the tunica media of pharyngeal arch artery-derived great vessels, the aorticopulmonary septum, and the outflow tract endocardial cushion as well as some noncardiac organs such as the thymus, parathyroid glands, and thyroid glands [4]. Ablation of the cardiac NC in chick embryos results in aortic arch anomalies and persistent truncus arteriosus [3, 5, 6]. In addition to direct contribution to the cardiovascular structure, cardiac NCCs affect the migration and alignment of myogenic precursors from the second heart field migrating into the outflow region.

Chromosome 22q11.2 deletion syndrome, formerly known as DiGeorge syndrome, velocardiofacial (Shprintzen) syndrome, and conotruncal anomaly face (Takao) syndrome, is a disease complex characterized by craniofacial, thymic, and parathyroid anomalies and cardiac manifestations including tetralogy of Fallot, persistent truncus arteriosus, and aortic arch anomalies [7]. This syndrome was formerly recognized as an NC disorder because of its resemblance to the avian phenotype of NC ablation. However, identification and analysis of the responsible genes in the 22q11.2 locus such as *TBX1* and *CRKL* and related factors have revealed that the pathogenesis is far more complex, involving interaction among NCCs, second heart field, endoderm, and other cell components.

30.3 Endothelin Signal and Neural Crest Development

Endothelin (Edn)-1 (Edn1), originally identified as a potent vasoconstrictor peptide, is a key regulator of craniofacial and cardiovascular development, acting on NCCs expressing Edn receptor type A (Ednra), a G protein-coupled receptor [8–10]. Inactivation of Edn1-Ednra signaling causes homeotic-like transformation of the lower jaw into an upper jaw structure and cardiovascular anomalies similar to chromosome 22q11.2 deletion syndrome. The craniofacial and cardiovascular anomalies are attributed to the disordered development of cranial (preotic) and cardiac (postotic) NCCs, respectively. In craniofacial development, the Edn1-Ednra signaling activates $G\alpha_q$-/$G\alpha_{11}$-dependent pathway, resulting in the induction of *Dlx5/Dlx6*, homeobox genes critical to ventral (mandibular) identity of the pharyngeal arches [10–12]. In cardiovascular development, the *Edn1-/Ednra*-null phenotype of aortic arch anomalies is independent of *Dlx5/Dlx6* [13], indicating that the Edn1-Ednra signaling pathway appears differently involved in craniofacial and cardiac development.

30.4 Preotic Neural Crest Contributing to Heart Development

Recently, we identified an additional cardiac phenotype of *Edn1-/Ednra*-null mice in the coronary artery [14]. The mutant mice exhibit marked dilatation of the septal branch and abnormalities of orifice and proximal branch formation. Labeling of NCCs using *Wnt1-Cre;Rosa26R* reporter mice revealed that NCCs contribute to coronary artery smooth muscle cells in the proximal region and septal branch, and NCC-derived smooth muscle cells are hardly detected in the smooth muscle layer in *Edn1-/Ednra*-null embryos. Correspondingly, NCC-specific knockout of $G\alpha_{12}$/$G\alpha_{13}$ rather than $G\alpha_q$/$G\alpha_{11}$ results in similar dilatation of the coronary artery septal branch [15], indicating that Edn1/Ednra signaling is necessary for NCC recruitment to coronary artery formation via $G\alpha_{12}$/$G\alpha_{13}$ and downstream Rho signaling.

Here, we faced to a conundrum where the NCCs came from. It had long been controversial whether and how NCCs contribute to coronary artery formation. Although NC-derived cell clusters are formed in association with the proximal portion of coronary arteries, quail-chick chimera experiments have shown that the cardiac NCCs do not differentiate into coronary smooth muscle cells [16, 17]. In contrast, *Wnt1-Cre* mice have indicated the possible direct involvement of NCCs as the source of coronary artery smooth muscle cells [18]. The apparent discrepancy was sometimes ascribed to differences in species, but no definite explanation had been given for it.

This controversy was settled by quail-chick chimera experiments, in which different regions of the chick neural folds were homotopically replaced by quail tissues. When the cardiac (postotic) NC at the level of r6-r8 (posterior to the mid-otic vesicle) was replaced, no contribution of quail NCCs to the wall of coronary arteries was observed. In contrast, replacement of the NC by exchanging the midbrain and preotic hindbrain (r1-r5) neural folds anterior to the otic vesicle resulted in a significant number of quail NCCs distributing into the heart and differentiating into coronary artery smooth muscle cells. The intracardiac migration of preotic NCCs and their contribution to the coronary artery smooth muscle layer were also confirmed by experiments using $R4$-$Cre;Z/AP$ reporter mice, in which r4-derived preotic NCCs were specifically and permanently labeled. Furthermore, ablation of the preotic NC in chick embryos caused abnormalities in coronary septal branch and orifice formation, reminiscent of the $Edn1$-/$Ednra$-null phenotype.

Are preotic and postotic NCCs spatially segregated within the heart region to play distinct roles? Double labeling with different dyes of premigratory NCCs at the levels of r3/4 (preotic) and somites 1/2 (postotic) in chick embryos revealed sequential migration of NCCs from preotic to postotic neural folds. Consequently, preotic and postotic NCCs distribute differently within the heart and great vessel-forming regions after migration, with anteroposterior order of NCCs corresponding to their proximodistal location within the heart. Preceding preotic NCCs are likely to differentiate into coronary artery smooth muscle cells, whereas subsequent postotic NCCs predominantly form the aorticopulmonary septum and the smooth muscle layer of the aorta and pulmonary artery (Fig. 30.1). In addition, both NCC populations differently distribute within semilunar valves, suggesting their distinct roles in valve formation (Fig. 30.1).

30.5   Future Direction and Clinical Implications

Identification of preotic NCCs as an origin of cardiac cellular components may provide a novel insight into cell lineage-based understanding of cardiac development, anatomy, and (patho-)physiology. The spatiotemporal pattern of preotic NCC migration and distribution suggests close interaction with second heart field-derived mesodermal cells. In coronary artery formation, interactions between preotic NCCs and other precursor cells from different origins such as the proepicardium and endocardium seem to be an important issue to be addressed. Considering endothelial and endocardial cells are major source of Edn1, the Edn signaling may play a role in these interactions.

From a clinical viewpoint, it is intriguing to pursue the relationship between the NC origin and susceptibility to atherosclerosis and calcification of the proximal coronary arteries. Preotic NCCs retain multipotent capacities including osteogenic and chondrogenic differentiation, leading us to speculate a possibility that these capacities may be related to the pathogenesis and progression of coronary artery diseases. Characterization of preotic NC-derived smooth muscle cells and other derivatives may open perspectives toward novel therapeutic strategies.

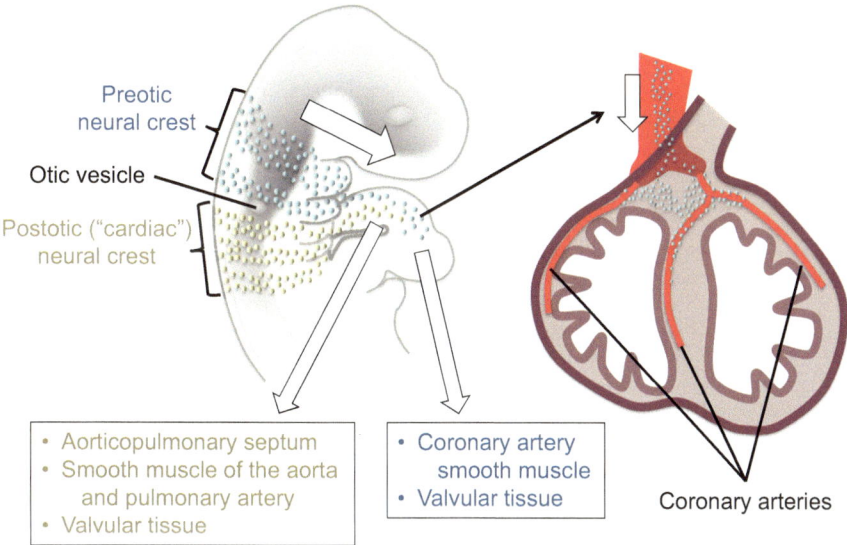

Fig. 30.1 Different contributions of preotic and postotic NCCs to craniofacial and cardiovascular development. Preotic NCCs migrate into the first and second pharyngeal arches to differentiate into the bone, cartilage, teeth, and connective tissue, a part of which further migrates into the heart to differentiate into the coronary artery smooth muscle and valvular tissues. Postotic NCCs follow preotic NCCs in migration and form the aorticopulmonary septum and the smooth muscle layer of the aorta and pulmonary artery with some contribution to the semilunar valves

References

1. His W. Untersuchungen über die erste Anlage des Wirbeltier- leibes. Die erste Entwickelung des Hühnchens im Ei. Leipzig: FCW Vogel; 1868.
2. Le Douarin N, Kalcheim C. The neural crest. 2nd ed. Cambridge: Cambridge University Press; 1999.
3. Kirby ML, Gale TF, Stewart DE. Neural crest cells contribute to normal aorticopulmonary septation. Science. 1983;220:1059–61.
4. Kirby ML, Hutson MR. Factors controlling cardiac neural crest cell migration. Cell Adh Migr. 2010;4:609–21.
5. Nishibatake M, Kirby ML, Van Mierop LH. Pathogenesis of persistent truncus arteriosus and dextroposed aorta in the chick embryo after neural crest ablation. Circulation. 1987;75:255–64.

6. Waldo K, Miyagawa-Tomita S, Kumiski D, Kirby ML. Cardiac neural crest cells provide new insight into septation of the cardiac outflow tract: aortic sac to ventricular septal closure. Dev Biol. 1998;196:129–44.

7. Momma K. Cardiovascular anomalies associated with chromosome 22q11.2 deletion syndrome. Am J Cardiol. 2010;105:1617–24.

8. Kurihara Y, Kurihara H, Suzuki H, et al. Elevated blood pressure and craniofacial abnormalities in mice deficient in endothelin-1. Nature. 1994;368:703–10.

9. Kurihara Y, Kurihara H, Oda H, et al. Aortic arch malformations and ventricular septal defect in mice deficient in endothelin-1. J Clin Invest. 1995;96:293–300.

10. Sato T, Kurihara Y, Asai R, et al. An endothelin-1 switch specifies maxillomandibular identity. Proc Natl Acad Sci U S A. 2008;105:18806–11.

11. Ozeki H, Kurihara Y, Tonami K, Watatani S, Kurihara H. Endothelin-1 regulates the dorsoventral branchial arch patterning in mice. Mech Dev. 2004;121:387–95.

12. Sato T, Kawamura Y, Asai R, et al. Recombinase-mediated cassette exchange reveals the selective use of G_q/G_{11}-dependent and -independent endothelin 1/endothelin type A receptor signaling in pharyngeal arch development. Development. 2008;135:755–65.

13. Kim K-S, Arima Y, Kitazawa T, et al. Endothelin regulates neural crest deployment and fate to form great vessels through Dlx5/Dlx6-independent mechanisms. Mech Dev. 2013;130:553–66.

14. Arima Y, Miyagawa-Tomita S, Maeda K, et al. Preotic neural crest cells contribute to coronary artery smooth muscle involving endothelin signalling. Nat Commun. 2012;3:1267.

15. Dettlaff-Swiercz DA, Wettschureck N, Moers A, Huber K, Offermanns S. Characteristic defects in neural crest cell-specific $G\alpha_q$/$G\alpha_{11}$- and $G\alpha_{12}$/$G\alpha_{13}$-deficient mice. Dev Biol. 2005;282:174–82.

16. Hood LC, Rosenquist TH. Coronary artery development in the chick: origin and deployment of smooth muscle cells, and the effects of neural crest ablation. Anat Rec. 1992;234:291–300.

17. Waldo KL, Kumiski DH, Kirby ML. Association of the cardiac neural crest with development of the coronary arteries in the chick embryo. Anat Rec. 1994;239:315–31.

18. Jiang X, Rowitch DH, Soriano P, McMahon AP, Sucov HM. Fate of the mammalian cardiac neural crest. Development. 2000;127:1607–16.

Roles of Endothelial Hrt Genes for Vascular Development

31

Masahide Sakabe, Takashi Morioka, Hiroshi Kimura, and Osamu Nakagawa

Keywords
Notch signaling • Vascular development • Endothelial cells

Various cellular signaling pathways play essential roles in regulating embryonic vascular development. Among them, Notch signaling is implicated in arterial endothelium differentiation and vascular morphogenesis. Mice that lack Notch receptors or other signaling components die in utero due to severe vascular abnormalities. We previously identified the Hairy-related transcription (Hrt) factor family, also called Hey, Hesr, CHF, Herp, and Gridlock, as downstream mediators of Notch signaling in the developing vasculature [1]. The Hrt family proteins, Hrt1/Hey1, Hrt2/Hey2, and Hrt3/HeyL, mainly act as transcriptional repressors, by binding to consensus DNA elements or by associating with other DNA-binding transcription factors. The mice deficient for *Hrt2* showed perinatal lethality due to ventricular septal defects and mitral valve insufficiency, and cardiomyocyte-specific deletion of *Hrt2* caused abnormal expression of atrial-specific genes in the ventricle and cardiac dysfunction in adulthood [2].

M. Sakabe • O. Nakagawa (✉)
Laboratory for Cardiovascular System Research, Nara Medical University Advanced Medical Research Center, 840 Shijo-cho, Kashihara, Nara 634-8521, Japan
e-mail: osamu.nakagawa@ncvc.go.jp

T. Morioka
Laboratory for Cardiovascular System Research, Nara Medical University Advanced Medical Research Center, 840 Shijo-cho, Kashihara, Nara 634-8521, Japan

Second Department of Internal Medicine, Nara Medical University, 840 Shijo-cho, Kashihara, Nara 634-8522, Japan

H. Kimura
Second Department of Internal Medicine, Nara Medical University, 840 Shijo-cho, Kashihara, Nara 634-8522, Japan

T. Nakanishi et al. (eds.), *Etiology and Morphogenesis of Congenital Heart Disease*, DOI 10.1007/978-4-431-54628-3_31

233

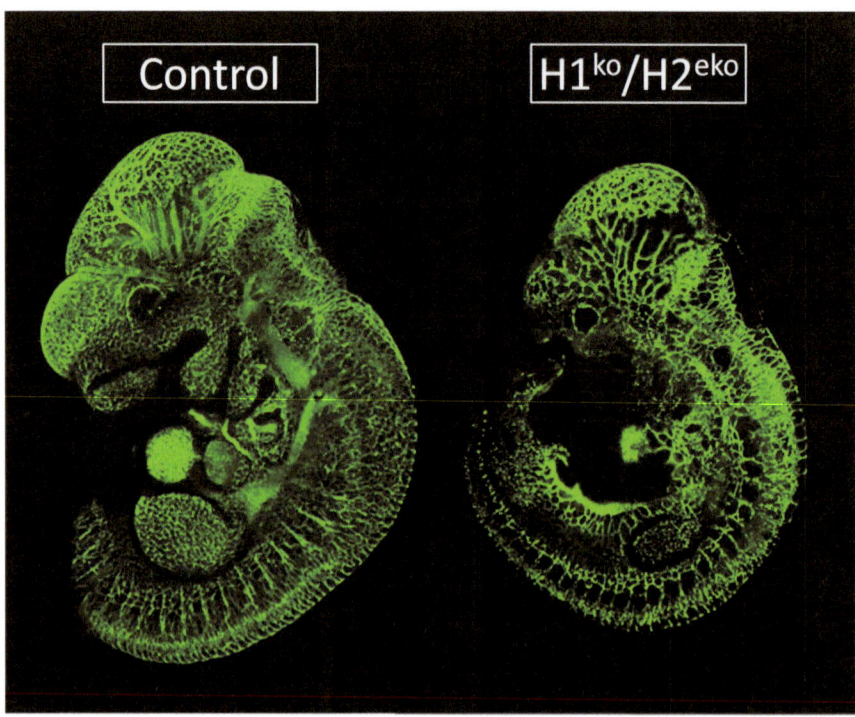

Fig. 31.1 The mice in which *Hrt2* was deleted specifically in endothelial cells with the global *Hrt1* null background (H1^{ko}/H2^{eko}) show embryonic lethality with severe defects of vascular morphogenesis. Whole mount PECAM1 immunostaining demonstrated impairment of vascular network formation in H1^{ko}/H2^{eko} embryos

It was also reported that combined loss of *Hrt1* and *Hrt2* resulted in early embryonic lethality due to vascular demise similar to that observed in Notch signal-deficient embryos. While *Hrt1* and *Hrt2* are expressed in endothelial cells as well as smooth muscle cells of embryonic vasculature, it remained unclear which vascular cell type requires *Hrt1*/*Hrt2* functions. In the present study, we generated the mice with endothelial-cell-specific deletion of *Hrt2* combined with global *Hrt1* null mutation and analyzed their vascular phenotypes during embryonic development. The loss of endothelial *Hrt1*/*Hrt2* caused early vascular abnormalities virtually identical to those observed in the global *Hrt1*/*Hrt2* knockout mouse embryos (Fig. 31.1), suggesting that *Hrt* functions in endothelial cells are indispensable for normal vascular development.

References

1. Nakagawa O, Nakagawa M, Richardson JA, Olson EN, Srivastava D. Hrt1, Hrt2, and Hrt3: a new subclass of bHLH transcription factors marking specific cardiac, somitic, and pharyngeal arch segments. Dev Biol. 1999;216:72–84.
2. Xin M, Small EM, van Rooij E, Qi X, Richardson JA, Srivastava D, Nakagawa O, Olson EN. Essential roles of the bHLH transcription factor Hrt2 in repression of atrial gene expression and maintenance of postnatal cardiac function. Proc Natl Acad Sci U S A. 2007;104:7975–80.

Inositol Trisphosphate Receptors in the Vascular Development

32

Keiko Uchida, Maki Nakazawa, Chihiro Yamagishi, Katsuhiko Mikoshiba, and Hiroyuki Yamagishi

Keywords

Intracellular calcium • Vascular development • Angiogenesis • Placenta

The placental circulation is crucial for the development of mammalian embryos [1]. The labyrinth layer in the placenta is created by extensive villous branching of the trophoblast and vascularization arising from the embryonic mesoderm. In the labyrinth, materials are exchanged between the maternal and embryonic circulation. Recently, we have found that inositol 1,4,5-trisphosphate (IP$_3$) receptors (IP$_3$Rs) may be required for the placental vascularization.

IP$_3$Rs are intracellular Ca^{2+} release channels that have three subtypes in mammals (IP$_3$R1, IP$_3$R2 and IP$_3$R3) [2]. We previously showed that IP$_3$R1 and IP$_3$R2 played an essential role in heart development from the analysis of mouse embryo double knockout for IP$_3$R1 and IP$_3$R2 [3]. A previous report on the

K. Uchida
Department of Pediatrics, Keio University School of Medicine, 35 Shinanomachi, Shinjuku-ku, Tokyo 160-8582, Japan

Health Center, Keio University, 4-1-1 Hiyoshi, Kohoku-ku, Yokohama, Kanagawa 223-8521, Japan

M. Nakazawa • C. Yamagishi • H. Yamagishi (✉)
Department of Pediatrics, Keio University School of Medicine, 35 Shinanomachi, Shinjuku-ku, Tokyo 160-8582, Japan
e-mail: hyamag@keio.jp

K. Mikoshiba
Laboratory for Developmental Neurobiology, RIKEN Brain Science Institute, Saitama 351-0198, Japan

Calcium Oscillation Project, International Cooperative Research Project and Solution-Oriented Research for Science and Technology, Japan Science and Technology Agency, Kawaguchi, Saitama 332-0012, Japan

© The Author(s) 2016
T. Nakanishi et al. (eds.), *Etiology and Morphogenesis of Congenital Heart Disease*, DOI 10.1007/978-4-431-54628-3_32

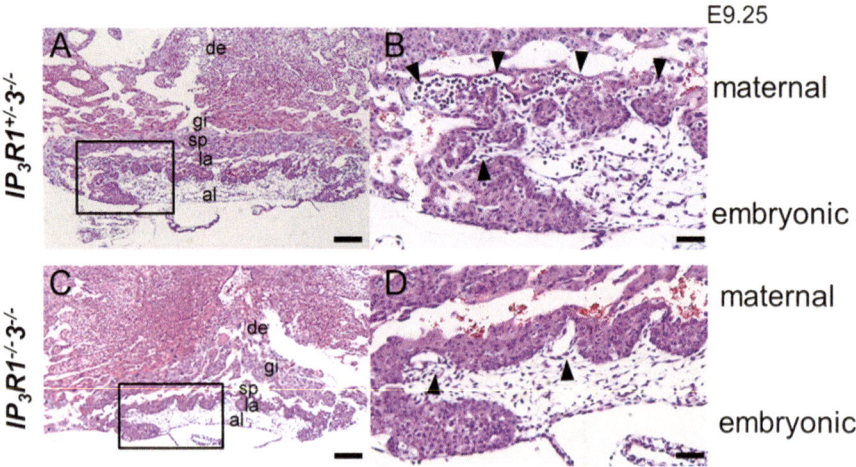

Fig. 32.1 Cross sections of E9.25 placentas from the $IP_3R1^{+/-}3^{-/-}$ (**a** and **b**) and $IP_3R1^{-/-}3^{-/-}$ (**c** and **d**) mice. (**b**) and (**d**) show higher-power fields of the *rectangular* areas of the labyrinth in (**a**) and (**c**), respectively. Embryonic vessels (*arrowheads*) fail to elongate to the maternal sinuses in the placenta of $IP_3R1^{-/-}3^{-/-}$ compared to that of $IP_3R1^{+/-}3^{-/-}$ (wild type). *al* allantois, *de* decidua, *gi* trophoblast giant cells, *la* labyrinth layer, *sp* spongiotrophoblast layer. Scale bars, 0.5 mm in (**a**) and (**c**) and 0.2 mm in (**b**) and (**d**)

requirement for phospholipase (PLC) $\delta1$ and $\delta3$ [4] that produce IP_3 for placentation led us to investigate the placental defects by deletion of any subtypes of IP_3Rs. Our preliminary result revealed that embryonic vasculature in the labyrinth was impaired in the placenta double knockout for IP_3R1 and IP_3R3 at E9.25 (Fig. 32.1). The detailed phenotype and the underlying mechanism how the intracellular Ca^{2+} signaling via IP_3Rs may be implicated in the development of extraembryonic vasculature are under investigation.

This work was supported by a Grant-in-Aid for Scientific Research from the Ministry of Education, Culture, Sports, Science and Technology, Japan (to K.U. and H.Y.).

References

1. Rossant J, Cross JC. Placental development: lessons from mouse mutants. Nat Rev Genet. 2001;2:538–48.
2. Berridge MJ, Lipp P, Bootman MD. The versatility and universality of calcium signalling. Nat Rev Mol Cell Biol. 2000;1:11–21.
3. Uchida K, Aramaki M, Nakazawa M, Yamagishi C, Makino S, et al. Gene knock-outs of inositol 1,4,5-trisphosphate receptors types 1 and 2 result in perturbation of cardiogenesis. PLoS One. 2010;5:e12500.
4. Nakamura Y, Hamada Y, Fujiwara T, Enomoto H, Hiroe T, et al. Phospholipase C-delta1 and -delta3 are essential in the trophoblast for placental development. Mol Cell Biol. 2005;25:10979–88.

Yukako Yoshikane, Mitsuhisa Koga, Tamaki Cho, Kyoko Imanaka-Yoshida, Yumi Yamamoto, Junichi Hashimoto, Hiroki Aoki, Koichi Yoshimura, and Shinichi Hirose

Keywords

Kawasaki disease • Tenascin-C • c-Jun N-terminal kinase • Aneurysm • Remodeling

Kawasaki disease is the most common acute systemic vasculitis of unknown etiology in children [1] and can cause inflammation of the coronary arteries leading to aneurysms. Tenascin-C, an extracellular matrix protein, and c-Jun N-terminal kinase (JNK), an intracellular signaling protein, are known to be associated with inflammation and tissue remodeling [2, 3]. The purpose of this study was to demonstrate tenascin-C and JNK might be involved in tissue remodeling in a *Candida albicans*-induced murine model of aneurysm.

Y. Yoshikane (✉) • K. Yoshimura • S. Hirose
Department of Pediatrics, Faculty of Medicine, Fukuoka University, Nanakuma 7-45-1, Jonan-ku, Fukuoka 814-0180, Japan
e-mail: yyoshika@fukuoka-u.ac.jp

M. Koga
Department of Pharmaceutical Care and Health Sciences, Fukuoka University, Fukuoka, Japan

T. Cho
Department of Functional Bioscience, Fukuoka Dental College, Fukuoka, Japan

K. Imanaka-Yoshida
Department of Pathology and Matrix Biology, Mie University Graduate School of Medicine, Tsu, Japan

Y. Yamamoto • J. Hashimoto
Department of Surgery and Clinical Science, Yamaguchi University Graduate School of Medicine, Yamaguchi, Japan

H. Aoki
Cardiovascular Research Institute, Kurume University, Kurume, Japan

© The Author(s) 2016
T. Nakanishi et al. (eds.), *Etiology and Morphogenesis of Congenital Heart Disease*,
DOI 10.1007/978-4-431-54628-3_33

1. More than 80 % of the mice showed the macroscopic features of aneurysms in the aorta and/or iliac and coronary arteries.
2. Marked inflammatory cell infiltration was observed in vascular wall and perivascular connective tissue, accompanied by fragmentation of elastic fibers.
3. Expression of tenascin-C was highly observed in vascular wall, accompanied by active degradation of elastic fibers.
4. Pharmacologic inhibition of JNK attenuated the aneurysm formation in the mice model.

In conclusion, these findings suggest that both tenascin-C and JNK are involved in abnormal tissue remodeling and inflammation in the *Candida albicans*-induced Kawasaki disease murine model of aneurysm and that JNK inhibition may represent a novel therapeutic target for preventing a Kawasaki disease-related aneurysm.

References

1. Kawasaki T, Kosaki F, Okawa S, Shigematsu I, Yanagawa H. A new infantile acute febrile mucocutaneous lymph node syndrome (MLNS) prevailing in Japan. Pediatrics. 1974;54:271–6.
2. Imanaka-Yoshida K. Tenascin-C in cardiovascular tissue remodeling: from development to inflammation and repair. Circ J. 2012;76:2513–20.
3. Yoshimura K, Aoki H, Ikeda Y, Fujii K, Akiyama N, Furutani A, Hoshii Y, Tanaka N, Ricci R, Ishihara T, Esato K, Hamano K, Matsuzaki M. Regression of abdominal aortic aneurysm by inhibition of c-Jun N-terminal kinase. Nat Med. 2005;11:1330–8.

Perspective

Susumu Minamisawa

The ductus arteriosus (DA), an important fetal artery that connects the main pulmonary artery and the aortic arch, closes immediately after birth. Therefore, closure of the DA is a symbolic event of the change from fetal to neonatal circulation. When the DA remains open after the first 3 days of life in humans, the condition is called patent DA (PDA), and it usually causes a left-to-right shunt. If the shunt volume significantly increases, neonates may exhibit pulmonary edema and respiratory distress. Especially in preterm infants, PDA could be a serious life-threatening risk. The incidence of PDA unaccompanied by any other cardiovascular abnormality has been estimated to be 0.06 % of term infants, and the incidence sharply increases in premature infants. Symptomatic PDA cases have been found in 28 % of infants with very low birth weight (<1500 g) and 55 % of infants with extremely low birth weight (<1000 g). On the other hand, PDA is lifesaving in some patients with congenital heart diseases such as hypoplastic left heart syndrome and pulmonary atresia. The DA must be open to maintain blood flow in the systemic or pulmonary circulation. Current pharmacological treatment for closing or maintaining PDA is limited to the agents that control the vasodilatory effect of prostaglandin E_2 (PGE_2), such as cyclooxygenase inhibitors or PGE_1 preparations.

DA closure is thought to occur in response to a combination of two different mechanisms. One mechanism is an acute response of smooth muscle constriction within the first several hours of life, known as functional closure of the DA. The other is a relatively chronic response of structural change in the DA during the

S. Minamisawa (✉)
Department of Cell Physiology, The Jikei University School of Medicine, 3-25-8 Nishishinbashi, Minatoku 105–8461, Tokyo, Japan
e-mail: sminamis@jikei.ac.jp

perinatal period, known as anatomical closure or vascular remodeling, such as progression of neointimal thickening and impaired elastic fiber formation. After shutting down blood flow, progressive apoptosis and fibrotic changes occur in the DA, resulting in permanent DA closure and a remnant structure known as the ligamentum arteriosum. In this part, four studies highlight the recent advances in molecular mechanisms underlying DA closure from the perspective of functional and anatomical closure.

The primary driving force behind functional DA closure is an increase in oxygen tension and a decrease in circulating PGE_2. The DA is an oxygen-sensitive vessel. Several potassium channels are known to play an important role in the oxygen-sensing system. Momma et al. show that ATP-sensitive potassium channels (K_{ATP} channels) are another oxygen sensor. They show that sulfonylureas, which are commonly used to treat diabetes, inhibit K_{ATP} channels to constrict the DA, suggesting that sulfonylureas can be used for patients with PDA. The response to oxygen is weaker in premature DA than in mature DA. Hayama et al. investigate the developmental changes in the contractile apparatus and sarcoplasmic reticulum in the DA and its connecting arteries that may contribute to the oxygen-sensitive mechanism.

Anatomical closure of the DA is associated with distinct differentiation of the vessel wall. Intimal thickening is the most prominent phenotypic change, and it involves several processes: (a) an area of subendothelial extracellular matrix deposition, (b) the migration of undifferentiated medial smooth muscle cells (SMCs) into the subendothelial space, and (c) the disassembly of the internal elastic lamina and the loss of elastic fibers in the medial layer. Yokoyama et al. demonstrate that chronic activation of the PGE_2 receptor EP4 plays a pivotal role in neointima formation and impaired elastic fiber formation in the DA. Therefore, PGE_2 signaling has dual roles in functional and anatomical DA closure. DA remodeling is known to resemble the vascular remodeling of aging arteries. Gittenberger-de Groot et al. demonstrate that progerin, an alternative splice variant of lamin A/C, is highly expressed in the closing DA as well as in premature aging of vessels in children with Hutchinson progeria syndrome, suggesting that progerin plays a role in vascular remodeling of the DA.

These studies show that development of a new strategy to treat DA closure or opening first requires understanding of the precise molecular mechanisms underlying both functional and anatomical DA closure.

Progerin Expression during Normal Closure of the Human Ductus Arteriosus: A Case of Premature Ageing?

34

Adriana C. Gittenberger-de Groot, Regina Bokenkamp, Vered Raz, Conny van Munsteren, Robert E. Poelmann, Nimrat Grewal, and Marco C. DeRuiter

Abstract

The ductus arteriosus (DA) is a fetal vessel bypassing the still nonfunctional lungs. Closure of the DA at birth is essential for the transition from a fetal to a neonatal circulation. This closing process begins with a physiological contraction followed by definitive anatomical closure. The latter process starts already before birth by development of intimal thickening followed after birth by degeneration of the inner media, including cytolytic necrosis and apoptosis. The DA will remain patent when there is insufficient maturation in prematurely born babies or when there is a structural abnormality as seen in persistent DA (PDA). The histological changes during normal DA closure resemble the features seen in the premature ageing vessels in children with the Hutchinson progeria syndrome. The latter syndrome is caused by a mutation in the lamin A/C gene resulting in accumulation of the progerin splice variant. We studied human DA biopsies from the fetal to the neonatal period to investigate whether lamin A/C and progerin might be involved in the DA closure process. The results

A.C. Gittenberger-de Groot (✉) • R.E. Poelmann
Department of Anatomy and Embryology, Leiden University Medical Center, Leiden, The Netherlands

Department of Cardiology, Leiden University Medical Center, Leiden, The Netherlands
e-mail: a.c.gittenberger_de-groot@lumc.nl

R. Bokenkamp
Department of Pediatric Cardiology, Leiden University Medical Center, Leiden, The Netherlands

V. Raz
Department of Genetics, Leiden University Medical Center, Leiden, The Netherlands

C. van Munsteren • M.C. DeRuiter
Department of Anatomy and Embryology, Leiden University Medical Center, Leiden, The Netherlands

N. Grewal
Department of Thoracic Surgery, Leiden University Medical Center, Leiden, The Netherlands

© The Author(s) 2016
T. Nakanishi et al. (eds.), *Etiology and Morphogenesis of Congenital Heart Disease*,
DOI 10.1007/978-4-431-54628-3_34

show an increase in the intima and inner media of progerin in the normal neonatal DA, while expression of lamin A/C is diminished. In the non-closing aorta, the fetal DA and the PDA, no or hardly any progerin expression was found. We postulate that the lamin A/C to progerin balance is important during normal anatomical closure of the DA presenting a unique case of physiological premature vascular ageing.

Keywords
Lamin A/C Progerin • Atherosclerosis • Apoptosis • Vascular biology • Persistent ductus arteriosus

34.1   Introduction

The ductus arteriosus (DA) is a fetal vessel that connects the pulmonary trunk to the distal aortic arch. The patent DA is functional before birth directing oxygen-rich blood from the placenta to the systemic circulation bypassing the still immature lung vascular bed. After birth, the DA contracts, and this physiological closure is followed by definitive anatomical closure. Ultimately, the DA will be remodeled into a fibrous strand [1, 2]. The physiological closure is regulated by many substances in which prostaglandins play an important role [3]. The onset of anatomical closure in the human fetus starts already in the second trimester of gestation with blebbing of the endothelium and formation of intimal thickening [2]. The degenerative changes in the media, in which cytolytic necrosis and apoptosis play an important role [4], start immediately after birth. By then, marked intimal thickening supports the closure of the DA upon contraction. This maturation process of ductal closure (Fig. 34.1) follows a temporo-spatial pattern. Upon premature delivery of the baby, the DA closure process may not be effective already resulting in an immature patent DA. Usually, this resolves spontaneously within a few days to weeks after birth. The process might be enhanced by providing prostaglandin inhibitors like indomethacin [2]. Another reason for non-closure resulting in a persistent DA (PDA) may be structural abnormalities, in most cases based on an elastin deposition problem [3]. The histological features of normal DA closure resemble atherosclerotic changes in the vessel wall albeit without the atheroma component [5]. The mature DA vessel wall structure is highly similar to the premature ageing histopathology seen in vessels of children with the Hutchinson-Gilford syndrome (HGPS) [6]. This syndrome is based on a genetic defect in the lamin A (*LMNA*) gene which leads to pathological accumulation of the splice variant progerin [7, 8]. Increased progerin expression has also been reported during normal vascular ageing [9, 10]. Based on the observed similarities, we investigated the role of an *LMNA*/progerin balance during normal perinatal closure of the DA.

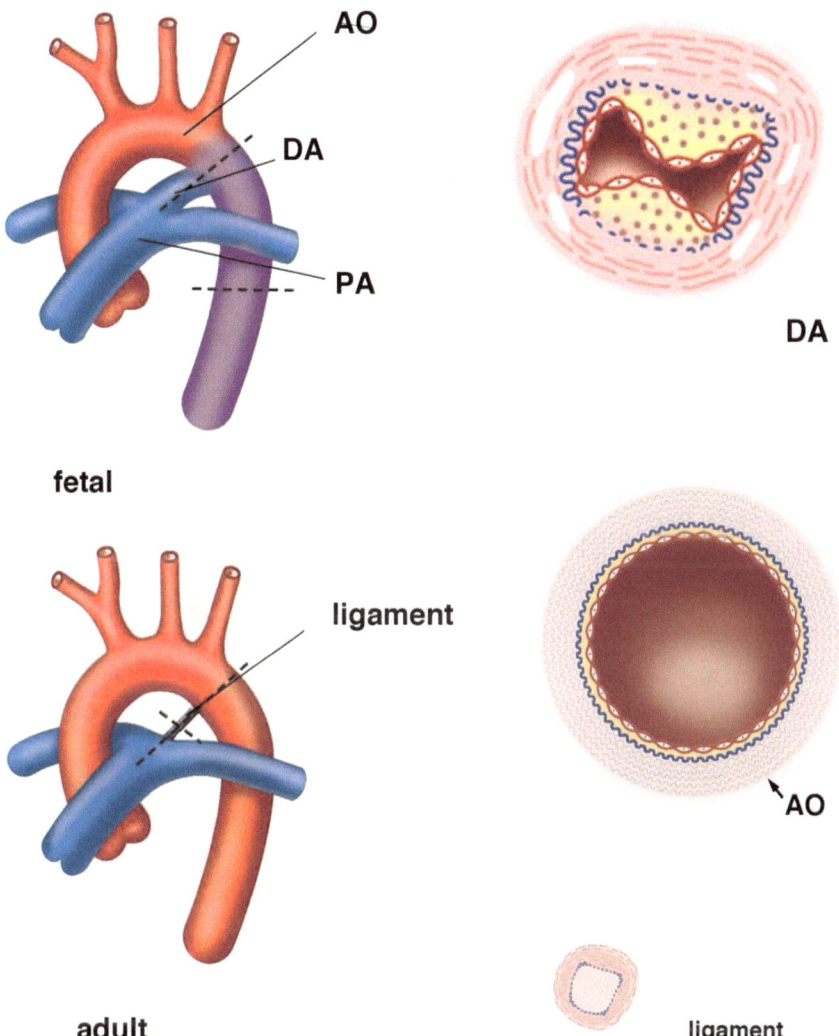

Fig. 34.1 Schematic representation of the aortic arch and the cross sections of the muscular ductus arteriosus with intimal thickening (DA) and the elastic aorta (AO) and pulmonary artery (PA), during the fetal, neonatal, and adult stage of DA, in which the latter is remodeled into a ligament

34.2   Material and Methods

Following aortic arch reconstruction and surgical closure of the DA, a biopsy specimen of human neonatal DA ($n = 16$) and adjoining descending aorta ($n = 2$) was obtained. The fetal DA (one of 14 weeks' and one of 18 weeks' gestation) was acquired from postmortem fetal specimen after legal or spontaneous abortion. One biopsy specimen was obtained from a 2-year-old child with PDA. Details on sectioning, fixation, and immunohistochemistry have been described as well as the technique for RNA isolation and RT-PCR reactions [11].

34.3   Results

The RT-PCR studies of the aorta and neonatal DA revealed that next to lamin A also progerin was detected. Lamin A was seen both in the aorta and neonatal DA, while low levels of progerin were only seen in the neonatal DA [11]. The elastin expression showed clearly that the DA is a muscular artery, while the aorta is an elastic artery (Fig. 34.2a, d, e, j). Immunohistochemistry of the tissue sections

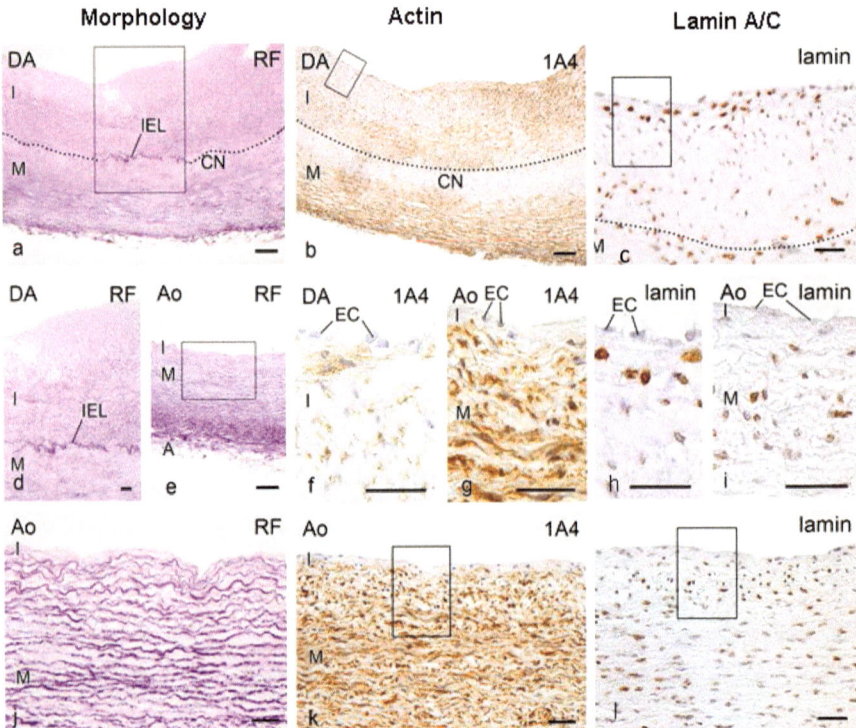

Fig. 34.2 Transverse histological sections of the ductus arteriosus (DA) and the aorta (AO). (**a**) Overview of the DA wall with a resorcin-fuchsin staining (RF) showing the morphology of a muscular artery with an internal elastic membrane (IEL) (**a, d**) between the intima (I) and the media (m). In the inner media, there is cytolytic necrosis (CN), which is better appreciated by the alpha smooth muscle actin loss (actin) in the smooth muscle cells (**b**). The endothelial cells (ECs) are negative for actin, also in the AO (**f, g**). (**e**) Overview of the elastic wall of the AO with a regular manifestation of elastic lamellae (**j**) and marked actin staining in the smooth muscle cells (**k**). There is no CN in the aorta. (**c, h, i, l**) In both the DA and AO lamin A/C, positive cells are found. ECs are negative for lamin A/C. Adventitia: A

revealed alpha smooth muscle actin to be present in all cases studied both in the intima (when thickened) and in the media. The expression level of alpha smooth muscle actin became less in the cases of a normal closing neonatal DA where inner media degeneration was seen with developing cytolytic necrosis (Fig. 34.2a, b). Furthermore, the expression of lamin A/C was found in all DA and the aorta with exception of the 14-week fetal DA. There was a clear difference with regard to progerin expression (Fig. 34.3). This was only found in the normal closing neonatal DA. If the latter was already fully developed including apoptosis and loss of nuclei, the progerin expression was diminished. This shows that progerin expression precedes apoptosis and degeneration. Progerin was not detected in the fetal 18-week DA, being also absent in the non-closing PDA and aorta.

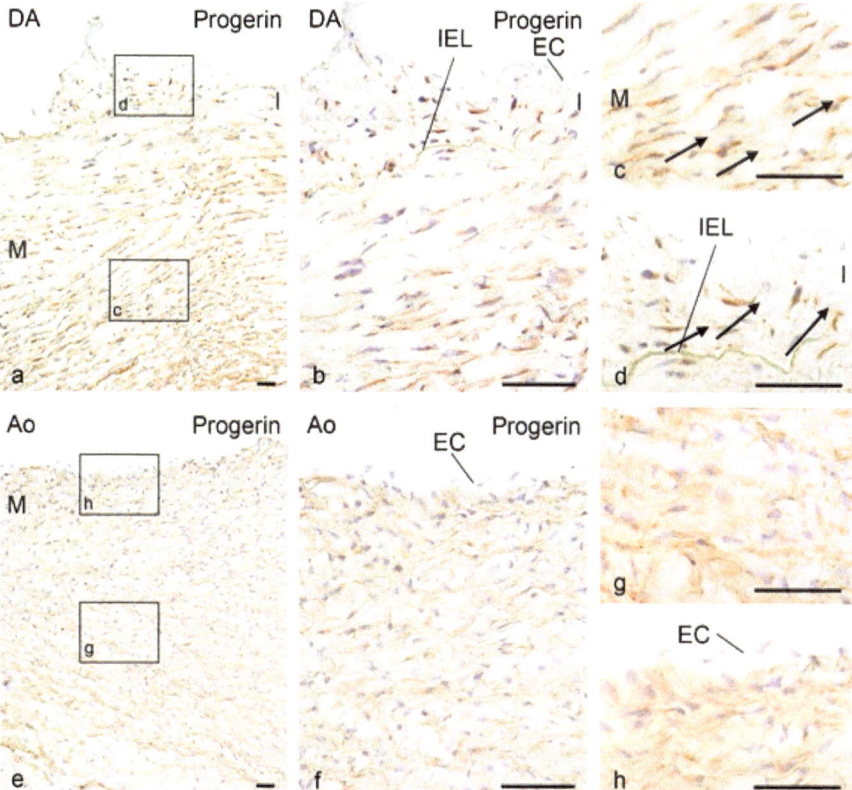

Fig. 34.3 (**a**) Overview of the intima (I) and media (M) of a 10-day-old neonatal ductus arteriosus (DA) with faint cytoplasmic and strong nuclear and perinuclear expression of progerin in the vascular smooth muscle cells of the media (**c**) and intima (**d**). The endothelial cells (ECs) are negative (**b, d**). (**e**) In the aortic wall (AO), no regions with strong progerin expression are found (**g, h**). ECs are also negative (**f, h**). *IEL* internal elastic lamina. *Bars* = 30 μm

34.4 Discussion

Normal ductal closure is regulated by many molecular pathways [3]. In our study, we introduced for the first time a role for lamina A/C and progerin [11]. This observation was triggered by the histopathology found in the vessel wall of young adolescents with the HGPS [8], in which premature atherosclerosis was observed. The physiological development of prenatal intimal thickening and the development of cytolytic necrosis in the DA showing contractions after birth mimic the degenerative changes seen in the HGPS. Alterations in the farnesylation process can be the cause of accumulation of progerin at the nuclear membrane. This process is associated with the initiation of apoptosis in these progerin-expressing cells [12], being characteristic for degeneration in the DA media [13]. The cause for progerin increase in the closing DA needs further investigation, but it is known that alternative splicing is affected by oxidative stress [14] which is activated during normal DA closure [15]. Accumulation of reactive oxygen species has also been observed during vascular ageing [10] in which an increase of progerin has been described in the smooth muscle cells of the coronary vessels of elderly persons as well as a gradual increase of progerin-expressing cells in other vessels with upclimbing age [7, 10]. It is noteworthy that the non-closing PDA, although only one case has been studied, and the aorta did not show progerin increase in the neonatal stage. This aspect needs further study as in families with bicuspid aortic valve (BAV) a correlation with PDA [16] has been reported. Unpublished data from our group show a diminished progerin expression in the dilated ascending aorta in BAV as compared to dilation of the aorta in tricuspid valves. The latter shows an increase in progerin indicative of advanced ageing.

Potentially, ductal closure based on increased progerin expression could lead to fetal demise in HGPS; this has, however, not been reported.

34.5 Future Directions and Clinical Applications

Further research is necessary to elucidate the role of the lamin A/C to progerin balance in vascular pathology, ageing, and the unique observation in selective DA closure at birth.

Clinical applications might be related to the use of farnesyltransferase inhibitors which may prevent onset and progression induced by the accumulation of progerin [17]. It cannot be excluded that inhibition of physiological progerin expression might prevent anatomical closure of the DA, which can be relevant in ductus-dependent congenital heart disease.

References

1. Gittenberger-de Groot AC. Persistent ductus arteriosus: most probably a primary congenital malformation. Br Heart J. 1977;39:610–8.
2. Gittenberger-de Groot AC, van Ertbruggen I, Moulaert AJMG, et al. The ductus arteriosus in the preterm infant: histological and clinical observation. J Pediatr. 1980;96:88–93.
3. Bokenkamp R, DeRuiter MC, van Munsteren C, et al. Insights into the pathogenesis and genetic background of patency of the ductus arteriosus. Neonatology. 2009;98:6–17.
4. Gittenberger-de Groot AC, Strengers JLM. Histopathology of the arterial duct (ductus arteriosus) with and without treatment with prostaglandin E1. Int J Cardiol. 1988;19:153–66.
5. Slomp J, van Munsteren JC, Poelmann RE, et al. Formation of intimal cushions in the ductus arteriosus as a model for vascular intimal thickening. An immunohistochemical study of changes in extracellular matrix components. Atherosclerosis. 1992;93:25–39.
6. Goldman RD, Shumaker DK, Erdos MR, et al. Accumulation of mutant lamin A causes progressive changes in nuclear architecture in Hutchinson-Gilford progeria syndrome. Proc Natl Acad Sci U S A. 2004;101:8963–8.
7. Olive M, Harten I, Mitchell R, Beers JK, Djabali K, et al. Cardiovascular pathology in Hutchinson-Gilford progeria: correlation with the vascular pathology of aging. Arterioscler Thromb Vasc Biol. 2010;30:2301–9.
8. Stehbens WE, Wakefield SJ, Gilbert-Barness E, et al. Histological and ultrastructural features of atherosclerosis in progeria. Cardiovasc Pathol. 1999;8:29–39.
9. McClintock D, Ratner D, Lokuge M, et al. The mutant form of lamin A that causes Hutchinson-Gilford progeria is a biomarker of cellular aging in human skin. PLoS One. 2007;2:e1269.
10. Ragnauth CD, Warren DT, Liu Y, et al. Prelamin A acts to accelerate smooth muscle cell senescence and is a novel biomarker of human vascular aging. Circulation. 2010;121:2200–10.
11. Bökenkamp R, Raz V, Venema A, et al. Differential temporal and spatial progerin expression during closure of the ductus arteriosus in neonates. PLoS One. 2011;6:e23975.
12. Bridger JM, Kill IR. Aging of Hutchinson-Gilford progeria syndrome fibroblasts is characterised by hyperproliferation and increased apoptosis. Exp Gerontol. 2004;39:717–24.
13. Tananari Y, Maeno Y, Takagishi T, et al. Role of apoptosis in the closure of neonatal ductus arteriosus. Jpn Circ J. 2000;64:684–8.
14. Kajimoto H, Hashimoto K, Bonnet SN, et al. Oxygen activates the Rho/Rho-kinase pathway and induces RhoB and ROCK-1 expression in human and rabbit ductus arteriosus by increasing mitochondria derived reactive oxygen species: a newly recognized mechanism for sustaining ductal constriction. Circulation. 2007;115:1777–88.
15. Slomp J, Gittenberger-de Groot AC, Glukhova MA, et al. Differentiation, dedifferentiation, and apoptosis of smooth muscle cells during the development of the human ductus arteriosus. Arterioscler Thromb Vasc Biol. 1997;17:1003–9.
16. Glancy DL, Wegmann M, Dhurandhar RW. Aortic dissection and patent ductus arteriosus in three generations. Am J Cardiol. 2001;87:813–5.
17. Capell BC, Olive M, Erdos MR, et al. A farnesyltransferase inhibitor prevents both the onset and late progression of cardiovascular disease in a progeria mouse model. Proc Natl Acad Sci U S A. 2008;105:15902–7.

Utako Yokoyama, Susumu Minamisawa, and Yoshihiro Ishikawa

Abstract

The ductus arteriosus (DA) is a shunt vessel between the aorta and the pulmonary artery during the fetal period. It is well recognized that prostaglandin E_2 (PGE_2) dilates the DA through activation of its receptor EP4 and subsequent cyclic AMP (cAMP) production during the fetal period and that oxygen constricts the DA by inhibiting potassium channels immediately after birth. In addition to the regulation of vascular tone, morphological remodeling of the DA throughout the perinatal period, such as prominent intimal thickening and poor elastogenesis, has been demonstrated.

We recently identified the molecular mechanisms of the acquisition of unique morphological remodeling in the DA during development. During the fetal period, PGE_2-EP4 signaling decreases elastic fiber formation through degradation of the cross-linking enzyme lysyl oxidase (LOX) and increases hyaluronan-mediated intimal thickening in the DA. This remodeling is mediated by activation of the EP4 receptor via diverse downstream intracellular signaling pathways. Hyaluronan-mediated intimal thickening was induced by the EP4-Gs protein-cyclic AMP-protein kinase A pathway. The attenuation of elastogenesis is mediated through a non-cyclic AMP signaling pathway, such as c-src-phospholipase C (PLC). These data suggest that placental PGE_2-mediated vascular remodeling via different signaling pathways orchestrates the subsequent luminal DA reorganization, leading to complete obliteration of the DA.

U. Yokoyama (✉) • Y. Ishikawa
Cardiovascular Research Institute, Graduate School of Medicine, Yokohama City University, Yokohama, Japan
e-mail: utako@yokohama-cu.ac.jp

S. Minamisawa
Department of Cell Physiology, School of Medicine, Jikei University, Tokyo, Japan

© The Author(s) 2016

253

T. Nakanishi et al. (eds.), *Etiology and Morphogenesis of Congenital Heart Disease*,
DOI 10.1007/978-4-431-54628-3_35

Keywords
Ductus arteriosus • Prostaglandin E • Intimal thickening • Smooth muscle • Elastic fiber

35.1   Introduction

The ductus arteriosus (DA) normally closes immediately after birth. Although the DA is a normal and essential fetal structure, it becomes abnormal if it remains patent after birth. DA closure occurs in two phases: functional closure of the lumen in the first hours after birth by smooth muscle constriction and anatomic occlusion of the lumen over the next several days due to extensive neointimal thickening in human DA [1–3]. There are several events that promote DA constriction immediately after birth. Increasing oxygen tension and a dramatic decrease in circulating PGE_2 promote muscular constriction of the DA. In addition, DA remodeling is also necessary for its complete closure. Remodeling is characterized by (a) an area of subendothelial deposition of extracellular matrix [4], (b) the disassembly of the internal elastic lamina and loss of elastic fiber in the medial layer [5], and (c) migration into the subendothelial space of undifferentiated medial smooth muscle cells (SMCs). Some of these changes begin about halfway through gestation, and some occur after functional closure of the DA in the neonate [3, 6]. In addition to the well-known vasodilatory role of PGE_2, our findings revealed the role of PGE_2 in the anatomical closure of the DA.

35.2   The Molecular Mechanisms of Intimal Thickening of the Ductus Arteriosus

35.2.1 Hyaluronan-Mediated Intimal Thickening

PGE_2 plays a primary role in maintaining the patency of the DA via its receptor EP4. However, previous studies have demonstrated that genetic disruption of the PGE receptor EP4 paradoxically results in fatal patent DA in mice [7, 8]. In addition, double mutant mice in which cyclooxygenase (COX)-1 and COX-2 are disrupted also exhibit patent DA [9]. We found that intimal thickening was completely absent in the DA of EP4-disrupted neonatal mice [3]. Moreover, a marked reduction in hyaluronan production was found in EP4-disrupted DA, whereas a thick layer of hyaluronan deposit was present in wild-type DA. PGE_2-EP4-cyclic AMP (cAMP)-protein kinase A (PKA) signaling upregulates hyaluronan synthase type 2 mRNA, which increases hyaluronan production in the DA. Accumulation of hyaluronan then promotes SMC migration into the subendothelial layer to form intimal thickening [3].

EP4 is a Gs protein-coupled receptor that increases intracellular cAMP by adenylyl cyclases (ACs) consisting of nine different isoforms of membrane-bound forms of ACs (AC1 through AC9). We found that AC2 and AC6 are more highly expressed in rat DA than in the aorta during the perinatal period [10]. Our data using AC subtype-targeted siRNAs and AC6-deficient mice suggest that AC6 is responsible for hyaluronan-mediated intimal thickening of the DA, whereas AC2 inhibits AC6-induced hyaluronan production. The activation of both AC2 and AC6 induces vasodilation.

35.2.2 Epac-Mediated SMC Migration

In addition to PKA, a new target of cAMP that is an exchange protein activated by cAMP has recently been discovered; it is called Epac [11]. Epac is a guanine nucleotide exchange protein that regulates the activity of small G proteins and has been known to exhibit a distinct cAMP signaling pathway that is independent of PKA [12]. There are two variants: Epac1 is expressed in most tissues, including the heart and blood vessels, whereas Epac2 is expressed in the adrenal gland and the brain. Although both Epac1 and Epac2 are upregulated during the perinatal period, Epac1, but not Epac2, acutely promotes SMC migration and thus intimal thickening in the DA [13]. Since Epac stimulation does not increase hyaluronan production, the effect of Epac1 on SMC migration is independent of that of hyaluronan accumulation, which operates through a mechanism different from that underlying PKA stimulation.

35.2.3 Regulation of Elastogenesis

Elastic fiber formation begins in mid-gestation and increases dramatically during the last trimester in the great arteries. However, the DA exhibits lower levels of elastic fiber formation [5], which may contribute to vascular collapse and subsequent closure of the DA after birth. We found that EP4 significantly inhibited elastogenesis and decreased lysyl oxidase (LOX) protein, which catalyzes elastin cross-links in DA SMCs but not in aortic SMCs. In EP4-knockout mice, electron microscopic examination showed that the DA acquired an elastic phenotype that was similar to the neighboring aorta. More importantly, human DA and aorta tissues from seven patients showed a negative correlation between elastic fiber formation and EP4 expression, as well as between EP4 and LOX expression. Together with in vitro experiments, these data suggest that PGE₂-EP4 signaling inhibits elastogenesis in the DA by degrading LOX protein. The EP4-cSrc-PLC-γ-signaling pathway, a signaling pathway that has not previously been recognized, most likely promoted the lysosomal degradation of LOX [14, 15].

35.3 Future Direction and Clinical Implications

The persistently patent DA after birth is a major cause of morbidity and mortality, especially in premature infants, that can lead to severe complications, including pulmonary hypertension, right ventricular dysfunction, postnatal infections, and respiratory failure [16]. The incidence of DA patency has been estimated to be 1 in 500 in term newborns [17]. In preterm babies with birth weights <1,500 g, the incidence of patent DA exceeds 30 % [18]. Therefore, it is important to improve current pharmacological therapy through understanding the precise mechanisms of the regulation of the DA. Since both vascular contraction and remodeling are required for complete DA closure, pharmacological therapies that promote vaso-constriction and remodeling would be ideal for premature infants with persistently patent DAs. On the other hand, vasodilation and inhibition of intimal thickening are required for DA-dependent congenital heart diseases.

Our data suggest that PGE_2-EP4-cAMP signaling promotes hyaluronan and Epac-mediated intimal thickening and that the EP4-PLC pathway attenuates elastogenesis in the DA. These cascades of events via different signaling pathways are thought to orchestrate the subsequent luminal DA reorganization (Fig. 35.1),

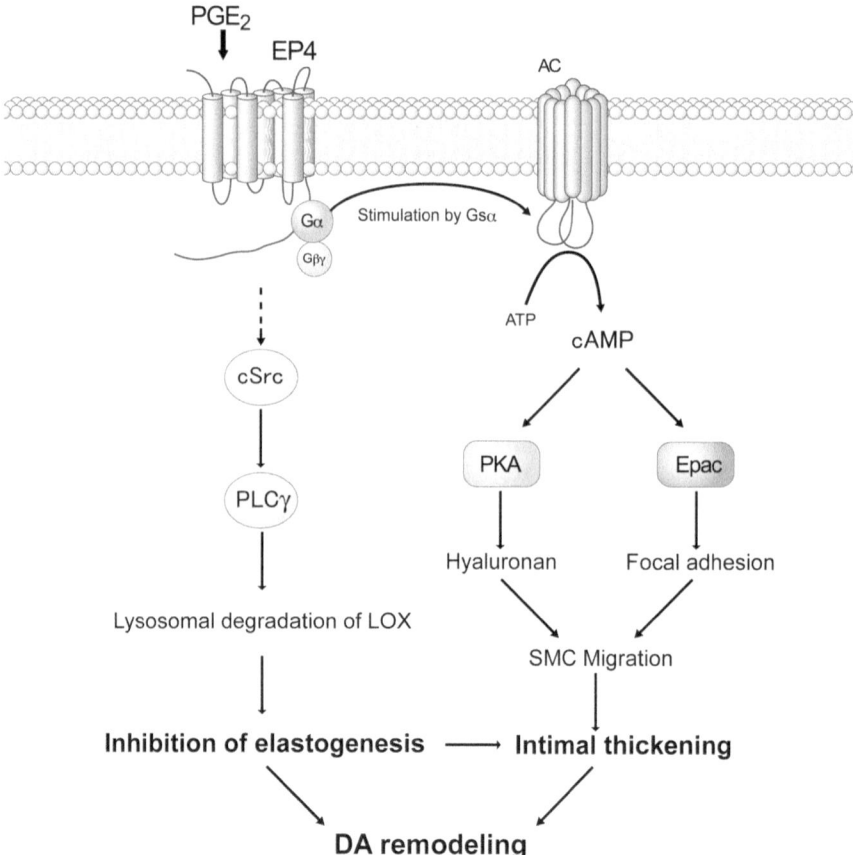

Fig. 35.1 The diverse EP4 signaling pathways. Both PKA and Epac synergistically promoted intimal cushion formation in the DA, but they work in two distinct ways. The cSrc-PLC pathway inhibited elastogenesis via degrading LOX proteins

leading to complete obliteration of the DA. In addition to its role in controlling vascular tone in the functional closure of the DA, the vascular remodeling of the DA is now attracting considerable attention as a target for novel therapeutic strategies for patients with persistently patent DA and DA-dependent cardiac anomalies.

Acknowledgments   This work was supported by grants from the Ministry of Health, Labor and Welfare of Japan (U.Y.), the Ministry of Education, Culture, Sports, Science and Technology of Japan (U.Y., S.M.), the Yokohama Foundation for Advanced Medical Science (U.Y., S.M.), the "High-Tech Research Center" Project for Private Universities: MEXT (S.M.), MEXT-Supported Program for the Strategic Research Foundation at Private Universities (S.M.), the Vehicle Racing Commemorative Foundation (U.Y., S.M.), Miyata Cardiology Research Promotion Funds (U.Y., S.M.), the Takeda Science Foundation (U.Y., S.M.), the Japan Heart Foundation Research Grant (U.Y.), the Kowa Life Science Foundation (U.Y.), the Sumitomo Foundation (U.Y.), and the Shimabara Science Promotion Foundation (S.M.).

References

1. Smith GC. The pharmacology of the ductus arteriosus. Pharmacol Rev. 1998;50:35–58.
2. Clyman RI. Mechanisms regulating the ductus arteriosus. Biol Neonate. 2006;89:330–5.
3. Yokoyama U, Minamisawa S, Quan H, Ghatak S, Akaike T, Segi-Nishida E, Iwasaki S, Iwamoto M, Misra S, Tamura K, Hori H, Yokota S, Toole BP, Sugimoto Y, Ishikawa Y. Chronic activation of the prostaglandin receptor EP4 promotes hyaluronan-mediated neointimal formation in the ductus arteriosus. J Clin Invest. 2006;116:3026–34.
4. Gittenberger-de Groot AC, Strengers JL, Mentink M, Poelmann RE, Patterson DF. Histologic studies on normal and persistent ductus arteriosus in the dog. J Am Coll Cardiol. 1985;6:394–404.
5. De Reeder EG, Girard N, Poelmann RE, Van Munsteren JC, Patterson DF, Gittenberger-De Groot AC. Hyaluronic acid accumulation and endothelial cell detachment in intimal thickening of the vessel wall. The normal and genetically defective ductus arteriosus. Am J Pathol. 1988;132:574–85.
6. Slomp J, Gittenberger-de Groot AC, Glukhova MA, Conny van Munsteren J, Kockx MM, Schwartz SM, Koteliansky VE. Differentiation, dedifferentiation, and apoptosis of smooth muscle cells during the development of the human ductus arteriosus. Arterioscler Thromb Vasc Biol. 1997;17:1003–9.
7. Segi E, Sugimoto Y, Yamasaki A, Aze Y, Oida H, Nishimura T, Murata T, Matsuoka T, Ushikubi F, Hirose M, Tanaka T, Yoshida N, Narumiya S, Ichikawa A. Patent ductus arteriosus and neonatal death in prostaglandin receptor EP4-deficient mice. Biochem Biophys Res Commun. 1998;246:7–12.

8. Nguyen M, Camenisch T, Snouwaert JN, Hicks E, Coffman TM, Anderson PA, Malouf NN, Koller BH. The prostaglandin receptor EP4 triggers remodelling of the cardiovascular system at birth. Nature. 1997;390:78–81.

9. Loftin CD, Trivedi DB, Tiano HF, Clark JA, Lee CA, Epstein JA, Morham SG, Breyer MD, Nguyen M, Hawkins BM, Goulet JL, Smithies O, Koller BH, Langenbach R. Failure of ductus arteriosus closure and remodeling in neonatal mice deficient in cyclooxygenase-1 and cyclooxygenase-2. Proc Natl Acad Sci U S A. 2001;98:1059–64.

10. Yokoyama U, Minamisawa S, Katayama A, Tang T, Suzuki S, Iwatsubo K, Iwasaki S, Kurotani R, Okumura S, Sato M, Yokota S, Hammond HK, Ishikawa Y. Differential regulation of vascular tone and remodeling via stimulation of type 2 and type 6 adenylyl cyclases in the ductus arteriosus. Circ Res. 2010;106:1882–92.

11. de Rooij J, Zwartkruis FJ, Verheijen MH, Cool RH, Nijman SM, Wittinghofer A, Bos JL. Epac is a Rap1 guanine-nucleotide-exchange factor directly activated by cyclic AMP. Nature. 1998;396:474–7.

12. Bos JL. Epac: a new cAMP target and new avenues in cAMP research. Nat Rev Mol Cell Biol. 2003;4:733–8.

13. Yokoyama U, Minamisawa S, Quan H, Akaike T, Suzuki S, Jin M, Jiao Q, Watanabe M, Otsu K, Iwasaki S, Nishimaki S, Sato M, Ishikawa Y. Prostaglandin E2-activated Epac promotes neointimal formation of the rat ductus arteriosus by a process distinct from that of cAMP-dependent protein kinase A. J Biol Chem. 2008;283:28702–9.

14. Yokoyama U, Minamisawa S, Shioda A, Ishiwata R, Jin MH, Masuda M, Asou T, Sugimoto Y, Aoki H, Nakamura T, Ishikawa Y. Prostaglandin E2 inhibits elastogenesis in the ductus arteriosus via EP4 signaling. Circulation. 2014;129:487–96.

15. Yokoyama U, Iwatsubo K, Umemura M, Fujita T, Ishikawa Y. The prostanoid EP4 receptor and its signaling pathway. Pharmacol Rev. 2013;65:1010–52.

16. Hermes-DeSantis ER, Clyman RI. Patent ductus arteriosus: pathophysiology and management. J Perinatol. 2006;26 Suppl 1:S14–8; discussion S22-13.

17. Mitchell SC, Korones SB, Berendes HW. Congenital heart disease in 56,109 births. Incidence and natural history. Circulation. 1971;43:323–32.

18. Van Overmeire B, Allegaert K, Casaer A, Debauche C, Decaluwe W, Jespers A, Weyler J, Harrewijn I, Langhendries JP. Prophylactic ibuprofen in premature infants: a multicentre, randomised, double-blind, placebo-controlled trial. Lancet. 2004;364:1945–9.

Developmental Differences in the Maturation of Sarcoplasmic Reticulum and Contractile Proteins in Large Blood Vessels Influence Their Contractility

36

Emiko Hayama and Toshio Nakanishi

Keywords
Ductus arteriosus • mRNA expression • Contractile system • Development

Developmental changes in the contractile system of blood vessels such as the ductus arteriosus (DA), pulmonary artery (PA), and aorta (Ao) have not been investigated extensively. We assessed the developmental changes in the expression of genes that regulate vasoconstriction of fetal blood vessels.

DA, PA, and Ao were taken from rabbit fetuses at 21, 27, and 30 days of gestation (full term = 31 days) as well as 2-day-old rabbits. Total DA, PA, and Ao RNA were isolated from pooled segments. Expression of target mRNAs was quantified using absolute quantitative real-time PCR:

1. *Contractile proteins* Contractile activity in smooth muscles is determined primarily by the phosphorylation state of myosin regulatory light chain (MRLC). During muscle contraction, intracellular Ca^{2+} levels increase substantially, and binding of Ca^{2+} to calmodulin activates MLC kinase, which then phosphorylates MRLC. Expression of calmodulin and MLC kinase was not significantly different among the vessels. Expression of tropomyosin 2 was higher in DA compared to PA.
2. *Sarcoplasmic reticulum (SR)* Cytosolic Ca^{2+} levels are increased through Ca^{2+} release from SR and Ca^{2+} entry from the extracellular space via Ca^{2+} channels. Expression of SR cardiac-type ryanodine receptor (RYR) increased throughout fetal maturation and was much higher than skeletal-type RYR. Expression of SR calcium storage protein calsequestrin-2 increased with development in Ao but

---

E. Hayama, Ph.D. • T. Nakanishi, M.D., Ph.D. (✉)
Department of Pediatric Cardiology, Tokyo Women's Medical University, 8-1 Kawada-cho, Shinjyuku-ku, Tokyo 162-8666, Japan
e-mail: pnakanis@hij.twmu.ac.jp

© The Author(s) 2016
T. Nakanishi et al. (eds.), *Etiology and Morphogenesis of Congenital Heart Disease*,
DOI 10.1007/978-4-431-54628-3_36

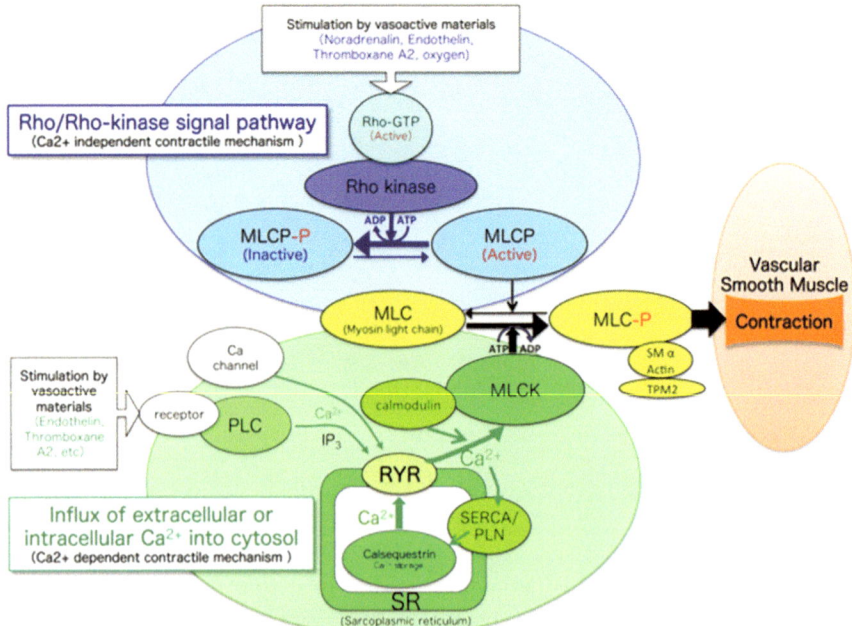

Fig. 36.1 Vascular smooth muscle contraction. Contractile activity in smooth muscle (SM) is determined primarily by the phosphorylation state of myosin regulatory light chain (MRLC). Increase in intracellular Ca^{2+} levels leads to Ca^{2+}-calmodulin binding, which activates MLC kinase (MLCK) to phosphorylate MRLC. Cytosolic Ca^{2+} is increased through Ca^{2+} release from the sarcoplasmic reticulum (SR) and Ca^{2+} entry from extracellular space via Ca^{2+} channels. The Ca^{2+}-independent Rho/Rho kinase pathway inhibits MLC phosphatase (MLCP) activity and promotes phosphorylation of MLC. *RYR* ryanodine receptor, *SERCA* sarcoplasmic/endoplasmic reticulum Ca^{2+}-ATPase, *PLN* phospholamban, *TPM2* tropomyosin 2

not in DA and PA, with expression levels remaining very low in the latter two. Expression of SR Ca^{2+} pump regulator phospholamban increased with development in PA and Ao but remained very low in DA. The expression of SR genes differs significantly at development stages and is vessel dependent, indicating differential maturity of SR in fetal vessels.

3. *Rho/Rho-kinase* The Ca^{2+}-independent Rho/Rho kinase pathway inhibits MLC phosphatase activity and promotes phosphorylation of MLC. Expression of small GTPase RhoB and Rho kinase-1 was higher than that of RhoA and Rho kinase-2. The expression levels of these Rho/Rho kinase pathway genes were similar in fetal and newborn vessels.

In conclusion, contraction of the premature DA, PA, and Ao may be regulated predominantly by the Rho/Rho kinase pathway, owing to the poor expression of the component protein genes in the immature SR. DA contractile systems may be well developed compared with those of the surrounding PA and Ao (Fig. 36.1).

Sources of Funding This work was supported by JSPS KAKENHI grant numbers, 20390303 and 21591399, and by a grant from the Japan Research Promotion Society for Cardiovascular Diseases.

Fetal and Neonatal Ductus Arteriosus Is Regulated with ATP-Sensitive Potassium Channel

37

Kazuo Momma, Mika Monma, Katsuaki Toyoshima, Emiko Hayama, and Toshio Nakanishi

Keywords

Ductus arteriosus • Fetus • ATP-sensitive potassium channel • Sulfonylurea • Potassium channel opener

The fetal patency and neonatal closure of the ductus arteriosus (DA) are regulated with oxygen and prostaglandins. The proposed oxygen sensors of fetal and neonatal DA include P450-endothelin and the Kv channel [1]. We hypothesized that the ATP-sensitive potassium channel (K_{ATP} channel) is another oxygen sensor [2].

Fetal and neonatal DA was studied with Wistar rats; sulfonylurea drugs including tolbutamide, chlorpropamide, gliclazide, glimepiride, and glibenclamide (K_{ATP} channel inhibitors); diazoxide and pinacidil (K_{ATP} channel openers, KCOs); and rapid whole-body freezing (Fig. 37.1).

Tolbutamide, chlorpropamide, and gliclazide easily passed across the placenta and constricted fetal DA dose-dependently following orogastric administration to near-term pregnant rats. The fetal DA constricted 30 % with clinical doses of sulfonylurea drugs and closed completely with larger doses.

Glimepiride and glibenclamide passed across the placenta minimally and only mildly constricted the fetal DA after maternal administration, but constricted and

K. Momma (✉) • E. Hayama • T. Nakanishi
Department of Pediatric Cardiology, Tokyo Women's Medical University, Kawadacho 8-1, Shinjyuku-ku, Tokyo 162-8666, Japan
e-mail: kmomma@iris.ocn.ne.jp

M. Monma
Department of Obstetrics and Gynecology, Shounan-Kamakura General Hospital, Okamoto 1370-1, Kamakura 247-8533, Japan

K. Toyoshima
Department of Neonatology, Kanagawa Children's Medical Center, Mutukawa 2-138-4 Minamiku, Yokohama 232-8555, Japan

T. Nakanishi et al. (eds.), *Etiology and Morphogenesis of Congenital Heart Disease*,
DOI 10.1007/978-4-431-54628-3_37

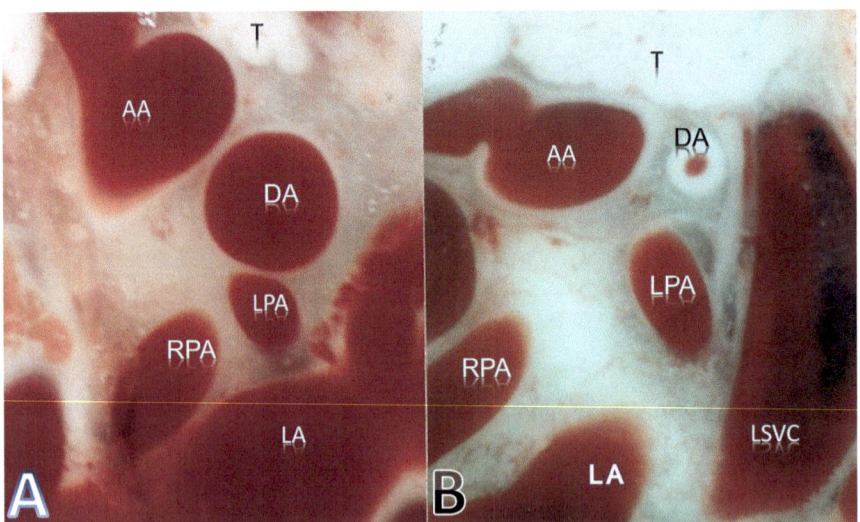

Fig. 37.1 The fetal ductus arteriosus was studied in the near-term fetus or newborn rat following rapid whole-body freezing, cutting on the freezing microtome, with a microscope and a micrometer. The control fetus shows a widely open ductus (**a**), and the fetus with glibenclamide (10 mg/kg; 10–100 times clinical dose injected at 1 h before) shows severely constricted ductus (**b**). *AA* aortic arch, *DA* ductus arteriosus, *LA* left atrium, *LPA* left pulmonary artery, *LSVC* left superior vena cava, *RPA* right pulmonary artery, *T* thymus

closed the fetal DA dose-dependently with direct fetal injection. Fetal DA closure was associated with hydrops and fetal death.

Diazoxide and pinacidil delayed DA closure following neonatal injection immediately postnatally and dilated the closing DA with injection at 60 min postnatally.

All tested sulfonylurea drugs constricted fetal DA dose-dependently and with complete closure at large doses. KCOs dilated the neonatal DA. These results indicate physiological regulation of fetal and neonatal DA with K_{ATP} channels.

This study has several clinical implications. Sulfonylurea-associated fetal death was first reported 50 years ago. The mechanism of death remained unclear prior to this study. Sulfonylureas may be useful for closing patent DA in premature neonates.

Recently reported neonatal DA reopening associated with the use of diazoxide for hyperinsulinemic hypoglycemia has been proved experimentally. DA-dilating effect of KCO drugs may be useful as a bridge to surgery in neonatal DA-dependent congenital heart diseases.

References

1. Coceani F, Baragatti B. Mechanisms for ductus arteriosus closure. Semin Perinatol. 2012;36:92–7.
2. Nakanishi T, Gu H, Momma K, et al. Mechanisms of oxygen-induced contraction of ductus arteriosus isolated from the fetal rabbit. Circ Res. 1993;72:1218–23.

Perspective

Michael Artman

One of the fascinating and essential characteristics of the cardiovascular system is that it functions automatically without requiring conscious input. In other words, the normal heart beats regularly and spontaneously regardless of whether or not the individual is thinking about it. Electrical impulses arise in the sinoatrial node, pass through the atria, pause at the atrioventricular node, and then are distributed in a tightly orchestrated spatial and temporal manner to the ventricles for coordinated and effective pumping activity. This elegantly synchronized process repeats itself over and over approximately three billion times during the average human life. Even minor disruptions in this normal process can have devastating consequences. Considerable morbidity and mortality result from abnormalities in intrinsic cardiac pacemaker activity and conduction of impulses. Congenital and acquired arrhythmias remain a significant health problem. Consequently, a clear understanding of the molecular and cellular processes involved in the formation and maintenance of normal electrical signaling in the heart has profound implications for human cardiovascular health and disease. Furthermore, it is likely that insights into the developmental processes involved in building a cardiac conduction system will likely have implications for understanding fundamental biology applicable to other organ systems.

This part reviews and summarizes the current state of understanding of the development of spontaneous pacemaker activity and the cardiac conduction system. Christoffels' group reviews the signaling pathways and molecules involved in the

M. Artman (✉)
Department of Pediatrics, Children's Mercy Hospital, University of Missouri–Kansas Medicine, Kansas City, MO, USA
e-mail: martman@cmh.edu

development of the cardiac conduction system. The authors emphasize the importance of understanding the three-dimensional architecture of genomic loci to provide insight into the regulation (normal and abnormal) of gene expression. Mikawa and colleagues identify and characterize a previously unrecognized source of cells that are destined to become pacemaker cells, even before cardiac morphogenesis begins. These results will likely lead to better understanding of the mechanisms involved in pacemaker cell differentiation, which in turn may ultimately have therapeutic implications. Additional information on the mechanisms, molecular signals, and pathways involved in specification of cell fate are provided by Asai et al. and by Morikawa et al. Taken together, the papers presented in Part VIII provide an interesting, informative, and elegant overview of the current understanding of pacemaker and conduction system development. Moreover, they provide a road map for future investigations into this essential and fundamental biological process.

Regulation of Vertebrate Conduction System Development

38

Jan Hendrik van Weerd and Vincent M. Christoffels

Abstract

The cardiac conduction system (CCS) consists of distinctive components that initiate and conduct the electrical impulse required for the coordinated contraction of the cardiac chambers. The development of the CCS involves complex regulatory networks of transcription factors that act in stage, tissue and dose-dependent manners. As disrupted function or expression of these factors may lead to disorders in the development or function of components of the CCS associated with heart failure and sudden death, it is crucial to understand the molecular and cellular mechanisms underlying their complex regulation. Here, we discuss the regulation of genes driving CCS-specific gene expression and demonstrate the complexity of the mechanisms governing their regulatory networks. The three-dimensional conformation of chromatin has recently been recognized as an important regulatory layer, shaping the genome in regulatory domains and physically wiring gene promoters to their regulatory sequences. Knowledge of the mechanisms by which distal-acting regulatory sequences exert their function to drive tissue-specific gene expression and understanding how the three-dimensional chromatin landscape is involved in this regulation will increase our understanding of how disease-associated genomic variation affects the function of such sequences.

Keywords

Heart development • Conduction system • Sinus node • Atrioventricular node • Transcriptional regulation • Functional genomics • Patterning • Pacemaker

J.H. van Weerd • V.M. Christoffels (✉)
Department of Anatomy, Embryology and Physiology, Academic Medical Center, University of Amsterdam, Meibergdreef 15 – L2-106, 1105 AZ Amsterdam, The Netherlands
e-mail: v.m.christoffels@amc.uva.nl

38.1   Introduction

The cardiac conduction system (CCS) initiates and propagates the electrical impulse that is required for the rhythmic and synchronized contraction of the heart. The impulse initiates in the sinoatrial node (SAN) and is rapidly propagated through the atria, thereby activating the contraction of the atrial myocardium. The impulse is then propagated through the atrioventricular node (AVN), the only electrical connection between the atria and ventricles. The AVN delays conduction of the impulse, allowing for the atrial contraction and ventricular filling to complete before the ventricles contract. Further propagation of the impulse to the fast-conducting atrioventricular bundle (AVB), bundle branches (BBs) and Purkinje network causes the depolarization of the ventricular myocardium, leading to ventricular contraction. The development of the CCS is regulated by transcription factors that act in strictly stage, tissue and dose-dependent manners [1, 2]. A disruption in the function or expression of these factors could lead to disorders in the development or function of the CCS that can lead to lethal arrhythmias and heart failure. Knowledge of the mechanisms underlying regulation of genes involved in CCS development is therefore crucial.

38.2   Genetic Pathways Controlling SAN and AVC Development

The heart is the first organ to form during embryonic development and starts as a primitive, linear tube with the inflow region at the caudal side and the outflow region at the cranial side. The slow conductive properties of the embryonic muscle cells within the primitive tube at this stage and dominant pacemaker activity at the caudal end cause a slow peristaltic pattern of contraction along the tube to propagate the blood. At this stage, the entire sinus venosus acts as pacemaker, characterized by the expression of the pacemaker channel Hcn4, a member of the family of channels responsible for the hyperpolarization-activated current *if* that is crucial for the pacemaker potential [3, 4]. The heart tube elongates by the addition of rapidly proliferating progenitor cells that differentiate to cardiac muscle. This implies that cells added to the inflow tract will acquire dominant pacemaker activity. With further development, specific regions in the heart tube start to divide rapidly and activate a working myocardial gene program, resulting in the ballooning of the primitive atrium and ventricle. Concomitant with the ballooning of the primitive atria is the formation of the sinus venosus including the SAN. Dominant pacemaker activity will gradually be confined to the SAN at the junction of the sinus venosus and atrium.

   The transcriptional activator Tbx5 is required for the sinus venosus expression of Shox2 [5], a homeobox transcription factor necessary for SAN formation and function [6, 7]. Shox2 represses cardiac homeobox transcription factor Nkx2-5 [6]. In the chambers, Nkx2-5 activates chamber-specific genes including *Nppa* and high-conductance gap junction subunit-encoding genes *Gja5* (Cx40) and *Gja1* (Cx43), whereas it represses SAN/CCS-specific genes *Hcn4* and T-box transcription factor *Tbx3* [8]. Tbx3 is required for the formation of the SAN by directly repressing atrial myocardial genes *Gja5*, *Gja1* and *Nppa* to prevent atrialization of the SAN and indirectly activating *Hcn4* and other SAN genes [9, 10].

During ballooning of the primitive cardiac chambers, the region in between the atria and ventricles does not proliferate and forms a constriction, the atrioventricular canal (AVC). Bmp2 expression in the AVC activates the expression of Tbx3 and Tbx2 [11]. Together with Msx2, these T-box factors repress the working myocardial gene program in the AVC and AVC-derived AVN [12, 13] and stimulate the pacemaker gene program and the program required for the formation of the AV cushions. Within the AVC, Tbx2 and Tbx3 interact with Nkx2-5 to repress genes that are activated by Nkx2-5 and Tbx5 in the working myocardium of the atria and ventricles [14–16]. Tbx2 and Tbx3 thus suppress working myocardial differentiation of the AVC, thereby causing the retention of the primitive phenotype of slow conduction and low rates of proliferation, providing a primitive morphological and functional constriction in between the atrial and ventricular chambers. Other factors that regulate the formation of the AVC and its border with the chamber myocardium include Wnts, acting upstream of Bmp2; Hey1 and Hey2, Notch target genes expressed in the chambers that suppress Tbx2 [17]; Tbx20, which represses BMP-mediated activation of Tbx2 in the chambers [18]; and Gata4/6, which act in complex with Smads and histone acetyltransferases (HATs) to activate AVC-specific enhancers in the AVC and with histone deacetylases (HDACs) and Hey1/2 to suppress these enhancers in the chambers [19]. The resulting pattern of conduction—fast in the atria, slow in the AVC and fast in the ventricles—results in the alternating contraction pattern of the chambers and an ECG that resembles the adult ECG (Fig. 38.1a).

38.3 Transcriptional Regulation of CCS Genes

Although the expression patterns and functions of genes involved in the development of the cardiac conduction system are relatively well studied, little is known about the molecular mechanisms underlying their regulation of expression. Tissue-specific gene expression often involves long-range regulatory elements, such as enhancers, which dictate the strictly time-, tissue- and dosage-dependent expression of their target genes. The identification and function of such enhancers is therefore highly relevant to fully understand the complex regulatory networks in CCS formation. However, to date only few studies have been carried out investigating in depth the regulation of genes driving CCS development.

The T-box transcription factor Tbx5 plays indispensable roles in the early patterning of the heart and CCS and is involved in limb development [15, 20, 21]. Mutations in *TBX5* are associated with Holt-Oram syndrome, a developmental disorder characterized by hand-heart defects [22, 23]. Using modified bacterial artificial chromosomes (BACs), the regulatory landscape of the *TBX5* locus was determined, and within this landscape, multiple cardiac-specific enhancers were identified by utilizing multiple genome-wide ChIP-seq datasets and evolutionary

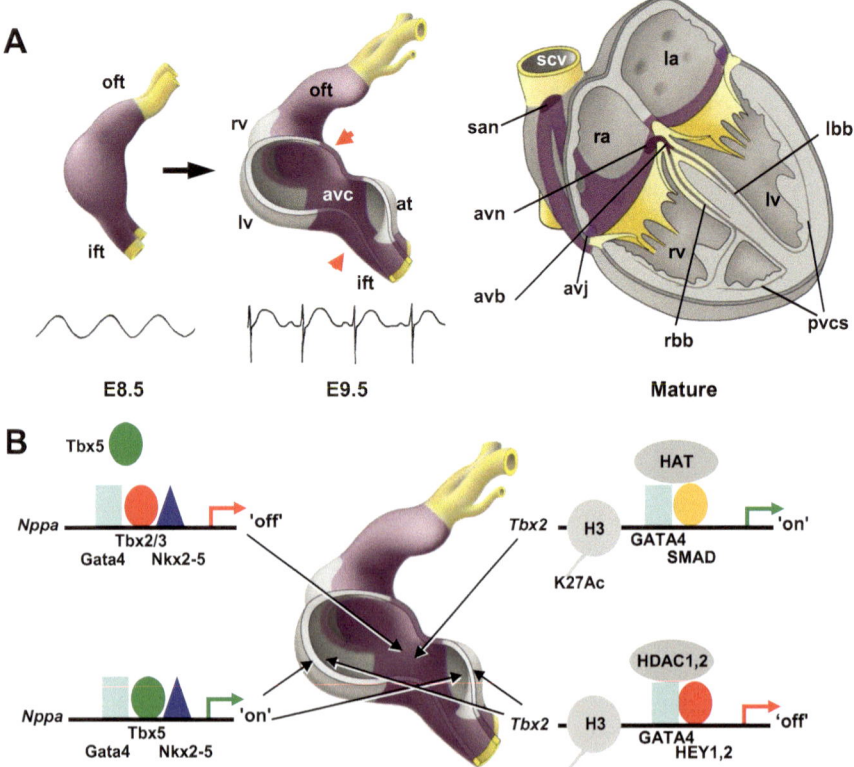

Fig. 38.1 Schematic overview of cardiac development. (**a**) The early heart tube has a primitive phenotype of slow conduction, represented by a sinusoidal ECG. With further development, regions at the outer curvatures of the primary heart tube expand and obtain a working myocardium phenotype of fast conduction (*grey*). The sinus venosus (sv), AVC, outflow tract (oft) and inner curvatures retain their primitive pacemaker-like phenotype and slow conductivity (*purple*). A more mature ECG can be derived from these hearts. Eventually, these non-chamber myocardial regions will give rise to the mature conduction system components. (**b**) The transcriptional repressors Tbx2 and Tbx3 compete with the transcriptional activator Tbx5 to regulate their target genes. In the AVC, expression of chamber myocardium genes like *Nppa* is actively repressed by Tbx2 and Tbx3, whereas in the developing chambers Tbx5 activates these genes. Transcriptional activation in the AVC is regulated by GATA binding site-dependent histone modifications which render the chromatin more (e.g. HATs) or less (e.g. HDACs) accessible for transcription factors to bind their target regulatory sequences, resulting in activation or repression of gene expression

conservation. These enhancers were shown to recapitulate part of the TBX5 expression pattern in the heart, but interestingly, none of these fragments drove reporter expression in the limbs, suggesting that the *cis*-regulation of TBX5 in the heart and limbs is compartmentalized [24]. Such knowledge can highly improve the understanding of the mechanisms underlying the development of congenital heart diseases by decoupling the heart and hand phenotypes seen with Holt-Oram syndrome, thereby presenting more compartmentalized phenotypes compared to disorders caused by protein-coding mutations.

Another example of how enhancer-mediated gene expression is involved in the tight regulation of the CCS is presented by the transcriptional repressor *Id2*. This factor was identified by serial analysis of gene expression (SAGE) as having CCS-specific expression. *Id2* is expressed throughout development in the AVB and BB and in non-CCS compartments such as the AV endocardial cushions and valves. The requirement of *Id2* for ventricular CCS structure and function was demonstrated as *Id2*-deficient mice exhibit structural and functional conduction system abnormalities, including left bundle branch block. *Id2* is cooperatively regulated by *Nkx2-5* and *Tbx5* in the developing ventricular conduction system by binding of both *Tbx5* and *Nkx2-5* to a 1052 bp fragment of the *Id2* promoter. Mutation of the *Tbx5* binding site within this promoter region completely abolished CCS expression whereas extracardiac expression was unaltered, illustrating the specificity of this transcriptional mechanism in the coordinated development of the ventricular conduction system [20].

The hyperpolarization-activated channel HCN4 is required for the generation of pacemaker action potentials in the embryonic heart. Using a transgenic BAC approach, it was shown that the regulatory regions sufficient to recapitulate the endogenous Hcn4 expression pattern in the SAN, AVN, His bundle, bundle branches and left ventricular Purkinje fibres reside within the region covered by one bacterial artificial chromosome (BAC) of 200 kbp [25]. Using transgenic mouse assays, multiple evolutionary conserved cis-acting regulatory sequences were identified to drive *Hcn4* expression in the AV conduction system. One of these regions drives reporter expression specifically in the non-chamber myocardium in a Mef2c-dependent manner. Furthermore, depletion of histone deacetylases resulted in ectopic expression of reporter activity in chamber myocardium, revealing a role for histone modifications in Mef2c-regulated enhancer-mediated expression of Hcn4 in components of the CCS [26].

More recently, Contactin-2 (*Cntn2*), a cell adhesion molecule critical for neuronal patterning and ion channel clustering, was described as a marker for the ventricular conduction system, with expression in the AVB, BBs and Purkinje fibres. Using a GFP-modified BAC, the boundaries of the regulatory domain involved in the control of *Cntn2* expression were identified, since reporter activity of the modified BAC completely recapitulates endogenous *Cntn2* expression [27]. Such knowledge facilitates in the identification of single, individual regulatory elements driving CCS development and will greatly add to our understanding of how genes involved in the complex development of CCS components are regulated.

Enhancer function is regulated by modifications of specific histone tails that mark active or poised enhancers. Active enhancers are associated with an open, accessible chromatin state, whereas poised enhancers are associated with dense, closed chromatin. Histone modifiers such as histone deactelyases (HDACs), histone methyltransferases (HMTs) and histone acetyltransferases (HATs) therefore regulate the accessibility of long-range regulatory sequences, allowing for the binding by cell type-specific transcription factors to activate transcription in a tissue-dependent manner. Specification of the AVC is regulated by Gata4, which activates AVC enhancers in synergy with Bmp2/Smad signaling to recruit HATs such as p300 [19, 28]. This leads to H3K27 acetylation, a marker of active enhancers. In contrast, in chamber myocardium, Gata4 cooperates with HDACs and chamber-specific genes Hey1 and Hey2, leading to H3K27 deacetylation and repression (Fig. 38.1b) [19].

38.4 Common Genomic Variants Influence CCS Function

The importance of the strict regulation of the spatial and temporal expression of CCS genes is illustrated by findings from recent genome-wide association studies, which revealed common genomic variation to be associated with conduction parameters like PR interval and QRS duration. Such variation was identified in non-coding regions flanking genes encoding ion channels like *SCN5A/10A*, *KCNQ1* and *KCNH2* and cardiac transcription factors like *NKX2-5*, *MEIS1* and *TBX3/5*, indicating they might affect the function of enhancers controlling the precise regulation of these genes [29–31]. Tbx5 is broadly expressed and acts as transcriptional activator, inducing transcription of genes involved in cardiac differentiation [15, 20]. The activity domain of Tbx3 is much more restricted and confined to the developing and mature CCS, where it acts as a transcriptional repressor, thereby imposing the pacemaker phenotype on cells within its expression domain [10, 32]. Tbx3 and Tbx5 both recognize the same regulatory sequences [33], suggesting that these factors compete for binding and implicating a fine balance between activation and repression of CS genes by these factors. The precise regulation of transcription and activity of both factors is therefore crucial for proper CCS patterning, and minor changes in regulatory elements controlling the regulation of expression of these factors could thus potentially have large consequences for CCS function and development. Knowledge of the mechanisms by which such developmental genes are regulated to exert their spatio-temporal transcriptional activity is therefore crucial in the understanding of how variation identified by GWAS influences development.

38.5 3D Architecture Regulates Transcription

Physical enhancer-promoter contacts are a requirement for enhancer-mediated cell type-specific gene expression, and as such, the three-dimensional topology of chromatin plays an indispensable role in gene regulation by physically wiring long-range regulatory sequences with their target promoters [34]. Several protein complexes, including CTCF, cohesin and mediator, have been proposed to be involved in the organization of these contacts [35]. Furthermore, recent data suggest that such genomic structural organizers not only mediate single enhancer-promoter contacts but also mediate the organization of the genome in relatively cell-type invariant topologically associated domains (TADs) within which sequences particularly contact each other. Genes located within the same TAD exhibit greater expression correlation than genes located in distinct ones, suggesting that such domains may act as a backbone for tissue-specific regulatory contacts [36]. The recent emergence of techniques aimed at capturing the 3D conformation of genomic loci [37] therefore provides valuable tools to elucidate regulatory mechanisms on the chromatin level.

38.6   Regulation of Tbx3 by a Large Regulatory Domain

The evolutionary conserved *Tbx3/5* genomic locus is one of the few genomic loci of which the 3D architecture has been studied and reveals an example of the complexity of gene regulation on the level of chromatin topology. *Tbx3* and *Tbx5* form an evolutionary conserved gene cluster derived from a primordial T-box gene [38] and, as mentioned above, play crucial roles in the formation and function of the cardiac conduction system. Using circular chromosome conformation capture sequencing (4C-seq), which captures all the genomic regions in close proximity to a chosen point of view [39], the 3D architecture of the Tbx3/5 locus was probed, and genomic regions contacting *Tbx3* or *Tbx5* were identified in different tissues (Fig. 38.2a). Interestingly, these data revealed that the regulatory landscape is in a preformed conformation that is similar in embryonic heart, brain and limb. Rather than the de novo formation of enhancer-promoter loops upon binding by cell-type relevant transcription factors to initiate transcription, the locus is in a fixed, permissive structure in which enhancer-promoter loops are pre-existing [40]. Such a permissive structure has previously been described for different loci, including the *Hox* and *Shh* gene loci. Long-range enhancer-promoter contacts in these loci were shown to be irrespective of cell type, revealing a preformed topology [36, 41]. The permissive, preformed nature of these loci was exemplified by the fact that even in the absence of a distal-acting *Shh* enhancer, contacts between the *Shh* promoter and the enhancer region still occur [41]. The benefit of such preformed regulatory landscapes is believed to lie in the ease by which tissue-specific transcription factors can utilize preformed contacts to target the gene of interest, involving only slight variations in internal contacts within an otherwise rigid and conserved structure. In agreement with this, small differences in contact profiles for the different tissue types were observed in the *Tbx3/5* locus despite the fact that the domain is largely preformed, most probably caused by cell type-specific transcription factor mediated enhancer activation. Among the multiple sites contacting *Tbx3* in the gene desert upstream of the gene, two evolutionary conserved enhancers have been identified that also contact each other. They are bound by cardiac-specific transcription factors Nkx2-5, Gata4, Tbx5 and Tbx3 and were shown to respond to a BMP-mediated signalling pathway to drive atrioventricular conduction system expression of *Tbx3* [40].

As mentioned before, Tbx3 and Tbx5 are expressed in overlapping patterns and have overlapping functions in CCS development. It could therefore be expected that both genes share common regulatory mechanisms on a genomic level. Studies on the transcriptional regulation of other clustered developmental genes, like the *Irx* and *Hox* clusters, revealed that regulatory sequences are not uniquely associated

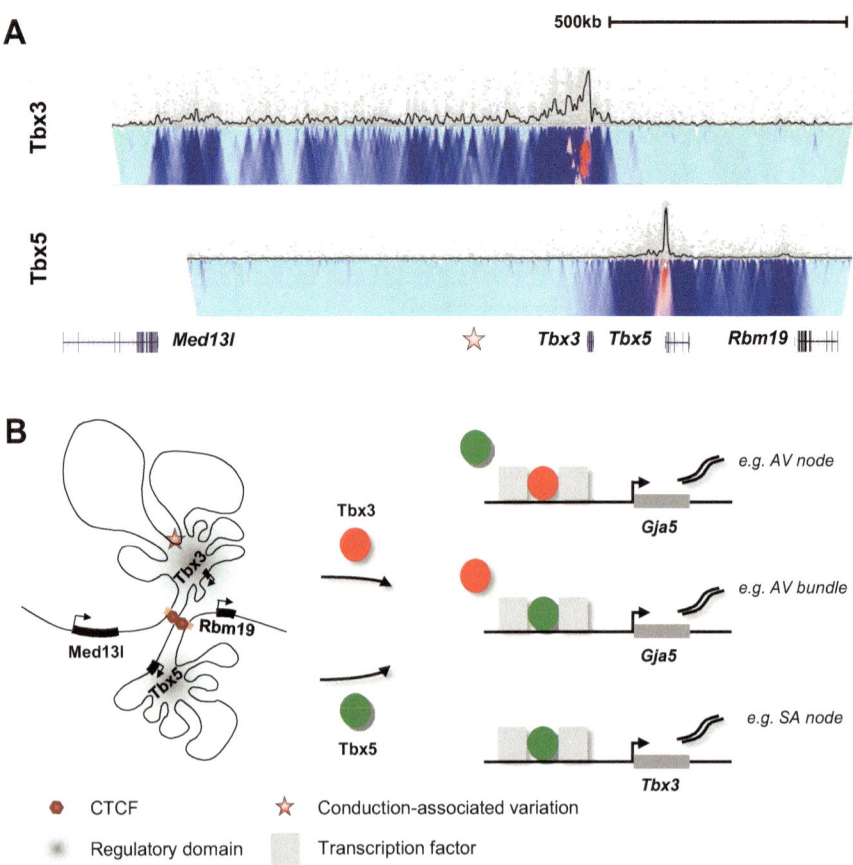

Fig. 38.2 The regulatory domains of *Tbx3* and *Tbx5* are physically separated. (**a**) Contact profiles of the *Tbx3* and *Tbx5* loci as determined by circular chromosome conformation capture (4C) reveal the genomic regions that physically contact *Tbx3* (*upper track*) or *Tbx5* (*lower track*) in mouse embryonic heart cells. *Red* and *dark blue* depict a high contact frequency, whereas *light blue* and *grey* depict a low contact frequency. The contact profiles of *Tbx3* and *Tbx5* hardly overlap, indicating that both genes do not share regulatory sequences and suggesting that common variants in humans upstream of *TBX3* as identified by GWAS (*star*) can be exclusively assigned to *TBX3* [40]. (**b**) Model of the 3D conformation of the *Tbx3/Tbx5* locus. The regulatory domains of *Tbx3* and *Tbx5* are physically separated; however, on the protein level, both genes recognize the same binding sequences and compete with each other to activate or repress their target genes, e.g. *Gja5* (Cx40). Despite the strict regulation on the chromatin level, Tbx5 also directly regulates Tbx3 by binding target sequences to activate transcription in, for example, the SAN

with single promoters, but rather are shared by multiple genes within the cluster. Such interplay between multiple enhancers coordinates the strict regulation of their expression patterns, and it has been proposed that such extensive enhancer sharing explains the conservation of the genomic organization throughout evolution [42, 43]. Interestingly, however, *Tbx3* and its flanking gene desert form a loop that is physically separated from that of the neighbouring *Tbx5* loop (Fig. 38.2b).

Genomic regions within the Tbx3 loop solely contact *Tbx3* but not *Tbx5* and vice versa, indicating enhancer sharing between these evolutionary conserved clusters is unlikely to occur [40]. The strict separation of the regulatory landscapes of *Tbx3* and *Tbx5* is not only cell type-independent, but also evolutionary conserved between mouse and human. Recent Hi-C data in human fibroblasts [36], revealing genome-wide contact profiles, reveal a similar separation of the *TBX3* and *TBX5* regulatory domains with hardly any overlap of the contact profiles. This organization of the Tbx3 locus in a ~1 Mb-scale self-regulatory domain corresponds well to the previously mentioned TADs. These domains are separated by boundary regions enriched for insulator binding protein CTCF, housekeeping genes, transfer RNA and short interspersed elements, hampering interactions of sequences within one TAD with regions exceeding the domain boundaries. Regions located within the regulatory domain of *Tbx3* are thus suggested to exclusively contact *Tbx3* and not *Tbx5* and vice versa.

38.7 Assigning Function to Genomic Variation

Understanding the 3D architecture of a genomic locus not only provides insight into the tight relationship between chromatin topology and the complex regulation of developmental gene expression, it could also provide valuable clues in the understanding of the role of functional variation as identified by genome-wide association studies on gene expression. Common genomic variation in the non-coding region upstream of *TBX3* and *TBX5* was found to influence PR interval and QRS duration in humans [29–31]. The fact that the regulatory domains of *TBX3* and *TBX5* are strictly separated indicates that the variation found in one of the domains can be exclusively assigned to its respective gene, facilitating our understanding of the functional effect of disease-associated variation.

A similar example of how knowledge on the chromatin conformation can increase our understanding of function of common variants is illustrated by studies on the *SCN5A/SCN10A* locus. *SCN5A* and *SCN10A* encode sodium channels important for conduction. GWAS implicated an intronic region in *SCN10A* as a major risk region for prolonged QRS duration [30]. The role of SCN10A in cardiac conduction however was not previously described, whereas mutations in the adjacent *SCN5A* are well established to cause several arrhythmogenic disorders, including Brugada and Long QT syndrome [44, 45]. It is therefore possible that the variation identified within the intron of *SCN10A* impacts the expression of *SCN5A*, rather than or in addition to that of *SCN10A*. Indeed, probing the 3D architecture with the promoters of both *SCN5A* and *SCN10A* and the site of the variation in the intron as point of view using 4C-seq revealed that this variant region contacts the *SCN5A* promoter and a strong enhancer downstream of SCN5A, suggesting that it might act as enhancer regulating the expression of *SCN5A* in the heart. Transgenic reporter assays revealed that indeed this enhancer is essential for cardiac *Scn5a* expression. In humans, the SNP located within the enhancer that correlates with slowed conduction is associated with lower *SCN5A* expression

[46]. Taken together, these results provide another example of how our understanding of the 3D architecture of a genomic locus harbouring functional variation can facilitate in the assignment of function to such variations, improving our understanding of the effect of disease-associated variation.

References

1. Christoffels VM, Smits GJ, Kispert A, et al. Development of the pacemaker tissues of the heart. Circ Res. 2010;106:240–54.
2. Munshi NV. Gene regulatory networks in cardiac conduction system development. Circ Res. 2012;110:1525–37.
3. Stieber J, Herrmann S, Feil S, et al. The hyperpolarization-activated channel HCN4 is required for the generation of pacemaker action potentials in the embryonic heart. Proc Natl Acad Sci U S A. 2003;100:15235–40.
4. Baruscotti M, Bucchi A, Viscomi C, et al. Deep bradycardia and heart block caused by inducible cardiac-specific knockout of the pacemaker channel gene Hcn4. Proc Natl Acad Sci U S A. 2011;108:1705–10.
5. Puskaric S, Schmitteckert S, Mori AD, et al. Shox2 mediates Tbx5 activity by regulating Bmp4 in the pacemaker region of the developing heart. Hum Mol Genet. 2010;19:4625–33.
6. Espinoza-Lewis RA, Yu L, He F, et al. Shox2 is essential for the differentiation of cardiac pacemaker cells by repressing Nkx2-5. Dev Biol. 2009;327(2):376–85.
7. Blaschke RJ, Hahurij ND, Kuijper S, et al. Targeted mutation reveals essential functions of the homeodomain transcription factor Shox2 in sinoatrial and pacemaking development. Circulation. 2007;115:1830–8.
8. Espinoza-Lewis RA, Liu H, Sun C, et al. Ectopic expression of Nkx2.5 suppresses the formation of the sinoatrial node in mice. Dev Biol. 2011;356:359–69.
9. Mommersteeg MTM, Hoogaars WMH, Prall OWJ, et al. Molecular pathway for the localized formation of the sinoatrial node. Circ Res. 2007;100:354–62.
10. Hoogaars WM, Engel A, Brons JF, et al. Tbx3 controls the sinoatrial node gene program and imposes pacemaker function on the atria. Genes Dev. 2007;21:1098–112.
11. Singh R, Hoogaars WM, Barnett, et al. Tbx2 and Tbx3 induce atrioventricular myocardial development and endocardial cushion formation. Cell Mol Life Sci. 2012;69:1377–89.
12. Boogerd KJ, Wong LYE, Christoffels VM, et al. Msx1 and Msx2 are functional interacting partners of T-box factors in the regulation of connexin 43. Cardiovasc Res. 2008;78:485–93.
13. Chen YH, Ishii M, Sucov HM, et al. Msx1 and Msx2 are required for endothelial-mesenchymal transformation of the atrioventricular cushions and patterning of the atrioventricular myocardium. BMC Dev Biol. 2008;8:75.
14. Habets PEMH, Moorman AFM, Clout DEW, et al. Cooperative action of Tbx2 and Nkx2.5 inhibits ANF expression in the atrioventricular canal: implications for cardiac chamber formation. Genes Dev. 2002;16:1234–46.

15. Bruneau BG, Nemer G, Schmitt JP, et al. A murine model of Holt-Oram syndrome defines roles of the T-box transcription factor Tbx5 in cardiogenesis and disease. Cell. 2001;106:709–21.
16. Lyons I, Parsons LM, Hartley L, et al. Myogenic and morphogenetic defects in the heart tubes of murine embryos lacking the homeo box gene Nkx2-5. Gene Dev. 1995;9:1654–66.
17. Kokubo H, Tomita-Miyagawa S, Hamada Y, et al. Hesr1 and Hesr2 regulate atrioventricular boundary formation in the developing heart through the repression of Tbx2. Development. 2007;134:747–55.
18. Singh R, Horsthuis T, Farin HF, et al. Tbx20 interacts with smads to confine tbx2 expression to the atrioventricular canal. Circ Res. 2009;105:442–52.
19. Stefanovic S, Barnett P, van Duijvenboden K, et al. GATA-dependent regulatory switches establish atrioventricular canal specificity during heart development. Nat Commun. 2014;5:3680.
20. Moskowitz IP, Kim JB, Moore ML, et al. A molecular pathway including id2, tbx5, and nkx2-5 required for cardiac conduction system development. Cell. 2007;129:1365–76.
21. Arnolds DE, Liu F, Fahrenbach JP, et al. TBX5 drives Scn5a expression to regulate cardiac conduction system function. J Clin Invest. 2012;122:2509–18.
22. Basson CT, Bachinsky DR, Lin RC, et al. Mutations in human TBX5 (corrected) cause limb and cardiac malformation in Holt-Oram syndrome. Nat Genet. 1997;15:30–5.
23. Li QY, Newbury-Ecob RA, Terrett JA, et al. Holt-Oram syndrome is caused by mutations in TBX5, a member of the Brachyury (T) gene family. Nat Genet. 1997;15:21–9.
24. Smemo S, Campos LC, Moskowitz IP, et al. Regulatory variation in a TBX5 enhancer leads to isolated congenital heart disease. Hum Mol Genet. 2012;21:3255–63.
25. Wu M, Peng S, Zhao Y. Inducible gene deletion in the entire cardiac conduction system using Hcn4-CreERT2 BAC transgenic mice. Genesis. 2014;52:134–40.
26. Vedantham V, Evangelista M, Huang Y, et al. Spatiotemporal regulation of an Hcn4 enhancer defines a role for Mef2c and HDACs in cardiac electrical patterning. Dev Biol. 2013;373:149–62.
27. Pallante BA, Giovannone S, Fang-Yu L, et al. Contactin-2 expression in the cardiac Purkinje fiber network. Circ Arrhythm Electrophysiol. 2010;3:186–94.
28. Dai YS, Markham BE. p300 Functions as a coactivator of transcription factor GATA-4. J Biol Chem. 2001;276:37178–85.
29. Pfeufer A, van Noord C, Marciante KD, et al. Genome-wide association study of PR interval. Nat Genet. 2010;42:153–9.
30. Sotoodehnia N, Isaacs A, de Bakker PI, et al. Common variants in 22 loci are associated with QRS duration and cardiac ventricular conduction. Nat Genet. 2010;42:1068–76.
31. Verweij N, Mateo Leach I, van den Boogaard M, et al. Genetic determinants of P wave duration and PR segment. Circ Cardiovasc Genet. 2014;7(4):475–81.
32. Bakker ML, Boukens BJ, Mommersteeg MTM, et al. Transcription factor Tbx3 is required for the specification of the atrioventricular conduction system. Circ Res. 2008;102:1340–9.
33. van den Boogaard M, Wong LY, Tessadori F, et al. Genetic variation in T-box binding element functionally affects SCN5A/SCN10A enhancer. J Clin Invest. 2012;122:2519–30.
34. de Laat W, Duboule D. Topology of mammalian developmental enhancers and their regulatory landscapes. Nature. 2013;502:499–506.
35. Phillips-Cremins JE, Sauria ME, Sanyal A, et al. Architectural protein subclasses shape 3D organization of genomes during lineage commitment. Cell. 2013;153:1281–95.
36. Dixon JR, Selvaraj S, Yue F, et al. Topological domains in mammalian genomes identified by analysis of chromatin interactions. Nature. 2012;485:376–80.
37. Dekker J, Marti-Renom MA, Mirny LA. Exploring the three-dimensional organization of genomes: interpreting chromatin interaction data. Nat Rev Genet. 2013;14:390–403.
38. Agulnik SI, Garvey N, Hancock S, et al. Evolution of mouse T-box genes by tandem duplication and cluster dispersion. Genetics. 1996;144:249–54.

39. van de Werken HJ, Landan G, Holwerda SJ, et al. Robust 4C-seq data analysis to screen for regulatory DNA interactions. Nat Methods. 2012;9:969–72.
40. van Weerd JH, Badi I, van den Boogaard M, et al. A large permissive regulatory domain exclusively controls Tbx3 expression in the cardiac conduction system. Circ Res. 2014;115 (4):432–41.
41. Amano T, Sagai T, Tanabe H, et al. Chromosomal dynamics at the Shh locus: limb bud-specific differential regulation of competence and active transcription. Dev Cell. 2009;16:47–57.
42. Tena JJ, Alonso ME, Calle-Mustienes E, et al. An evolutionarily conserved three-dimensional structure in the vertebrate Irx clusters facilitates enhancer sharing and coregulation. Nat Commun. 2011;2:310.
43. Duboule D. Vertebrate hox gene regulation: clustering and/or colinearity? Curr Opin Genet Dev. 1998;8:514–8.
44. Wilde AA, Brugada R. Phenotypical manifestations of mutations in the genes encoding subunits of the cardiac sodium channel. Circ Res. 2011;108:884–97.
45. Bezzina CR, Barc J, Mizusawa Y, et al. Common variants at SCN5A-SCN10A and HEY2 are associated with Brugada syndrome, a rare disease with high risk of sudden cardiac death. Nat Genet. 2013;45(9):1044–9.
46. van den Boogaard M, Smemo S, Burnicka-Turek O, et al. A common genetic variant within SCN10A modulates cardiac SCN5A expression. J Clin Invest. 2014;124:1844–52.

Cardiac Pacemaker Development from a Tertiary Heart Field

39

Michael Bressan, Gary Liu, Jonathan D. Louie, and Takashi Mikawa

Abstract

Rhythmic heartbeats are paced by electrical impulses that are autonomously generated by cardiac pacemaker cells. This chapter briefly summarizes our recent findings regarding the embryonic origin of and molecular mechanism delineating cardiac pacemaker cells, showing that pacemaker cells are physically segregated and molecularly programmed, in a tertiary heart field, prior to the onset of cardiac morphogenesis.

Keywords

Differentiation • Fate determination • Fate mapping • Lateral mesoderm patterning • Pacemaker • Pacemaker cell fate specification • Tertiary heart field • Wnt

39.1 Introduction

Rhythmic heartbeat is initiated by electrical impulses evoked at the sinoatrial node (SAN) that are then conducted to atrial muscle and converge to the atrioventricular node. After a brief delay, pacemaker-initiated action potentials (APs) rapidly pass down the conduction system network and finally spread into ventricular muscle (Fig. 39.1a). The SAN was first described more than a century ago [1], and its anatomical, physiological, and molecular characteristics have been thoroughly investigated [2]. While significant progress has been made in our understanding of the mechanism responsible for the differentiation and patterning of the distal

M. Bressan • G. Liu • J.D. Louie • T. Mikawa (✉)
Cardiovascular Research Institute, University of California San Francisco, San Francisco, CA 94143-3120, USA
e-mail: takashi.mikawa@ucsf.edu

T. Nakanishi et al. (eds.), *Etiology and Morphogenesis of Congenital Heart Disease*, DOI 10.1007/978-4-431-54628-3_39

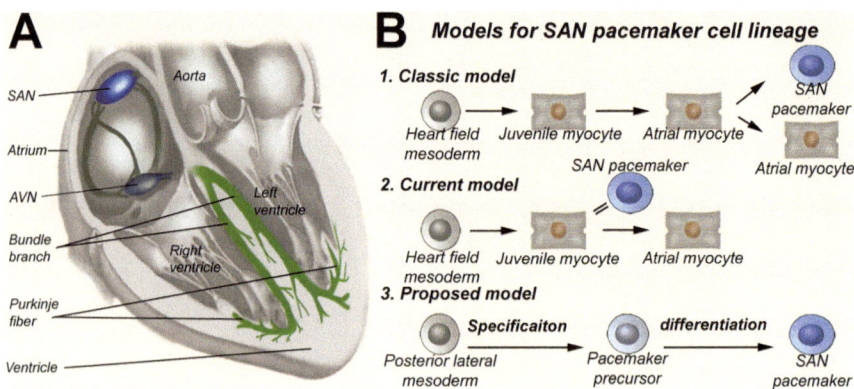

Fig. 39.1 (**a**) Diagram of the pacemaking and conduction system network consisting of distinct subcomponents. (**b**) Controversies in the developmental pathway of SAN pacemaker cells. *The classic model* assumes that this cell population differentiates from atrial myocytes. *A current dogma* suggests that SAN pacemaker cells arise from a part of the heart field mesoderm which remains as immature myocytes. These models are based on phenotypic similarities between embryonic myocytes and SAN cells in action potential shape and the expression of unique genes in common. The only way to reliably establish the origins of a cell type is to tag its antecedents. Our direct cell fate mapping studies have identified an origin and novel developmental pathway for SAN pacemaker cells (*proposed model*)

conduction components, the developmental pathway of the SAN remains controversial (Fig. 39.1b).

The identification of the definitive origin of and inductive mechanisms that define SAN pacemaker cells (PCs) will be critical for systematic investigation of developmental mechanisms, such as cell fate specification and differentiation, of this specialized cell population essential for cardiac function. We have recently identified a novel role of Wnt signaling in promoting pacemaker cell induction and differentiation, which is completely contrary to its well-documented inhibitory role in heart field induction [3].

39.2 Pacemaking Site Transitions From Left To Right During Heart Looping

Classic studies in the chick embryo have shown that as soon as the primitive heart tube forms, myocytes in the posterior inflow tract become electrically active and predominantly evoke pacemaking impulses [4]. While this population has been a priori thought as the progenitor of the SAN pacemaker, no direct cell lineage-tracing study has tested this dogma. Therefore, we have revisited this critical issue. Consistent with previous studies, our optical mapping analysis has detected that the primitive heart tube evokes APs preferentially at the left inflow. Importantly, however, our fate mapping studies have revealed that cells of the left inflow later differentiate into AV junction myocytes rather than the right side pacing cells at the

SA [5]. The above findings are consistent with a hypothesis that there are a successive series of pacemaker zones present in the early developing heart [6], indicating that the true origin of SAN pacemaker cells remains to be identified.

39.3   A Novel Cell Population That Juxtaposes the Right Atrium Takes Over Pacing Function by Mid-heart Looping Stage

While classic and current models assume that SAN PCs arise from a subpopulation of atrial precursors (Fig. 39.1b), to our surprise, our optical mapping data showed that a small region juxtaposing the right atrium preferentially evokes pacemaking action potentials [5]. This pacemaking site has often been neglected in previous studies as this region is routinely dissected away during isolation of the embryonic heart. Our optical mapping and in situ hybridization analyses [5] have identified that this pacemaker site preferentially expresses *HCN4*, a major member of the HCN gene family expressed in the heart responsible for the hyperpolarization-activated inward "funny" current, and an atrial-type myosin *AMHC1*, but is negative for a cardiac transcription factor *Nkx2-5* [9]. These are consistent with previous reports on the adult mouse SAN. Other tissues of the conduction system, such as AVN and Purkinje fibers, co-express *HCNs* and *Nkx2-5*. Thus, differentiated SAN pacemaker cells can be distinguished by these physiological, pharmacological, and molecular characteristics from other myocytes and conduction cells. Unfortunately, expression of these marker genes is dynamic and no *HCN4+/AMHC1+/Nkx2-5-* cells can be found in earlier stage embryos. It is therefore unclear when and where SAN pacemaker cell fate is induced and specified. As the heart matures through the processes of looping, heart primordium continually expands with cells being added to both the inflow and outflow segments [7, 8]. The origin of this novel pacemaking cell population needs to be determined.

39.4   The Right-Sided Pacemaking Cells Indeed Differentiate into SAN Pacemaker Cells

Our in ovo cell-tracing studies have mapped the fate of these extracardiac right-sided pacing cells to the physiologically correct SAN region of the resulting heart at stage 35 (E9) [5]. Optical mapping of these labeled hearts has shown that action potentials are predominantly evoked from these specific cells and propagate into the atrium [5]. Thus, these studies have identified a novel extracardiac cell population as SAN pacemaker precursors. However, until recently it was unknown where this cell population came from. Indeed, the fate of cells in the inflow has been controversial. No systematic fate mapping data was available about the pacemaker field posterior to the heart field at pre-heart tube formation stages. Genetic lineage tracing critically depends upon a gene that is exclusively expressed in pacemaker precursors but not in daughter cells. To our knowledge, no such gene had been identified to date. To explore the origin of SAN pacemaker cells, we performed fate

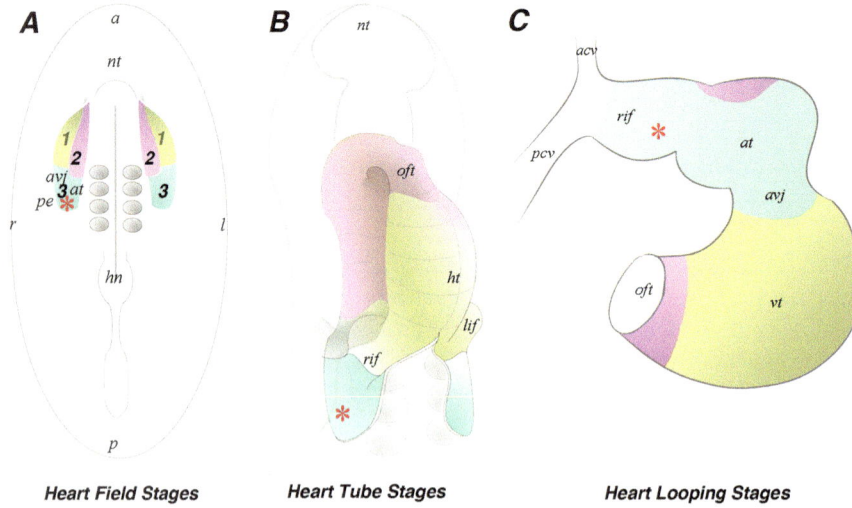

Fig. 39.2 Fate map of pacemaker field. (**a**) At early somite stages, heart precursors occupy bilateral fields within the lateral plate mesoderm. The primary heart field is indicated in *yellow*, the secondary heart field is indicated in *pink*, and the posterior tertiary heart field [5] is indicated in *blue*. Fate mapping studies indicate the progenitors of the SAN pacemaker cells reside within the tertiary heart field (*asterisk*). (**b**) At heart tube stages, the primary heart fields have fused along the midline, while the secondary and tertiary heart fields have not yet been incorporated into the heart. The pacemaker precursors maintain their position within the tertiary heart field mesoderm (*blue*). (**c**) At looping stages, the pacemaker cells have incorporated into/can be seen attaching the right inflow of the heart and begin to pace the heartbeat. *at* atria, *avj* atrioventricular junction, *ht* heart tube, *vt* ventricle, *a* anterior, *p* posterior, *r* right, *l* left

mapping for earlier embryonic stages [5; Fig. 39.2]. For example, in stage 7–8 embryos, the PCs were mapped to a small region in the right lateral plate at somite level 3. Importantly, "the pacemaker field (PF)" does not overlap with "the heart field (HF)" defined by expression of either *Nkx2-5* or *Isl1* [9, 10]. The data show for the first time that SAN precursors arise from an unpredicted small area of the lateral mesoderm immediately posterior to the known HF. It should be noted that the PF is embedded in the zone that also generates other cardiac-related tissues, such as the proepicardium, right atrium, and *vena cava* (Bressan 2013). We therefore tentatively termed this previously unrecognized mesodermal area "tertiary heart field," distinguishing it from the primary heart field and the secondary (anterior) heart field (Fig. 39.3).

Fig. 39.3 Model for distinct roles of Wnt signaling in HF induction vs. PF induction

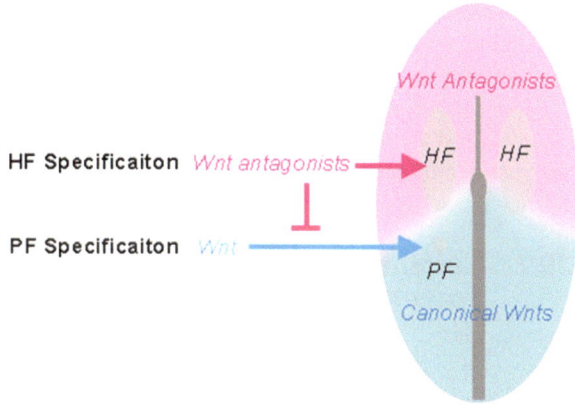

HF Specificaiton *Wnt antagonists*

PF Specificaiton *Wnt*

39.5 Pacemaker Cell Fate Specification Has Already Completed Prior to Heart Morphogenesis

The fate mapping data alone however do not provide any information as to where and when the pacemaker cell fate is specified. A protocol is needed to detect timing and location of cell fate specification apart from marker gene expression. "The cell fate specification" is defined by Harrison [11] and Slack [12] as "when a cell or a tissue becomes capable of autonomously differentiating under a neutral environment, such as in petri dish or test tube."

Using an established culture model that was previously used for studying classic HF specification, we examined autonomous pacemaker differentiation from PF, which has been mapped in stage 8 embryos (see Fig. 39.2). Under the neutral culture condition, HF explants from stage 8 embryos initiated spontaneous contractions within 24 h. Our data are consistent with previous studies on HF specification. In striking contrast, PF explants did not show any contractions during the first 24 h, but by 48 h many of them started rhythmic contractions with a higher beat rate and exhibited AP waveform characteristic of PCs. PF explants continued to rhythmically beat at a similar rate throughout the extended culture period for at least up to 120 h. The data show that cells posterior to the HF autonomously differentiate to initiate and maintain spontaneous rhythmic contractions, which are distinct from those of the HF.

39.6 PF Explants Are Sensitive to Blockers Specific for Pacemaking Ion Channel

Our pharmacological tests interrogated whether PF explants use ion channel characteristic of pacemaker cells. Our data show that a HCN channel blocker, ZD7288, induced PF explants to increase the interval between contractions by approximately

~50 % without detectable loss of rhythmicity. An L-type Ca^{2+} channel blocker, nifedipine, completely diminished spontaneous contraction of PF explants. These pharmacological responses of PF explants are characteristic of SAN pacemaker cells which evoke action potentials mainly through HCN channels and L-type Ca^{2+} channels and are distinct from cardiomyocytes which mainly use Na^+ channels for the upstroke of action potentials.

39.7 Current Models for Molecular Regulation of SAN Pacemaker Differentiation

It was once postulated that the SAN pacemaker arises from a part of the HF mesoderm which co-expresses *Isl1* and *Tbx18*. *Pitx2c*, *Shox2*, *Nkx2-5*, *Tbx3*, and *Tbx5* are believed to play a role in SAN development. In mice, gene deletion of a laterality gene, *Pitx2c*, leads to bilateral SANs. S*hox2* deficiency results in upregulation of *Nkx2-5* and *Cx43* in the SAN domain. Both *Shox2* and *Tbx3* expression require *Tbx5*. Because *Nkx2-5* is absent in the SAN, *Nkx2-5* is suggested to suppress *HCN4* and *Tbx3*. *Tbx3* gene knockouts result in expression of atrial genes in the SAN and partial loss of SAN-specific gene expression. Ectopic expression of *Tbx3* in the atria causes arrhythmia. Taken together, *Tbx3* and *Tbx18* have been proposed as key transcription factors for SAN pacemaker differentiation. Contradictory to these models, deletion of *Tbx3*, *Tbx18*, or *Shox2* results in no or only modest pacemaker defects. Further *Tbx3* expression occurs at the AV junction and AV node where *Nkx2-5* is highly expressed. Our data have revealed, however, that *Tbx3*, *Tbx18*, and *Isl1* are absent from SAN pacemaker cells during the early stages that correspond with cell fate specification and differentiation. Thus, further elucidation was needed to better understand the molecular mechanisms that regulate the specification and formation of SAN pacemaker cells.

39.8 A Novel Role of Wnt Signaling for Pacemaker Cell Fate Specification

Heart field induction and specification are promoted by BMP signaling and are restricted to the anterior region by inhibition of Wnt signaling [3, 13]. In chick, Wnt from the neural plate ectoderm and the posterior mesoderm inhibits myocardial differentiation in paraxial and posterior mesoderm, while the secreted Wnt antagonist, crescent, produced by the anterior endoderm supports myocardial development. Noncanonical Wnts that block canonical Wnt/β-catenin signaling can act positively on HF establishment and myocardial electro gradient. Isl1-Cre/β-catenin mutants show defects in outflow tract formation with decreased expression of *Tbx2*, *Tbx3*, *Shh*, and *Wnt11*. Canonical Wnt signaling is necessary for growth of second HF-derived right ventricular myocytes during heart looping and onward but its role for early specification of the second HF remains obscure. Taken together, inhibition of canonical Wnt/β-catenin signaling is critical for restricting heart field induction to the anterior mesoderm. In contrast to extensive studies of heart field formation, little is known about Wnt signaling in pacemaker development.

Activation of Wnt signaling in the HF diminishes the expression of several cardiac genes, including *Nkx2-5*. However, it remains largely unknown what HF fate becomes once "HF identity" is lost. Our fate mapping data indicated a surprisingly posterior origin for pacemaker cell progenitors within the *Nkx2.5* negative mesoderm, as well as an earlier timing of specification than previously believed. These results led us to reexamine the question of heart field identity following loss of *Nkx2.5*. Therefore, we investigated this untouched area of Wnt signaling on heart field cells using physiological approaches. Our data have revealed that Wnt-treated heart field cells do not lose contractility either in vivo or in vitro. Instead, they developed a rhythmic, high-rate contraction pattern similar to PF explants, including AP waveforms reminiscent of PCs [5]. A transient exposure to Wnt for only 8 h was sufficient to obtain the pacemaker-like beating pattern. Our in vivo and in vitro data have further demonstrated that inhibition of Wnt signaling for pacemaker progenitors results in a conversion of their fate to the ordinary cardiomyocyte type.

39.9 Concluding Remarks

An elucidation of the origin and molecular mechanisms that specify the pacemaker cell fate is fundamental to dissecting the earliest steps critical for SAN pacemaker development. Our data have revealed that the pacemaker cell fate is specified in a previously unconsidered embryonic region at very early embryonic stages even before heart morphogenesis begins. The work has also revealed that differential Wnt-mediated signaling cues in the lateral plate mesoderm are sufficient to induce pacemaker-like versus working myocardial fates, and that these fates are maintained throughout early cardiac morphogenesis. These results will significantly increase our understanding of the basis for the mechanisms that regulate pacemaker cell specification and differentiation.

Acknowledgments This work has been funded in part by NIH R01078921, R01HL093566, and R01HL112268.

References

1. Keith A, Flack M. The form and nature of the muscular connections between the primary divisions of the vertebrate heart. J Anat Physiol. 1907;41(Pt 3):172–89.
2. Boyett MR, Honjo H, Kodama I. The sinoatrial node, a heterogeneous pacemaker structure. Cardiovasc Res. 2000;47(4):658–87.
3. Marvin MJ, Di Rocco G, Gardiner A, et al. Inhibition of Wnt activity induces heart formation from posterior mesoderm. Genes Dev. 2001;15(3):316–27.
4. Kamino K, Hirota A, Fujii S. Localization of pacemaking activity in early embryonic heart monitored using voltage-sensitive dye. Nature. 1981;290(5807):595–7.
5. Bressan M, Liu G, Mikawa T. Early mesodermal cues assign avian cardiac pacemaker fate potential in a tertiary heart field. Science. 2013;340(6133):744–8.
6. Patten BM. Initiation and early changes in the character of the heart beat in vertebrate embryos. Physiol Rev. 1949;29:31–47.
7. Jeter Jr JR, Cameron IL. Cell proliferation patterns during cytodifferentiation in embryonic chick tissues: liver, heart and erythrocytes. J Embryol Exp Morphol. 1971;25(3):405–22.
8. Kirby ML, Gale TF, Stewart DE. Neural crest cells contribute to normal aorticopulmonary septation. Science. 1983;220(4601):1059–61.
9. Lints TJ, Parsons LM, Hartley L, et al. Nkx-2.5: a novel murine homeobox gene expressed in early heart progenitor cells and their myogenic descendants. Development. 1993;119:419–31.
10. Cai CL, Liang X, Shi Y, et al. Isl1 identifies a cardiac progenitor population that proliferates prior to differentiation and contributes a majority of cells to the heart. Dev Cell. 2003;5(6):877–89.
11. Harrison RG. Some difficulties of the determination problem. Am Nat. 1933;67:306–21.
12. Slack JMW. From egg to embryo: regional specification in early development. New York: Cambridge University Press; 1991.
13. Schultheiss TM, Burch JB, Lassar AB. A role for bone morphogenetic proteins in the induction of cardiac myogenesis. Genes Dev. 1997;11:451–62.

Endothelin Receptor Type A-Expressing Cell Population in the Inflow Tract Contributes to Chamber Formation

40

Rieko Asai, Yuichiro Arima, Daiki Seya, Ki-Sung Kim, Yumiko Kawamura, Yukiko Kurihara, Sachiko Miyagawa-Tomita, and Hiroki Kurihara

Keywords
Endothelin • First heart field • Chamber formation

The avian and mammalian heart mainly originates from two distinct embryonic regions: an early differentiating first heart field and a dorsomedially located second heart field. It remains largely unknown when and how these subpopulations of the heart field are established as regions with different fates.

Endothelin-1 (Edn-1) acts on cardiac neural crest cells through endothelin receptor type A (Ednra) and is involved in the normal formation of pharyngeal artery-derived great vessels and ventricular septum [1]. Previously, we identified a distinct cell population defined by the expression of *Ednra* in the mouse inflow region [2]. These cells are derived from a part of the first heart field, and largely confined to the inflow region at E8.25.

R. Asai (✉)
Division of Cardiovascular Development and Differentiation, Medical Research Institute, Department of Pediatric Cardiology, Tokyo Women's Medical University, Tokyo 162-8666, Japan

Department of Physiological Chemistry and Metabolism, Graduate School of Medicine, The University of Tokyo, Tokyo 113-0033, Japan
e-mail: shallow@m.u-tokyo.ac.jp

Y. Arima • D. Seya • K.-S. Kim • Y. Kawamura • Y. Kurihara • H. Kurihara
Department of Physiological Chemistry and Metabolism, Graduate School of Medicine, The University of Tokyo, Tokyo 113-0033, Japan

S. Miyagawa-Tomita
Division of Cardiovascular Development and Differentiation, Medical Research Institute, Department of Pediatric Cardiology, Tokyo Women's Medical University, Tokyo 162-8666, Japan

© The Author(s) 2016
T. Nakanishi et al. (eds.), *Etiology and Morphogenesis of Congenital Heart Disease*, DOI 10.1007/978-4-431-54628-3_40

From the expression patterns of β-gal in the *Ednra-lacZ* knock-in mice, we thought a possibility that this *Ednra*-positive cell population might move into cardiac chambers from the inflow region. By dye injection and transplantation experiments, we showed that the *Ednra*-positive cell population moved toward not only the left and right atria but also the left ventricle.

Then, to perform lineage analysis, we have generated an *Ednra-CreERT2* mouse line (unpublished). We activated Cre recombinase by tamoxifen at E7.25 and E8.25 and analyzed the distribution of β-gal-labeled cells in the *Ednra$^{CreERT2/+}$;R26R* embryo hearts. As a result, the β-gal-labeled cells were found to contribute to the left ventricle and both atria.

To facilitate this lineage analysis, we established a mouse-chick chimera model. When we transplanted the inflow region of the *Ednra$^{CreERT2/+}$;R26R* mouse embryos (tamoxifen i.p. at E7.25) to orthotopically into chick embryos, β-gal-positive cells were detected in the right atrium and left ventricles 7 days after transplantation.

From these results, we conclude that the *Ednra*-positive cell lineage in the early inflow tract certainly contribute to chamber myocardial formation.

References

1. Kurihara Y, Kurihara H, Suzuki H, Kodama T, Maemura K, Nagai R, Oda H, Kuwaki T, Cao WH, Kamada N, et al. Elevated blood pressure and craniofacial abnormalities in mice deficient in endothelin-1. Nature. 1994;368:703–10.
2. Asai R, Kurihara Y, Fujisawa K, Sato T, Kawamura Y, Kokubo H, Tonami K, Nishiyama K, Uchijima Y, Miyagawa-Tomita S, Kurihara H. Endothelin receptor type A expression defines a distinct cardiac subdomain within the heart field and is later implicated in chamber myocardium formation. Development. 2010;137:3823–33.

Specific Isolation of HCN4-Positive Cardiac Pacemaking Cells Derived from Embryonic Stem Cells

41

Kumi Morikawa, Yasuaki Shirayoshi, and Ichiro Hisatome

Keywords

Pacemaker • Embryonic stem cells • HCN4

Bradycardia causes slow heart beating, which has high risk for heart failure or stroke. The only available treatment for bradycardia is implantation of electronic pacemakers. However, this treatment for bradycardia has several shortcomings: requirement of operation for implantation and for exchange of battery and the lack of response to autonomic nerve regulation. The purpose of this study was to develop a biological pacemaker, which could be used for replacement of electronic pacemakers.

1. In differentiating mouse embryonic stem (mES) cells, cardiac pacemaker cells can be specifically visualized with green fluorescent protein (GFP) on the basis of their specific expression of hyperpolarization-activated cyclic nucleotide-gated (HCN) channel 4 at sinoatrial node (SAN) [1]. GFP knock-in ES cells at HCN4 locus (H7 clone) were established. The expression of GFP was specifically restricted at their contracting region in differentiating H7 ES cells.
2. Cell sorting revealed that a few cells (0.1–0.5 %) of H7 embryoid bodies (EBs) were GFP positive, of which approximately 80 % of cells showed the spontaneous beating activity with cesium-sensitive action potential. Sorted GFP+ cells expressed endogenous HCN4 and had essentially the same properties as the cardiac pacemaker cells at SAN.

K. Morikawa (✉)

Center for Promoting Next-Generation Highly Advanced Medicine, Tottori University Hospital, 36-1 Nishicho, Yonago 683-8504, Japan

e-mail: kumi@med.tottori-u.ac.jp

Y. Shirayoshi • I. Hisatome

Division of Regenerative Medicine and Therapeutics, Graduate School of Medical Science, Tottori University, 86 Nishicho, Yonago 683-8503, Japan

© The Author(s) 2016

T. Nakanishi et al. (eds.), *Etiology and Morphogenesis of Congenital Heart Disease*, DOI 10.1007/978-4-431-54628-3_41

291

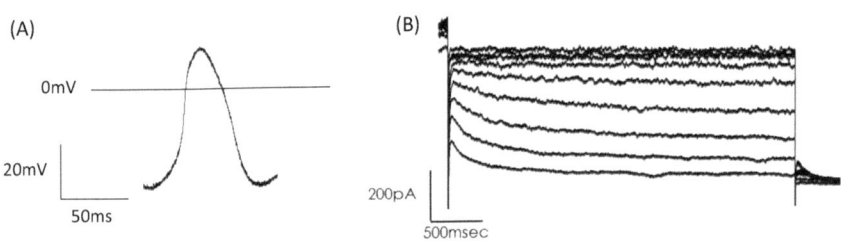

Fig. 41.1 Automaticity and I_f current from HCN4+ pacemaking cells. HCN4-GFP+ pacemaking cells derived from mESCs show typical automaticity (**a**) and I_f current (**b**)

(i) GFP+ cells expressed cardiac pacemaker specific markers such as HCN4, Cav3.1, and Connexin43 as well as cardiomyocyte-specific marker, Tropomyosin C.

(ii) Patch-clamp analysis revealed that GFP+ cells expressed pacemaker current I_f with spontaneously oscillating action potentials (automaticity) (Fig. 41.1).

(iii) GFP+ cells were capable of actively responding to adrenergic stimulation and cholinergic repression.

3. We further investigated whether mESC-derived HCN4+ cells can restore myocardial electromechanical properties. Using imaging techniques, we demonstrated that HCN4+ cells established electrical coupling with HL-1 cells, a cardiac muscle cell line derived from the mouse atrial cardiomyocytes tumor, to induce rhythmic electrical and contractile activities in vitro. Similarly, transplanted HCN4+ cells paced the hearts of rats with complete atrioventricular block, indicating that HCN4+ cells could substitute for pacemaker cells and elicit an ectopic rhythm.

These results demonstrated the potential of HCN4+ pacemaking cells derived from ES cells to act as a rate-responsive biological pacemaker and for future myocardial regenerative medicine to bradycardia.

Reference

1. Morikawa K, Bahrudin U, Miake J, et al. Identification, isolation and characterization of HCN4-positive pacemaking cells derived from murine embryonic stem cells during cardiac differentiation. PACE. 2010;33:290–303.

Part IX

Current Molecular Mechanism in Cardiovascular Development

Perspective

Osamu Nakagawa

Progenitor cell populations of multiple origins, including those from the mesodermal primary heart field, secondary heart field, and neural crest, participate in the formation of complex structures of the heart and great vessels. A variety of differentiated cell types, such as cardiomyocytes, fibroblasts, endocardial cells, and vascular cells, are coordinated to fulfill the mature functions of the heart and circulatory system. Sequential and combinatorial functions of numerous genes in these cells are necessary for proper cardiovascular development. Phenotype analyses of mutant mice and genetic studies of human patients have revealed the genes essential for cardiovascular cell differentiation, migration, proliferation, and alignment in developing embryos. A significant fraction of those genes encodes DNA-binding transcription factors and their cofactors, indicating that transcriptional regulation plays central roles in cardiovascular differentiation and morphogenesis. Mutations of the genes for inter-/intracellular signaling molecules often cause perturbation of downstream gene expression patterns. Genes associated with the laterality control and those encoding the cardiac sarcomere components were also found responsible for the congenital cardiovascular defects.

Emerging evidence indicates that additional molecules involved in the epigenetic regulation of gene expression are indispensable for cardiovascular development. Chromatin-remodeling protein complexes, SWI/SNF, ISWI, NuRD, and INO80, control the chromatin structure and accessibility of transcriptional machinery including DNA-binding proteins and RNA polymerase II. Gene deletion of a

O. Nakagawa (✉)
Department of Molecular Physiology, National Cerebral and Cardiovascular Center Research Institute, Osaka, Japan
e-mail: osamu.nakagawa@ncvc.go.jp

SWI/SNF complex factor Brg1/Smarca4 or its partner protein Baf60c/Smarcd3 caused various defects of cardiac structures and growth in mice. *CHD7*, a gene responsible for CHARGE syndrome, functions as a component of the NuRD transcriptional repressor complex. Histone modification was also shown to be important for cardiovascular development. Mutations of *MLL2/KMT2D*, which encodes a histone methyltransferase, are a major cause of Kabuki (Niikawa-Kuroki) syndrome. Mutations in the genes involved in histone methylation are significantly enriched in those identified in sporadic cases of congenital heart defects. Importance of histone methylation by polycomb group proteins, histone demethylation by jumonji family proteins, and histone acetylation/deacetylation by HAT/HDAC proteins was also documented using mouse models. In addition, noncoding functional transcripts, namely, microRNAs and long noncoding RNAs, possess crucial functions in the regulation of gene expression during cardiovascular development. MicroRNAs repress gene expression through the degradation of target mRNAs and/or the inhibition of their translation. Long noncoding RNAs modulate gene expression by various mechanisms including the association with epigenetic factors, transcriptional machinery, and microRNAs. Lack of miR-1, miR-133, and many other miRNAs as well as long noncoding RNAs such as Braveheart and Fendrr markedly affects differentiation and growth of cardiomyocytes in experimental models.

In this part, among these epigenetic regulators of gene expression, Shirai et al. describe the significance of polycomb group proteins in embryonic heart development, while Kataoka and Wang provide an overview on the roles of microRNAs and long noncoding RNAs in the developing heart and vasculature. Studies of new molecular mechanisms of cardiovascular development will lead to a better understanding of the etiologies of human congenital heart defects.

Combinatorial Functions of Transcription Factors and Epigenetic Factors in Heart Development and Disease

Kazuko Koshiba-Takeuchi, Yuika Morita, Ryo Nakamura, and Jun K. Takeuchi

Abstract

Heart malformations are the most common type of birth defect, affecting more than 2 % of newborns and causing significant morbidity and mortality. In the past two decades, studies have revealed the function and importance of cardiac transcription factors during heart development and in congenital heart disease. Transcription factors generally form complexes with other transcription factors and/or with chromatin factors to perform specific functions. This review focuses on how chromatin factors modify cardiac transcription factors during cardiovascular development and disease.

Keywords

Cardiac development • Cardiac disease • T-box genes • Epigenetic factors • SWI/SNF-type chromatin remodeling factors

K. Koshiba-Takeuchi
Department of Integrated Biosciences, Graduate School of Frontier Science, Tokyo, Japan

Department of Biological Sciences, Graduate School of Sciences, The University of Tokyo, Tokyo, Japan

Y. Morita • R. Nakamura
Department of Integrated Biosciences, Graduate School of Frontier Science, Tokyo, Japan

JST PRESTO, Tokyo, Japan

J.K. Takeuchi (✉)
Department of Integrated Biosciences, Graduate School of Frontier Science, Tokyo, Japan

Department of Biological Sciences, Graduate School of Sciences, The University of Tokyo, Tokyo, Japan

JST PRESTO, Tokyo, Japan
e-mail: junktakeuchi@iam.u-tokyp.ac.jp

© The Author(s) 2016
T. Nakanishi et al. (eds.), *Etiology and Morphogenesis of Congenital Heart Disease*,
DOI 10.1007/978-4-431-54628-3_42

42.1 Transcription Factors in Heart Development

The heart is an organ that pumps blood to and from the body's tissues through the blood vessels. Cardiac muscle contains cardiomyocytes, which ensure the heart's contractile ability; however, cardiomyocytes alone are not sufficient for the heart to function. Other components, such as the conduction cells, fibroblasts, blood vessels, and endocardial cells, are important for maintaining the heart's systemic pumping ability. Each cell type can be identified by its expression of specific transcription factors, signaling molecules, and/or function-specific proteins (Fig. 42.1). The heart is the first organ to form in vertebrates, and it performs a vital role in distributing oxygen and nutrients throughout the embryo. The primordial heart is derived from cardiovascular mesodermal cells that transiently express *T* (*Brachyury*), *Mesp1*, and *Flk1* during gastrulation. A subset of these cardiac mesodermal cells gives rise to cardiac progenitor cells, which can differentiate into any type of cardiac cell.

T-box transcription factors compose a conserved family of genes that are important for heart development and patterning. In humans, disruption of the cardiac *T-box* genes leads to various congenital heart defects [1]. Mutations in *Tbx5* are associated with Holt-Oram syndrome [2, 3], whereas mutations in the *Tbx20* gene result in atrial septal defect [4, 5]. Interestingly, in the developing heart, Tbx5 and Tbx20 are complementarily expressed in the left and right ventricle, respectively (Fig. 42.2) [6]. The regions in which the *T-box* genes are expressed and the regions that are defective in a given disease are very similar. *Tbx5* is expressed in the inflow tract, atria, AV cushion, and left ventricle but not in the outflow tract or

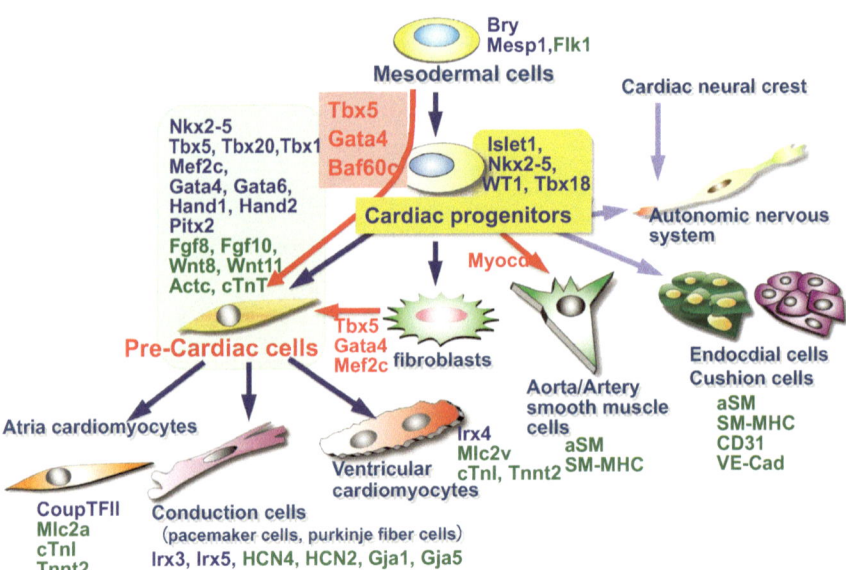

Fig. 42.1 Multiple cardiac cell types. Each cell type differentiates from mesoderm-derived cardiac progenitor cells. The major molecules associated with each cell type are indicated

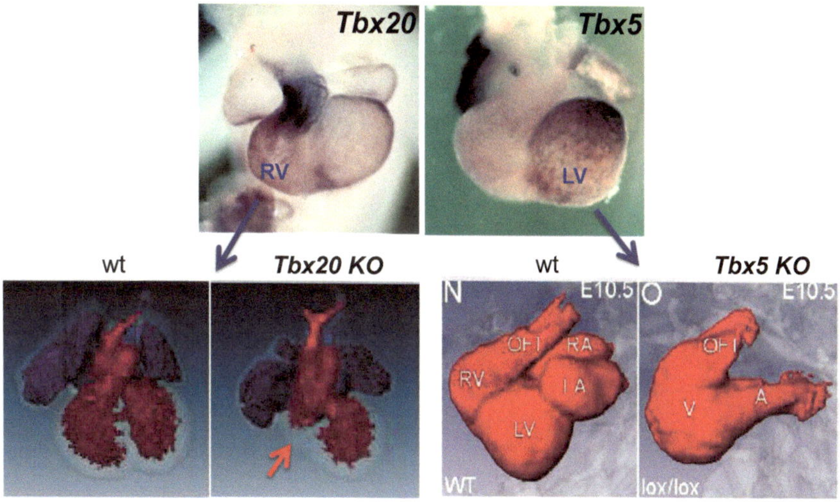

Fig. 42.2 The expression patterns of *Tbx5* and *Tbx20* in the mouse heart. These genes show complementary expression patterns in the ventricles. The knockout mice for each gene show hypoplasia in the same region in which the gene is normally expressed (Adapted from Bruneau et al. [8] and Takeuchi et al. [9])

the right ventricle; *Tbx5* expression appears to be restricted to the first heart field (FHF)-derived region [7]. *Tbx5* knockout mice experience severe left ventricle hypoplasia and die at approximately E9 without their hearts ever beating (Fig. 42.2) [8]. By contrast, *Tbx20* is primarily expressed in the outflow tract and the right ventricle, which are derived from the second heart field (SHF). *Tbx20* knockdown mice develop a single ventricle and show severe hypoplasia of the right ventricle (Fig. 42.2) [9]. These facts indicate that Tbx5 and Tbx20 may specify the identity of each ventricle. Tbx5 also acts in association with Sall4 in ventricular septum formation. Sall4 is a zinc-finger transcription factor that, when mutated, causes Okihiro syndrome (Duane-radial ray syndrome, DRRS) in humans [10, 11]. The heart and limb phenotypes of Okihiro syndrome are very similar to those of Holt-Oram syndrome. In fact, some Holt-Oram patients lack mutations in *TBX5* and instead have mutations in *SALL4* [12]. Tbx5 and Sall4 participate in protein-protein interactions and synergistically regulate downstream gene expression [13]. Furthermore, Tbx5 is a key gene involved in the acquisition of the ventricular septum during vertebrate evolution [14]. During vertebrate evolution from aquatic to terrestrial life, the morphology of the heart has changed. As a result, avian and mammalian hearts contain four chambers—two atria and two ventricles—and their circulatory systems contain two loops, the pulmonary and systemic loops, that separate the oxygen-rich and oxygen-poor blood. In vertebrates with four-chambered hearts, *Tbx5* expression is restricted to the left ventricle, whereas in animals with a single ventricle, *Tbx5* expression is observed throughout the ventricle. Reptiles show a unique *Tbx5* expression pattern that is associated with

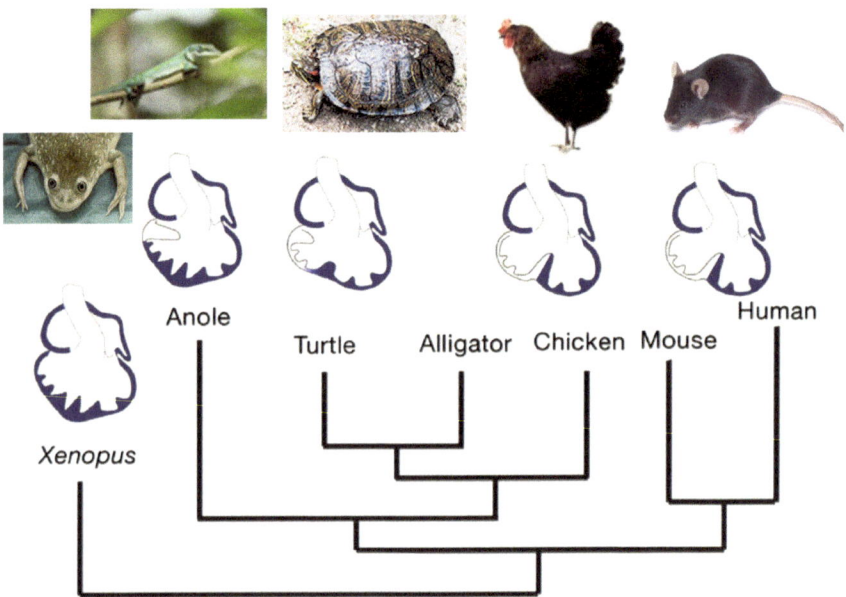

Fig. 42.3 *Tbx5* expression and heart morphology in vertebrates. Note that animals with two ventricles express *Tbx5* on the left side (Adapted from Koshiba-Takeuchi et al. [14])

their ventricular morphology. *Anolis*, which are a type of squamate, have a single ventricle that expresses *Tbx5* throughout this chamber and throughout development. Interestingly, turtles show a left-high to right-low gradient of *Tbx5* expression during late developmental stages, and a septum-like structure forms in the middle of the ventricle (Fig. 42.3). To confirm the precise interaction between *Tbx5* expression patterns and ventricular septum formation, we performed Tbx5 mis-expression experiments using transgenic mice. Transgenic mice that express Tbx5 throughout the ventricle fail to form a ventricular septum. These results strongly indicate that *Tbx5* expression in the left ventricle is important for the development of two-chambered ventricles. We hypothesized that the regulatory region of *Tbx5* might have been modified during vertebrate evolution, thereby changing the *Tbx5* expression pattern and the ventricular morphology, as shown in Fig. 42.3.

42.2 Chromatin Factors and Cardiac Differentiation

Recent studies have shown that chromatin factors are essential for determining cell fate in several organs. In heart development, SWI/SNF-type chromatin remodeling factors play key roles in the differentiation of cardiomyocytes by interacting with heart-specific transcription factors [15, 16]. Cardiac transcription factors alone are not sufficient to induce cardiomyocyte differentiation in vivo or in vitro (Figs. 42.4 and 42.5). This result suggests that chromatin accessibility is important for

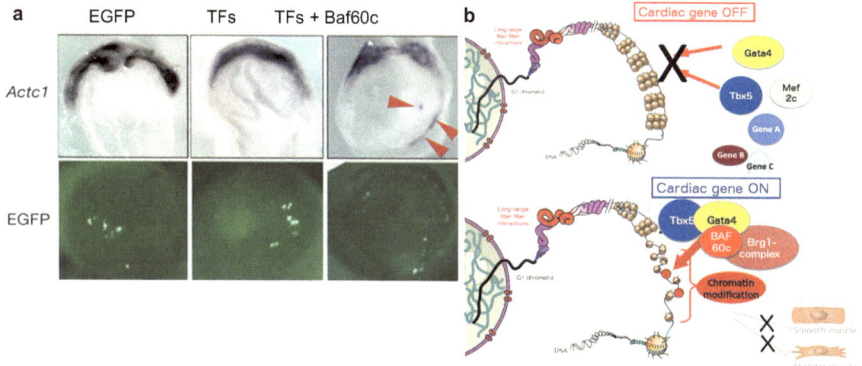

Fig. 42.4 (**a**) A mixture of transcription factors (*TFs*), Baf60c and EGFP, but not the control (EGFP) or TFs + EGFP, ectopically induce *Actc1* in the lateral plate mesodermal region. (**b**) A schematic of cardiac gene regulation. The SWI/SNF complex-mediated change in chromatin conformation is important for the activation of cardiac gene transcription (Adapted from Takeuchi and Bruneau [18] and Van Weerd et al. [15])

Tbx5	Gata4	Gata1	Nkx2-5	Baf60c	Baf60b	Actc1,Myl7 expression	Beating
+	−	−	−	−	−	X	X
−	+	−	−	−	−	X	X
−	−	+	−	−	−	X	X
+	+	−	−	−	−	X	X
+	−	+	−	−	−	X	X
−	−	−	+	−	−	X	X
+	−	−	+	−	−	X	X
−	+	−	+	−	−	X	X
+	+	−	+	−	−	X	X
−	+	−	−	+	−	O	X
−	+	−	−	−	+	O	X
−	−	+	−	+	−	O	X
−	−	+	−	−	+	X	X
+	+	−	+	+	−	O	O
+	+	−	−	+	−	O	O

Fig. 42.5 Schematic diagram shows that ectopic expression of major cardiac contracted genes (Actc1 and Myl7) are observed by combinatorial transfection of cTF and Baf60c into ex vivo mouse

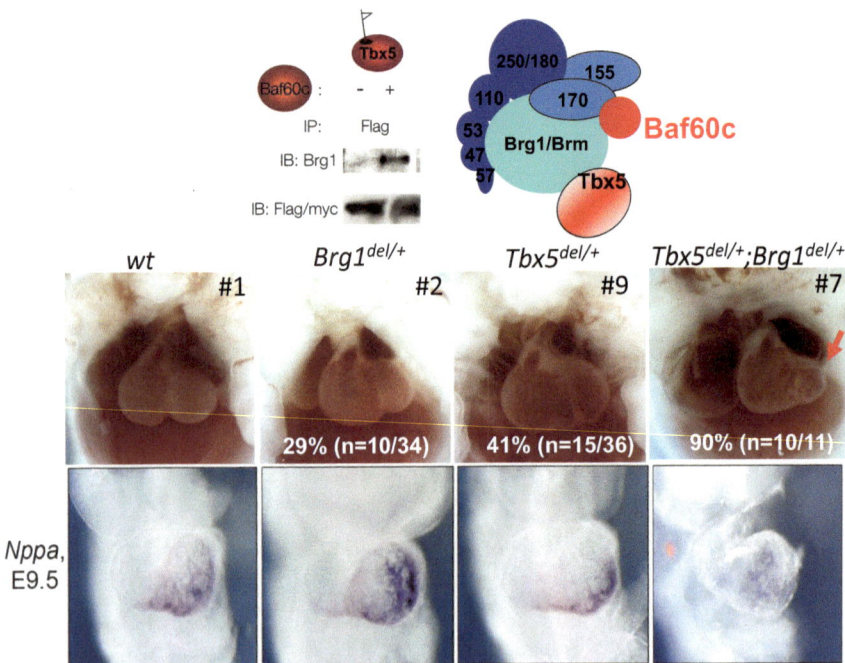

Fig. 42.6 The combinatorial functions of cardiac transcription factors and Baf chromatin remodeling factors. Only the transcription factors can induce cardiac markers, but the differentiation of beating cardiomyocytes requires both transcription factors and chromatin remodelers

transcription factors to bind to their target sites. Therefore, we searched for chromatin remodeling factors that are expressed in the cardiac region at early stages of heart development. A previous study showed that a component of the SWI/SNF-type chromatin remodeling complex, Baf60c (also known as *Smarcd3*), has specific roles in heart development [17]. When a mixture of cardiac transcription factors and Baf60c was injected into the lateral plate mesoderm of mouse embryos, alpha cardiac actin-positive cells were ectopically induced (Fig. 42.4) [18]. These ectopically induced cardiac cells could beat, which showed that they were functional cardiomyocytes. Baf60c can directly associate with the Tbx5, Nkx2-5, Gata4, and RBPjk proteins and regulate the transcription of downstream genes [17, 19]. Mutations in chromatin factors cause abnormal cardiac function in both mice and humans. Baf60c determines a cell's fate by not only loosening the chromatin structure but also synergistically interacting with specific factors. When associated with Tbx5, Brg1, a core protein of the SWI/SNF-type chromatin remodeling complex, synergistically regulates cardiac differentiation in the presence of Baf60c [20]. Mice heterozygous for both *Brg1* and *Tbx5* had more severe defects than did single mutants, particularly in the left ventricle (Fig. 42.6). These results indicate that the dosage of epigenetic factors affects the severity of the *Tbx5* mutant phenotype (i.e., left ventricular hypoplasia). The severity of congenital heart disease in humans may be related to the level of expression of epigenetic factors and/or of partner factors. To address this possibility, we need to elucidate the relationship between the expression level of epigenetic factors and the penetration of heart failure.

42.3   Future Directions and Clinical Implications

We have analyzed the functions of Baf60c, a component of the SWI/SNF-type chromatin remodeling complex, in heart development in vivo and in vitro, but the mechanism by which Baf60c is regulated is still unknown. An important question is whether the functions of Brg1 or Baf60c are altered in each type of tissue. If their functions do not vary, they may be regulated in a partner-dependent manner. One approach to confirm this hypothesis is to use ChIP-sequencing to compare Baf60c and Brg1 target genes in different tissues. Another question that must be addressed is how Baf60c's expression pattern and dosage are regulated and which molecule (s) participate in this regulation. We found a candidate transcription factor that directly regulates Baf60c expression, but this molecule alone could not explain the dynamic change in Baf60c's expression pattern. Further analysis is required to determine the molecular mechanisms of Baf60c regulation.

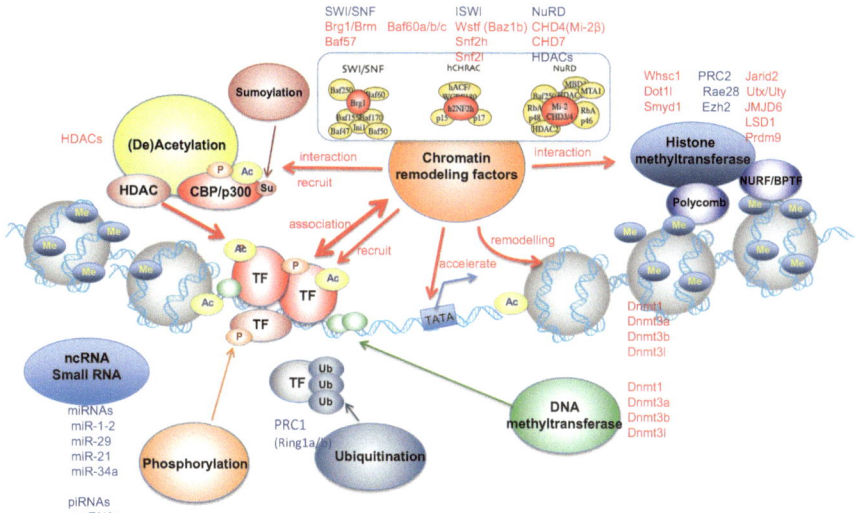

Fig. 42.7 Immunoprecipitation experiments indicate that Brg1 can strongly bind to Tbx5 with Baf60c. The lower panel shows the morphology of *Brg1del/+*, *Tbx5del/+*, and double heterozygote mouse hearts. The heart of the double heterozygote shows severe hypoplasia of the *left* ventricle (Adapted from Takeuchi et al. [20])

Over the last 20 years, cardiac researchers have elucidated many causes of congenital heart disease and have identified many genes that are involved. However, it is not sufficient to only understand the diversity or severity of the disease. In the future, we must also determine the role of epigenetic factors in heart failure because these factors regulate cardiac gene transcription (Fig. 42.7).

References

1. Greulich F, Rudat C, Kispert A. Mechanisms of T- box gene function in the developing heart. Cardiovasc Res. 2011;91:212–22.
2. Li QY, Newbury-Ecob RA, Terrett JA, et al. Holt-Oram syndrome is caused by mutations in TBX5, a member of the Brachyury (T) gene family. Nat Genet. 1997;15:21–9.
3. Basson CT, Bachinsky DR, Lin RC, et al. Mutations in human TBX5 cause limb and cardiac malformation in Holt-Oram syndrome. Nat Genet. 1997;15:30–5.
4. Kirk EP, Sunde M, Costa MW, et al. Mutations in cardiac T-box factor gene TBX20 are associated with diverse cardiac pathologies, including defects of septation and valvulogenesis and cardiomyopathy. Am J Hum Genet. 2007;81:280–91.
5. Posch MG, Gramlich M, Sunde M, et al. A gain-of-function TBX20 mutation causes congenital atrial septal defects, patent foramen ovale and cardiac valve defects. J Med Genet. 2010;47:230–5.
6. Takeuchi JK, Ohgi M, Koshiba–Takeuchi K, et al. Tbx5 specifies the left/right ventricles and ventricular septum position during cardiogenesis. Development. 2003;130:5953–64.
7. Bruneau BG, Logan M, Davis N, et al. Chamber-specific cardiac expression of Tbx5 and heart defects in Holt-Oram syndrome. Dev Biol. 1999;211:100–8.
8. Bruneau BG, Nemer G, Schmitt JP, et al. A murine model of Holt-Oram syndrome defines roles of the T– box transcription factor Tbx5 in cardiogenesis and disease. Cell. 2001;106:709–21.
9. Takeuchi JK, Mileikovskaia M, Koshiba-Takeuchi K, et al. Tbx20 dose-dependently regulates transcription factor networks required for mouse heart and motoneuron development. Development. 2005;132:2463–74.
10. Al-Baradie R, Yamada K, St Hilaire C, et al. Duane radial ray syndrome (Okihiro syndrome) maps to 20q13 and results from mutations in SALL4, a new member of the SAL family. Am J Hum Genet. 2002;71:1195–9.
11. Kohlhase J, Heinrich M, Schubert L, et al. Okihiro syndrome is caused by SALL4 mutations. Hum Mol Genet. 2002;11:2979–87.
12. Brassington A-ME, Sung SS, Toydemir RM, et al. Expressivity of Holt-Oram syndrome is not predicted by TBX5 genotype. Am J Hum Genet. 2003;73:74–85.
13. Koshiba–Takeuchi K, Takeuchi JK, Arruda EP, et al. Cooperative and antagonistic interactions between Sall4 and Tbx5 pattern the mouse limb and heart. Nat Genet. 2006;38:175–83.

14. Koshiba-Takeuchi K, Mori AD, Kaynak BL, et al. Reptilian heart development and the molecular basis of cardiac chamber evolution. Nature. 2009;461:95–8.
15. van Weerd JH, Koshiba-Takeuchi K, Kwon C, Takeuchi JK. Epigenetic factors and cardiac development. Cardiovasc Res. 2011;91:203–11.
16. Hang CT, Yang J, Han P, et al. Chromatin regulation by Brg1 underlies heart muscle development and disease. Nature. 2010;466:62–7.
17. Lickert H, Takeuchi JK, Von Both I, et al. Baf60c is essential for function of BAF chromatin remodelling complexes in heart development. Nature. 2004;432:107–12.
18. Takeuchi JK, Bruneau BG. Directed transdifferentiation of mouse mesoderm to heart tissue by defined factors. Nature. 2009;459:708–11.
19. Takeuchi JK, Lickert H, Bisgrove BW, et al. Baf60c is a nuclear notch signaling component required for the establishment of left–right asymmetry. Proc Natl Acad Sci U S A. 2007;104:846–51.
20. Takeuchi JK, Lou X, Alexander JM, et al. Chromatin remodelling complex dosage modulates transcription factor function in heart development. Nat Commun. 2011;2:187–97.

Pcgf5 Contributes to PRC1 (Polycomb Repressive Complex 1) in Developing Cardiac Cells

Manabu Shirai, Yoshihiro Takihara, and Takayuki Morisaki

Abstract

Polycomb-group (PcG) proteins maintain transcriptional silencing through specific histone modification and are essential for cell-fate transition and proper development of embryonic and adult stem cells. Recent advances in molecular analysis of PcG proteins have revealed that the distinct subunit composition of PRC1 confers specific and nonoverlapping functions for regulation of embryonic and adult stem cells. Here, we provide an overview of recent findings regarding the role of PcG proteins in cardiac development, with focus on the diversity of PcG complexes.

Keywords
Polycomb-group protein • Cardiac development • Transcriptional silencing • Histone modification

M. Shirai
Department of Bioscience and Genetics, National Cerebral and Cardiovascular Center Research Institute, 5-7-1 Fujishirodai, Suita, Osaka 565-8565, Japan

Y. Takihara
Department of Stem Cell Biology, Research Institute for Radiation Biology and Medicine, Hiroshima University, 1-2-3 Kasumi, Minami-ku, Hiroshima 734-8551, Japan

T. Morisaki (✉)
Department of Bioscience and Genetics, National Cerebral and Cardiovascular Center Research Institute, 5-7-1 Fujishirodai, Suita, Osaka 565-8565, Japan

Department of Molecular Pathophysiology, Graduate School of Pharmaceutical Sciences, Osaka University, 1-6 Yamadaoka, Suita, Osaka 565-0871, Japan
e-mail: morisaki@ri.ncvc.go.jp

T. Nakanishi et al. (eds.), *Etiology and Morphogenesis of Congenital Heart Disease*, DOI 10.1007/978-4-431-54628-3_43

43.1   Introduction

Cardiac development is a complex and ordered process that requires cellular specification, proliferation, and differentiation, as well as further migration of cell populations from diverse sites. The primary heart field (PHF) originates in the anterior splanchnic mesoderm, then gives rise first to the cardiac crescent, later to the linear heart tube, and ultimately contributes to parts of the left ventricular (LV) region. The second cardiogenic region, known as the second heart field (SHF), lies in the anterior, posterior, and dorsal to the linear heart tube and is derived from the pharyngeal mesoderm located medial and anterior to the cardiac crescent. Cells from the SHF are added to the developing heart tube and give rise to the outflow tract (OFT), right ventricular (RV) region, and main parts of atrial tissues [1].

Congenital heart defects (CHDs) represent the most common anomaly seen in human newborns, with a prevalence of approximately 1 % of all births [1]. Traditionally, focus on causes of CHD has involved transcriptional networks during cardiogenesis, because correct alignment and septation of cardiac structures regulated by cardiac specific transcriptional factors, such as *Tbx1*, *Tbx5*, *Tbx20*, *Gata4*, and *Nkx2-5*, are essential for cardiac morphogenesis [1]. In addition to these multiple genetic factors, recent studies have shown that some chromatin remodeling factors moderate gene expression to control cardiogenesis and are also involved in the molecular pathogenesis of CHD [2–7].

Polycomb-group (PcG) proteins maintain transcriptional silencing by regulating chromatin configuration [8, 9]. There are two principal PcG repressive complexes (PRCs), PRC1 and PRC2. Mammalian PRC2 contains four core proteins (Ezh1/2, Eed, Suz12, Rbbp4/7) and trimethylates histone H3 at lysine 23 (H3K27), while the other complex, PRC1, consists of a combination of several protein families, including chromobox (Cbx), Ring, polyhomeotic (Ph), and posterior sex combs (Psc), and induces mono-ubiquitination of histone H2A at lysine 119 (Fig. 43.1). In recent studies, PRC1 have been divided to Cbx-PRC1 (canonical PRC1) and Rybp-PRC1 (noncanonical PRC1) [10, 11]. Since development of high-throughput techniques for analyzing the genome in the past decade, PcG-mediated transcriptional repression has resulted in increased molecular information regarding its role in a number of important biological activities such as cell cycle progression, differentiation, and cell-fate transition in multiple cell types and tissue contexts, including embryonic, adult, and cancer stem cells. This PcG-mediated transcriptional repression can vary during development and among cell types. However, the precise role in the context of cell conditions remains unclear.

43.2   PcG Functions in Cardiac Development

In this review, we primarily focused on the functions of PcG proteins in cardiac development. Their roles have been elucidated via generation of knockout (KO) mice for each of the PcG components. Among the PRC2 components, loss

Fig. 43.1 PcG complexes in mammals. (**A**) Molecular functions of PcG complexes (PRCs). PRC2 trimethylates lysine 27 of histone H3. Canonical PRC1 binds to the H3K27me3 mark and mediates the mono-ubiquitination of histone H2A at lysine 119. (**B**) Canonical PRC1 components and results of KO of each in mice

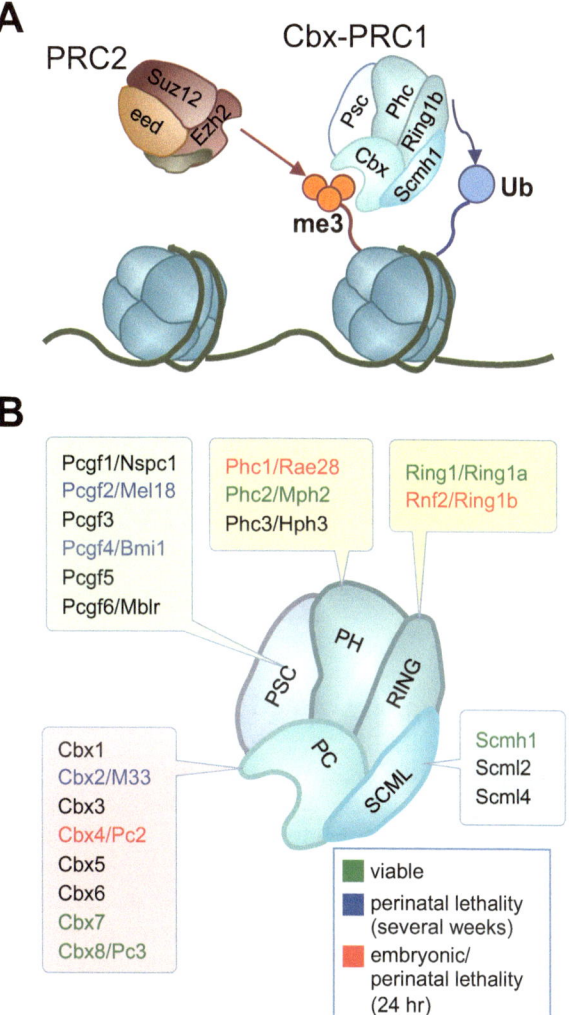

of Suz12, Ezh2, or Eed results in embryonic lethality during the early postimplantation stage [12–14]. To address the role of PRC2 in cardiac development, Ezh2 and Eed were conditionally inactivated in specified cardiac cells using Nkx2-5:Cre or TnT:Cre [3, 4]. Inactivation of Ezh2 by Nkx2-5:Cre ($Ezh2^{NK}$) and Eed by TnT:Cre (Eed^{TnT}) led to embryonic lethality and several cardiac defects including compact myocardial hypoplasia, whereas inactivation of Ezh2 by TnT:Cre ($Ezh2^{TnT}$) did not result in severe defects in cardiogenesis despite a modest upregulation of some cardiac genes, probably because of the redundant functions of Ezh1 and Ezh2.

Embryos deficient of *Ring1b*, a core component of PRC1, also displayed early embryonic lethality caused by gastrulation arrest [15]. Although early developmental arrest in *Ring1b* KO embryos was partially restored by inactivation of Cdkn2a

(Ink4a/ARF), cardiac tissue did not develop in double-KO embryos. Unlike early developmental defects seen in KO mice lacking some of the core PRC1 and PRC2 components, deficiency of other components has been shown to give rise to restricted effects. For example, loss of Rae28/Phc1 resulted in perinatal lethality with cardiac anomalies, double outlet right ventricle, and tetralogy of Fallot [6, 7]. In addition to cardiac defects, *Rae28/Phc1* deficient mice also showed craniofacial developmental defects, as well as thymus and parathyroid gland defects as seen in human DiGeorge syndrome.

Among Cbx proteins, Cbx4 may play an important role in cardiogenesis. SUMO-specific protease 2 (SENP2) was reported to regulate transcription of *Gata4* and *Gata6*, mainly through alteration of the occupancy of Cbx4 on their promoters [5]. In *SENP2*-deficient embryos, sumoylated Cbx4 accumulates on the promoters of target genes, leading to transcriptional repression of *Gata4* and *Gata6*. Furthermore, *Cbx4* mutant mice displayed postnatal lethality with severe hypoplasia of the developing thymus as a result of reduced thymocyte proliferation. However, the function of Cbx4 in cardiogenesis has not been clearly elucidated [16]. Thus, PcG proteins are essential for molecular regulation of the expression of several cardiac genes during embryogenesis and important for cardiac morphogenesis.

43.3 Diversity of PcG Proteins

In our recent studies, expression of the PRC1 components Ring1b, Bmi1/Pcgf4, and Rae28/Phc1 was detected in embryonic hearts containing both the PHF and SHF embryonic cardiac fields (Fig. 43.2), though their expression patterns were not restricted in cardiac cells. Despite the ubiquitous expression of PRC1 components, mice with single PcG KO except for the core proteins show more distinct and restricted phenotypes. Among Psc proteins, genetic deletion of *Mel18/Pcgf2* or *Bmi1/Pcgf4* resulted in postnatal lethality caused defects in anterior-posterior specification, while there was no effect on cardiogenesis [17, 18]. Also, *Mel18/ Bmi1* double-KO mice died around E9.5 and exhibited more severe developmental defects than those with KO of either alone, suggesting that Mel18/Pcgf2 and Bmi1/ Pcgf4 have partially redundant functions [18]. In addition to functional redundancy, recent studies have shown that distinct Cbx and Pcgf proteins confer specific and nonoverlapping functions of PRC1 in embryonic and adult stem cells [10, 11]. Exchanging Cbx protein in canonical PRC1 is involved in the switch from self-renewal to a differentiation state, whereas Pcgf and other noncanonical components, such as Rybp, Kdm2b, and L3mbtl2, also confer restricted functions to PRC1.

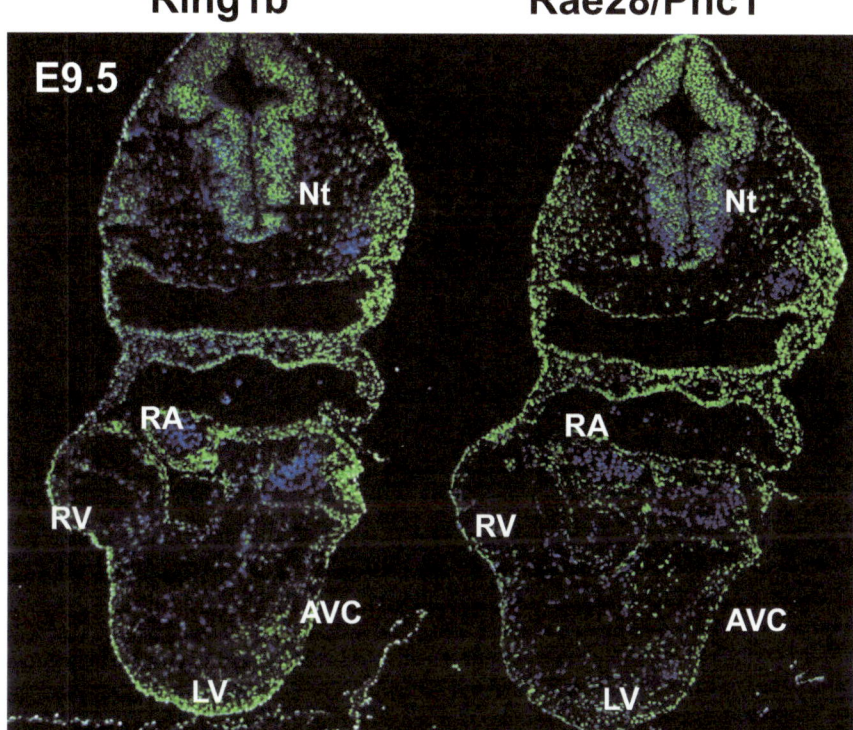

Fig. 43.2 Expression patterns of Ring1b and Rae28/Phc1 in E9.5 mouse embryos. Anti-Ring1b and anti-Rae28/Phc1 antibody-positive cells are ubiquitously distributed on E9.5. AVC, atrioventricular canal, LV, left ventricle, Nt, neural tube, RA, right atrium, RV, right ventricle

43.4 Pcgf5 Expression in the Developing Heart

Recently, we identified *Pcgf5* as an upregulated gene during cardiomyocyte differentiation of ES cells as well as strong expression of *Pcgf5* in cardiac fields during the early embryonic stages as compared to other embryonic tissues [19] (Fig. 43.3). Unlike *Bmi1/Pcgf4*, *Pcgf5* expression, patterns were more restricted to early mouse embryos. Our hypothesis states that exchanging Pcgf components within PRC1 determines the specificity of their function in the developing heart.

43.5 Conclusions

Recent increasing evidence obtained in experiments with embryonic and adult stem cells indicates that the diversity of PRC components, particularly PRC1, contributes to specification of their function (Fig. 43.4). Furthermore, transcriptional repression

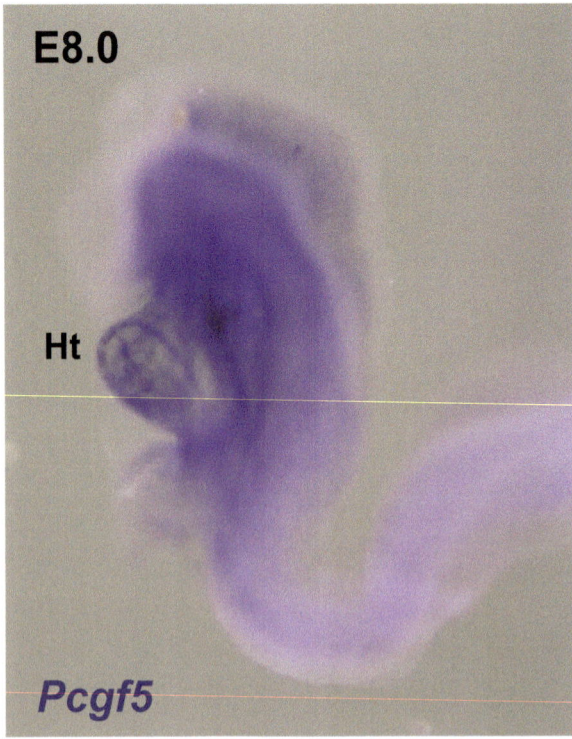

Fig. 43.3 Expression pattern of *Pcgf5* in E8.0 mouse embryos. Whole-mount in situ hybridization (WISH) showing expression of *Pcgf5* at E8.0. High Pcgf5 expression was observed in developing hearts. Ht, heart

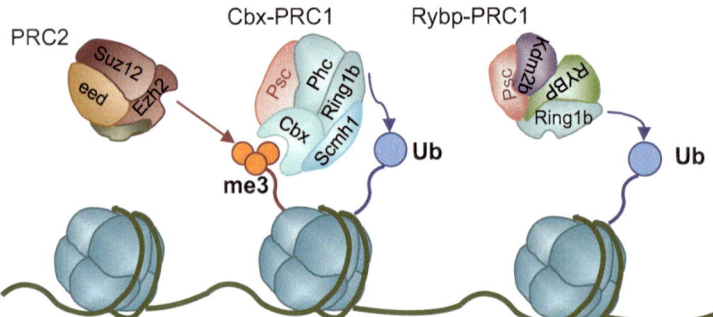

Fig. 43.4 Recent insight regarding PRC1 complexes. Canonical PRC1 (Cbx-PRC1) is recruited to H3K27me3 and causes mono-ubiquitination of lysine 119 of histone H2A. Noncanonical PRC1 (Rybp-PRC1) mediates the mono-ubiquitination of H2A in a PRC2-independent manner

by PcG is essential for cardiogenesis. Understanding PcG functions and more detailed characterization of the PcG complex during cardiogenesis may be crucial for elucidating the molecular mechanisms regulating cardiac development.

Acknowledgments This work was supported by a JSPS KAKENHI Grant (24591595).

References

1. Buckingham M, Meilhac S, Zaffran S. Building the mammalian heart from two sources of myocardial cells. Nat Rev Genet. 2005;6:826–35. doi:10.1038/nrg1710.
2. Hang CT, Yang J, Han P, et al. Chromatin regulation by Brg1 underlies heart muscle development and disease. Nature. 2010;466:62–7. doi:10.1038/nature09130.
3. Delgado-Olguin P, Huang Y, Li X, et al. Epigenetic repression of cardiac progenitor gene expression by Ezh2 is required for postnatal cardiac homeostasis. Nat Genet. 2012;44:343–7. doi:10.1038/ng.1068.
4. He A, Ma Q, Cao J, et al. Polycomb repressive complex 2 regulates normal development of the mouse heart. Circ Res. 2012;110:406–15. doi:10.1161/CIRCRESAHA.111.252205.
5. Kang X, Qi Y, Zuo Y, et al. SUMO-specific protease 2 is essential for suppression of polycomb group protein-mediated gene silencing during embryonic development. Mol Cell. 2010;38:191–201. doi:10.1016/j.molcel.2010.03.005.
6. Shirai M, Osugi T, Koga H, et al. The polycomb-group gene Rae28 sustains Nkx2.5/Csx expression and is essential for cardiac morphogenesis. J Clin Invest. 2002;110:177–84. doi:10.1172/JCI14839.
7. Takihara Y, Tomotsune D, Shirai M, et al. Targeted disruption of the mouse homologue of the drosophila polyhomeotic gene leads to altered anteroposterior patterning and neural crest defects. Development. 1997;124:3673–82.
8. Aloia L, Di Stefano B, Di Croce L. Polycomb complexes in stem cells and embryonic development. Development. 2013;140:2525–34. doi:10.1242/dev.091553.
9. Schwartz YB, Pirrotta V. A new world of polycombs: unexpected partnerships and emerging functions. Nat Rev Genet. 2013;14:853–64. doi:10.1038/nrg3603.
10. Gao Z, Zhang J, Bonasio R, et al. PCGF homologs, CBX proteins, and RYBP define functionally distinct PRC1 family complexes. Mol Cell. 2012;45:344–56. doi:10.1016/j.molcel.2012.01.002.
11. Morey L, Pascual G, Cozzuto L, et al. Nonoverlapping functions of the polycomb group Cbx family of proteins in embryonic stem cells. Cell Stem Cell. 2012;10:47–62. doi:10.1016/j.stem.2011.12.006.
12. Faust C, Schumacher A, Holdener B, Magnuson T. The eed mutation disrupts anterior mesoderm production in mice. Development. 1995;121:273–85.

13. O'Carroll D, Erhardt S, Pagani M, et al. The polycomb-group gene Ezh2 is required for early mouse development. Mol Cell Biol. 2001;21:4330–6. doi:10.1128/MCB.21.13.4330-4336. 2001.

14. Pasini D, Bracken AP, Jensen MR, et al. Suz12 is essential for mouse development and for EZH2 histone methyltransferase activity. EMBO J. 2004;23:4061–71. doi:10.1038/sj.emboj. 7600402.

15. Voncken JW, Roelen BAJ, Roefs M, et al. Rnf2 (Ring1b) deficiency causes gastrulation arrest and cell cycle inhibition. Proc Natl Acad Sci U S A. 2003;100:2468–73. doi:10.1073/pnas. 0434312100.

16. Liu B, Liu Y-F, Du Y-R, et al. Cbx4 regulates the proliferation of thymic epithelial cells and thymus function. Development. 2013;140:780–8. doi:10.1242/dev.085035.

17. van der Lugt NM, Domen J, Linders K, et al. Posterior transformation, neurological abnormalities, and severe hematopoietic defects in mice with a targeted deletion of the bmi-1 proto-oncogene. Genes Dev. 1994;8:757–69. doi:10.1101/gad.8.7.757.

18. Akasaka T, van Lohuizen M, van der Lugt N, et al. Mice doubly deficient for the polycomb group genes Mel18 and Bmi1 reveal synergy and requirement for maintenance but not initiation of Hox gene expression. Development. 2001;128:1587–97.

19. Narumiya H, Hidaka K, Shirai M, et al. Endocardiogenesis in embryoid bodies: novel markers identified by gene expression profiling. Biochem Biophys Res Commun. 2007;357:896–902. doi:10.1016/j.bbrc.2007.04.030.

Noncoding RNAs in Cardiovascular Disease

44

Masaharu Kataoka and Da-Zhi Wang

Abstract

For decades, it has been recognized that proteins, which are encoded by our genomes via transcription and translation, are building blocks that play vital roles in almost all biological processes. Mutations identified in many protein-coding genes are linked to various human diseases. However, this "protein-centered" dogma has been challenged in recent years with the discovery that majority of our genome is "noncoding" yet transcribed. Noncoding RNA has become the focus of "next generation" biology. Here, we review the emerging field of noncoding RNAs, including microRNAs (miRNAs) and long noncoding RNAs (lncRNAs), and their function in cardiovascular biology and disease.

Keywords

Cardiac disease • Heart development • Long noncoding RNAs • MicroRNAs

44.1 Introduction

When the human genome project was completed, it was surprising that only about 20,000–25,000 protein-coding genes exist in our species, with less than 2 % of the human genome used for coding proteins. What are the functions of noncoding sequences, which make up more than 98 % of our genome? We are now finding answers with the recognition that the majority of the genome is actively transcribed to produce thousands of noncoding transcripts, including microRNAs (miRNAs) and long noncoding RNAs (lncRNAs), in many cell types and tissues. miRNAs are a class of small noncoding RNAs (~22 nucleotides) and were first discovered in

M. Kataoka, M.D., Ph.D. • D.-Z. Wang, Ph.D. (✉)

Department of Cardiology, Harvard Medical School, Boston Children's Hospital, 320 Longwood Ave, Boston, MA, USA

e-mail: dwang@enders.tch.harvard.edu

T. Nakanishi et al. (eds.), *Etiology and Morphogenesis of Congenital Heart Disease*, DOI 10.1007/978-4-431-54628-3_44

C. elegans two decades ago. More than 2,000 miRNAs have been found in humans, and many of them are evolutionarily conserved. By imperfect base pairing with mRNAs in a sequence-dependent manner, miRNAs repress gene expression by degrading target mRNAs and/or inhibiting their translation. Roles for miRNAs have been demonstrated in the regulation of a broad range of biological activities and diseases [1]. More recently, thousands of lncRNAs, which are transcribed noncoding RNAs greater than 200 nucleotides, were discovered and implicated in a variety of biological processes [2]. Clearly, investigating and understanding of how miRNAs and lncRNAs regulate gene expression during cardiovascular development and function will greatly facilitate therapeutic treatment of cardiovascular disease.

44.2   miRNAs in Cardiac Development

Global disruption of all miRNA expression in the heart is the first step to understanding the function of miRNAs in cardiac development and physiology. Dicer, a RNase III endoribonuclease, is a critical enzyme for the maturation of most miRNAs. Conventional deletion of Dicer causes early embryonic lethality in mice [3]. Disrupting miRNA expression in the early embryonic stage using Nkx2.5-Cre leads to improperly compacted ventricular myocardium in mutant embryos [4], and α-MHC-Cre-mediated conditional deletion of Dicer causes postnatal lethality due to dilated cardiomyopathy and heart failure [5]. These studies suggest that many miRNAs have crucial roles in cardiac development. miR-1 is tissue-specifically expressed in the heart and skeletal muscle, and genetic deletion of both miR-1-1 and miR-1-2 indicated that miR-1 is required for cardiomorphogenesis and the expression of many cardiac contractile proteins [6].

44.3   Cardiac Regeneration, Remodeling, and Ischemia Regulated by miRNAs

Mammalian adult cardiomyocytes are terminally differentiated cells with very limited regenerative ability. A recent report identified about 40 miRNAs that strongly increased cell proliferation in neonatal mouse and rat cardiomyocytes. Two of these miRNAs, miR-590 and miR-199a, were further demonstrated to induce cardiomyocyte proliferation both in vitro and in vivo [7]. Using both gain- and loss-of-function approaches in transgenic and knockout mice models, we demonstrated that the miR-17-92 cluster is required for and sufficient to induce cardiomyocyte proliferation. More specifically, we identified miR-19a/b as the major contributors among the miR-17-92 cluster to the regulation of the cardiomyocyte proliferation [8]. These studies suggest that miRNAs are key regulators of cardiomyocyte proliferation and heart regeneration, suggesting their significant therapeutic potential to treat cardiac-degeneration-associated heart disease.

Cardiac remodeling, which is defined as an alteration in the structure (dimensions, mass, shape) of the heart, is one of the major responses of the heart to biomechanical stress and pathological stimuli. Numerous studies have demonstrated the functional involvement of many miRNAs during cardiac remodeling [9]. Recently, we and others demonstrated that miR-22, a miRNA enriched in cardiomyocytes but only mildly upregulated during cardiac hypertrophy, significantly promotes cardiac hypertrophy in vitro and in vivo [10, 11].

Ischemia is an independent risk factor of cardiovascular events, which leads to myocardial infarction (MI) and ischemia-reperfusion (I/R) injury. Several miRNAs participate in the regulation of these pathologic processes, especially cardiomyocyte apoptosis following MI and I/R injury. miR-92a, a member of the miR-17-92 cluster involved in cardiomyocyte proliferation, also participates in the control of cardiomyocyte survival by targeting integrin subunit α5 and eNOS. Inhibition of miR-92a by antagomir has improved cardiac function and reduced cardiomyocyte apoptosis after MI in mice [12]. miR-21 serves as an anti-apoptotic factor in MI animal models by targeting PDCD4 and repressing its expression. Interestingly, miR-21 seems to target cardiac fibroblasts, not cardiomyocytes, in the heart [13]. Conversely, miR-320 is downregulated after I/R injury. Gain- and loss-of-function studies demonstrated that miR-320 promotes cardiomyocyte apoptosis via maintaining HSP20 levels [14]. Together, these studies establish miRNAs as key regulators of cardiomyocyte survival and cardiac remodeling in response to pathophysiological stresses.

44.4 LncRNAs in Cardiac Development

While many lncRNAs have recently been discovered, relatively little is known about their function. A novel lncRNA, *Braveheart*, has been defined as a critical regulator of cardiovascular commitment from embryonic stem cells (ESCs) [15]. *Braveheart* activates a cardiovascular gene network and functions upstream of mesoderm posterior 1, a master regulator of a common multipotent cardiovascular progenitor. *Braveheart* mediates the epigenetic regulation of cardiac commitment by interacting with SUZ12, a component of the polycomb repressive complex 2 (PRC2). *Braveheart* therefore represents the first lncRNA that defines cardiac cell fate and lineage specificity, linking lncRNAs to cardiac development and disease. It remains to be seen if *Braveheart* is required for normal heart development in vivo. Equally critically, it will be important to determine whether genetic mutation of the *Braveheart* gene is linked to human cardiovascular disorders. Nevertheless, the discovery of *Braveheart* will significantly impact the cardiovascular research field and link lncRNAs to human cardiovascular disease.

Fendrr, another novel lncRNA, has been defined as an essential regulator of heart and body wall development. *Fendrr* is expressed in the mouse lateral plate mesoderm, from which precursors for the heart and body wall are derived, and the knockout of *Fendrr* resulted in defects in heart development [16]. Like *Braveheart*, *Fendrr* interacts with the PRC2 complex to regulate gene expression. It is expected that many more lncRNAs will be found to play important roles in cardiovascular development and function.

44.5   Noncoding RNAs in Cardiac Disease

The expression and function of multiple miRNAs have been associated with human cardiovascular disease. Recent studies also linked several lncRNAs to heart disease. *ANRIL*, a lncRNA, was identified as a risk factor for coronary disease [17]. Though it is still not fully understood how ANRIL functions, evidence suggests that this lncRNA may participate in the regulation of histone methylation [18]. Another lncRNA MIAT (myocardial infarction-associated transcript) (or *Gomafu/RNCR2*) was identified as a risk factor associated with patients with myocardial infarction [19]. However, how MIAT controls MI status remains largely unknown. Intriguingly, the genetic loci that encode MYH6 and MYH7, the main myosin heavy chain genes in cardiac muscle, appear to produce a noncoding antisense transcript (*Myh7-as*). *Myh7-as* transcription may regulate the ratio of *Myh6* and *Myh7*, altering the function of muscle contraction [20].

We have just started the era of "noncoding." We are looking forward to see more and more reports on the roles of noncoding RNAs in the regulation of a variety of essential biological processes. Furthermore, with efficient strategies for gain- and loss-of-function investigations, more fruitful work about the molecular mechanism and therapeutic application of noncoding RNAs in cardiovascular disease will emerge.

Acknowledgments Work in Dr. Wang's laboratory was supported by the March of Dimes Foundation and National Institutes of Health. We thank FeiFei Wang for critical reading of this commentary. Masaharu Kataoka was supported by Banyu Life Science Foundation International. DZ Wang is an established investigator of the American Heart Association.

References

1. Espinoza-Lewis RA, Wang DZ. MicroRNAs in heart development. Curr Top Dev Biol. 2012;100:279–317. doi:10.1016/B978-0-12-387786-4.00009-9.
2. Ulitsky I, Bartel DP. lincRNAs: genomics, evolution, and mechanisms. Cell. 2013;154:26–46. doi:10.1016/j.cell.2013.06.020.

3. Bernstein E, Kim SY, Carmell MA, et al. Dicer is essential for mouse development. Nat Genet. 2003;35:215–7.
4. Zhao Y, Ransom JF, Li A, et al. Dysregulation of cardiogenesis, cardiac conduction, and cell cycle in mice lacking miRNA-1-2. Cell. 2007;129:303–17.
5. Chen JF, Murchison EP, Tang R, et al. Targeted deletion of Dicer in the heart leads to dilated cardiomyopathy and heart failure. Proc Natl Acad Sci U S A. 2008;105:2111–6. doi:10.1073/pnas.0710228105.
6. Heidersbach A, Saxby C, Carver-Moore K, et al. MicroRNA-1 regulates sarcomere formation and suppresses smooth muscle gene expression in the mammalian heart. Elife. 2013. doi:10.7554/eLife.01323.
7. Eulalio A, Mano M, Dal Ferro M, et al. Functional screening identifies miRNAs inducing cardiac regeneration. Nature. 2012;492:376–81. doi:10.1038/nature11739.
8. Chen J, Huang ZP, Seok HY, et al. mir-17-92 cluster is required for and sufficient to induce cardiomyocyte proliferation in postnatal and adult hearts. Circ Res. 2013;112:1557–66.
9. Small EM, Olson EN. Pervasive roles of microRNAs in cardiovascular biology. Nature. 2011;469:336–42. doi:10.1038/nature09783.
10. Huang ZP, Chen J, Seok HY, et al. MicroRNA-22 regulates cardiac hypertrophy and remodeling in response to stress. Circ Res. 2013;112:1234–43. doi:10.1161/CIRCRESAHA.112.300658.
11. Gurha P, Abreu-Goodger C, Wang T, et al. Targeted deletion of microRNA-22 promotes stress-induced cardiac dilation and contractile dysfunction. Circulation. 2012;125:2751–61. doi:10.1161/CIRCULATIONAHA.111.044354.
12. Bonauer A, Carmona G, Iwasaki M, et al. MicroRNA-92a controls angiogenesis and functional recovery of ischemic tissues in mice. Science. 2009;324:1710–3. doi:10.1126/science.1174381.
13. Dong S, Cheng Y, Yang J, et al. MicroRNA expression signature and the role of microRNA-21 in the early phase of acute myocardial infarction. J Biol Chem. 2009;284:29514–25. doi:10.1074/jbc.M109.027896.
14. Ren XP, Wu J, Wang X, et al. MicroRNA-320 is involved in the regulation of cardiac ischemia/reperfusion injury by targeting heat-shock protein 20. Circulation. 2009;119:2357–66. doi:10.1161/CIRCULATIONAHA.108.814145.
15. Klattenhoff CA, Scheuermann JC, Surface LE, et al. Braveheart, a long noncoding RNA required for cardiovascular lineage commitment. Cell. 2013;152:570–83. doi:10.1016/j.cell.2013.01.003.
16. Grote P, Wittler L, Hendrix D, et al. The tissue-specific lncRNA Fendrr is an essential regulator of heart and body wall development in the mouse. Dev Cell. 2013;24:206–14. doi:10.1016/j.devcel.2012.12.012.
17. Broadbent HM, Peden JF, Lorkowski S, et al. Susceptibility to coronary artery disease and diabetes is encoded by distinct, tightly linked SNPs in the ANRIL locus on chromosome 9p. Hum Mol Genet. 2008;17:806–14.
18. Yap KL, Li S, Muñoz-Cabello AM, et al. Molecular interplay of the noncoding RNA ANRIL and methylated histone H3 lysine 27 by polycomb CBX7 in transcriptional silencing of INK4a. Mol Cell. 2010;38:662–74. doi:10.1016/j.molcel.2010.03.021.
19. Ishii N, Ozaki K, Sato H, et al. Identification of a novel non-coding RNA, MIAT, that confers risk of myocardial infarction. J Hum Genet. 2006;51:1087–99.
20. Pandya K, Smithies O. β-MyHC and cardiac hypertrophy: size does matter. Circ Res. 2011;109:609–10. doi:10.1161/CIRCRESAHA.111.252619.

iPS Cells and Regeneration in Congenital Heart Diseases

Perspective

Deepak Srivastava

Human adult somatic cells can be reprogrammed to induced pluripotent stem (iPS) cells upon the introduction of four transcription factors that are part of the pluripotency network in embryonic stem cells. The ability to readily from patients with disease has ushered in new opportunities to understand disease mechanisms, screen for therapeutics, and consider regenerative approaches using personalized human cells. In this session, several examples of the use of iPS cells in novel ways were presented.

By making iPS cells from patients with genetically defined disease, the investigators were able to differentiate the pluripotent cells into the cell type affected by the congenital cardiovascular disorder. These cells carried the disease-causing mutation and provided a platform for understanding the cellular and molecular consequences of the mutation in the most relevant human cells. Deep interrogation of such cells promises to reveal fundamental mechanisms of disease and should point to new targets to intervene in the disease process. This is being done for diseases involving cardiomyocytes, smooth muscle cells, and endothelial cells, each of which can be easily differentiated from human iPS cells with good efficiency and purity. Once new targets for disease pathology are discovered in such cells, small molecule or biologic screens can be performed to identify lead

D. Srivastava, M.D. (✉)
Gladstone Institute of Cardiovascular Disease, San Francisco, CA 94158, USA

Department of Pediatrics and Department of Biochemistry and Biophysics, University of California, San Francisco, San Francisco, CA 94158, USA
e-mail: dsrivastava@gladstone.ucsf.edu

candidates for new therapeutics. For those congenital diseases that have ongoing consequences after birth, there is potential to intervene postnatally in the disease evolution.

In addition to the use of iPS cells for disease modeling and drug discovery, there are robust efforts to use pluripotent stem cells for regenerative medicine. Such efforts often involve bioengineering approaches to assemble stem cell-derived cardiomyocytes into a three-dimensional structure. This can be useful for cardiomyocytes, valves, or vessels. The use of iPS cells may allow personalized tissues to be developed, as tissue could be generated with one's own cells. New approaches using efficient gene-editing techniques may allow correction of abnormal genes and subsequent use of corrected cells for transplant. Other types of progenitor cells are also being studied for their regenerative capacity and are discussed in this section.

While there is great hope that the use of iPS cells will lead to new therapeutic approaches, many hurdles must be overcome. For disease modeling, purifying specific subtypes of cells that are affected by disease will be important, as will the ability to generate more mature, adult-like cells from the iPS cells. For regenerative medicine approaches, the ability to generate mature cells that can survive and integrate upon transplantation will be critical and will likely require clever engineering strategies. Nevertheless, it is likely that iPS-based technologies will provide us a better understanding of human disease and lead to new interventions.

Human Pluripotent Stem Cells to Model Congenital Heart Disease

Seema Mital

Abstract

Congenital heart disease (CHD) is the most common cause of neonatal mortality related to birth defects. Etiology is multifactorial including genetic and/or environmental causes. The genetic etiology is known in less than 20 % cases. Animal studies have identified genes involved in cardiac development. However, generating cardiac phenotypes usually requires complete gene knockdown in animal models which does not reflect the haplo-insufficient model commonly seen in human CHD. Human pluripotent stem cells which include human embryonic stem cells (hESC) and human-induced pluripotent stem cells (hiPSC) provide a unique in vitro platform to study human "disease in a dish" by providing a renewable resource of cells that can be differentiated into virtually any somatic cell type in the body. This chapter will discuss the use of human pluripotent stem cells to model human CHD.

Keywords

Human embryonic stem cells • Induced pluripotent stem cells • Williams syndrome • Hypoplastic left heart syndrome • Fetal reprogramming

45.1 Introduction

Human embryonic stem cells (hESC) can give rise to all three germ layers – ectoderm, endoderm, and mesoderm – and can be used to generate differentiated cells of different lineages [1]. The Nobel prize-winning discovery by Yamanaka of the ability to reprogram somatic cells to induced pluripotent stem cells (iPSC) using

S. Mital, M.D. (✉)
Division of Cardiology, Department of Pediatrics, Hospital for Sick Children, University of Toronto, 555 University Avenue, Toronto, Ontario M5G 1X8, Canada
e-mail: seema.mital@sickkids.ca

© The Author(s) 2016
T. Nakanishi et al. (eds.), *Etiology and Morphogenesis of Congenital Heart Disease*,
DOI 10.1007/978-4-431-54628-3_45

specific reprogramming factors [2] uncovered a whole new field of research focused on the use of iPSCs to model human disease, perform drug screens, and explore strategies for autologous cell-based therapies in the future. Reprogramming protocols include ectopic expression of four transcription factors [2, 3] that induce reprogramming of somatic cells into an embryonic state. Viral integration-free protocols are also used albeit are less efficient. These cells can then be expanded and differentiated into several somatic cell types including cardiac lineages such as cardiomyocytes, vascular smooth muscles cells (SMCs), and endothelial cells. The process of cardiac differentiation of hESCs and hiPSCs recapitulates cardiac embryogenesis thereby providing a unique opportunity to explore the impact of gene or environmental defects on early cardiac development and gain novel insights into disease mechanisms [4]. Strategies for modeling cardiac malformations are discussed.

45.2 Modeling Fetal Cardiac Reprogramming in Hypoplastic Left Heart Syndrome (HLHS)

HLHS is one of the most severe cardiac malformations characterized by poor growth of left-sided cardiac structures. This is commonly associated with endocardial fibroelastosis (EFE). The mechanism of LV growth failure and fibrosis is poorly understood. We studied 29 normal and 30 HLHS fetal hearts during second trimester [5]. We found increased nuclear expression of hypoxia-inducible factor 1α (HIF1α) in fetal HLHS compared to normal LVs, a central hypoxia-responsive gene that promotes activation of angiogenic, metabolic, and other genes to facilitate cardiac adaptation to hypoxia. However, expression of vascular endothelial growth factor (VEGF) was downregulated. The failure of hypoxia-induced angiogenesis was likely related to cell senescence as shown by DNA damage (nuclear γH2AX activation and p53 upregulation) and of cell senescence (β-galactosidase upregulation). Senescent cells, although functional, do not produce growth factors essential for the survival and proliferation of stem/progenitor cells thereby compromising tissue renewal capacity. Not surprisingly, HLHS hearts showed fewer cardiac progenitor markers, as well as reduced differentiated cardiomyocyte and endothelial cells. DNA damage was most prominent in endothelial cells followed by myocytes, with SMCs being least susceptible. Additionally, there was increased transforming growth factor (TGFβ1) expression, increased myofibroblast transformation, and increased interstitial and perivascular fibrosis in fetal HLHS compared to controls. Together this suggested that the fetal LV may be susceptible to chronic hypoxia or reduced blood flow (a phenomenon that occurs in HLHS due to reduced antegrade flow through the diminutive ascending aorta) resulting in DNA damage and cell senescence and consequent loss of cell replication and growth capacity as well as fibrosis.

To clarify the role of hypoxia in fetal cardiac differentiation, we exposed hESC-derived cardiac lineages to 1 % hypoxia for 72 h. This was associated with recapitulation of the fetal HLHS phenotype including increased HIF1α; reduced

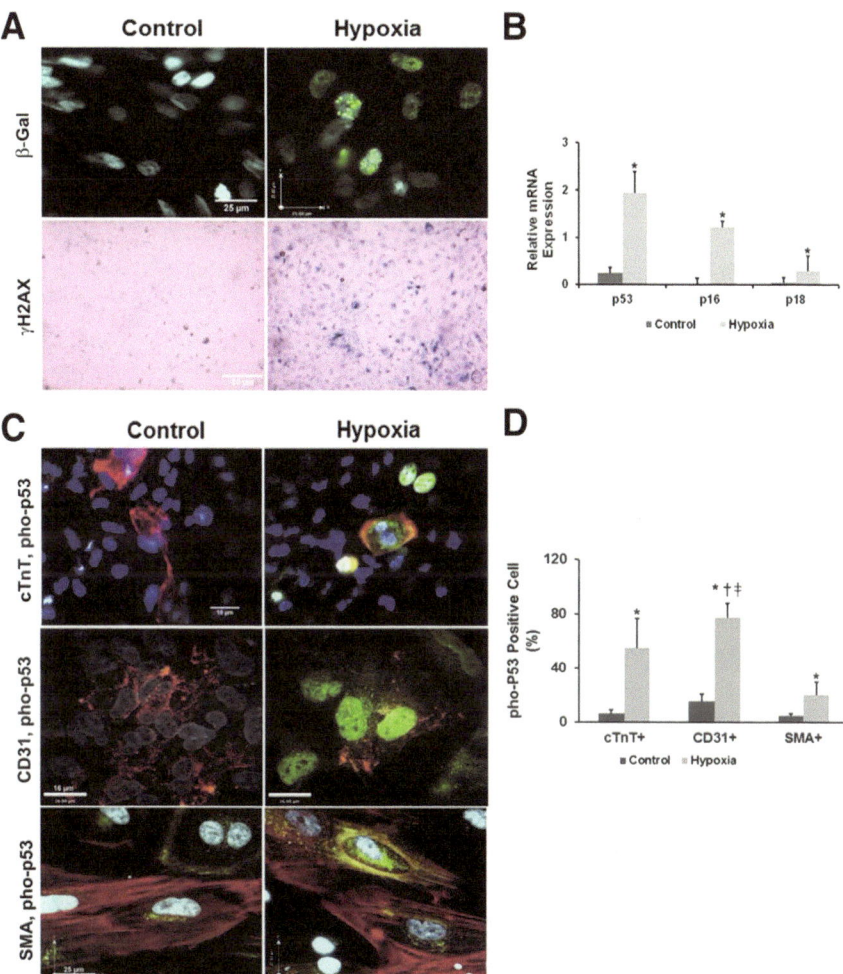

Fig. 45.1 Effect of hypoxia on DNA damage and oncogene upregulation in hPSC-derived cardiac lineages. (**a**) DNA damage-related marker γH2AX (*green nuclear foci*) and senescence marker β-gal (*blue*) are increased in hypoxic cells. *Blue* represents nuclear staining with DAPI. (**b**) qPCR results revealed higher mRNA expression of the tumor suppressor oncogene p53 and the G1 cell cycle inhibitors p16 and p18 in hypoxic (*gray bars*) compared with control cells (*black bars*). (**c**) Double immunostaining revealed co-localization of ph-p53 (*green*) with cTnT+ myocytes (*red*), CD31+ endothelial cells (*red*), and SMA+ SMCs (*red*), indicating DNA damage in all three lineages. *Blue* represents nuclear staining with TO-PRO-3. (**d**) Cellomics quantification confirmed the higher number of phospho-p53+ cardiac lineages in hypoxic cells (*gray bars*) compared with controls (*black bars*), with most severe injury in ECs followed by myocytes and then SMCs. ∗$P < 0.01$ versus controls; †$P < 0.05$ versus SMA+ cells; ‡$P < 0.05$ versus cTnT+ cells ($n = 3$ experiments in each group). Original magnification: ×1,000 (γH2AX); ×600 (β-gal) (**a**); ×1,000 (**c**) (Reproduced with permission) [5]. Reprinted from Gaber et al. [5], Copyright (2013), with permission from Elsevier [5])

VEGF; TGFβ1 upregulation; DNA damage (highest in endothelial cells followed by myocytes followed by SMCs); cell senescence; reduced cell proliferation, resulting in a reduction in myocyte and endothelial lineages but increase in SMC lineages; and reduced contractility (Fig. 45.1). Treatment with TGFβ1 inhibitor reversed this abnormal phenotype. This suggests that immature cardiac lineages may be susceptible to hypoxic injury and that this may be mediated in part by TGFβ1 activation. This may contribute to the phenotype of LV growth failure and fibrosis in cardiac malformations like fetal HLHS.

These findings have several implications. They suggest that antenatal intervention for HLHS may be more effective in promoting LV growth if performed before irreversible tissue injury. However, complementary strategies to provide missing growth factors and/or inhibit TGFβ1 either pre- or postnatally may be needed to promote LV growth and ameliorate progressive fibrosis.

45.3   hiPSCs to Model Williams-Beuren Syndrome (WBS)

WBS is a genetic disorder caused by deletion of 26–28 genes in the 7q11.23 region. Cardiac manifestations are common and are related primarily to haploinsufficiency of the elastin gene in the deleted region. Elastin insufficiency causes vascular SMC proliferation resulting in either generalized arteriopathy or discrete arterial stenoses including supravalvar aortic stenosis, coronary stenosis, pulmonary stenosis, and renal artery stenosis [6]. Surgical correction is often associated with recurrence of stenosis, and there are no medical therapies to prevent or reduce vascular stenoses. Mouse models require complete elastin gene knockdown to reproduce supravalvar aortic stenosis. We therefore generated iPSCs from a patient with WBS with supravalvar aortic stenosis to provide a more human-relevant model for study. Skin fibroblasts obtained at the time of surgery were reprogrammed using four factor retroviral reprogramming. Four iPSC lines were characterized for pluripotency and subjected to SMC differentiation using a published protocol [7]. SMCs generated from iPSCs from normal BJ fibroblasts showed high elastin expression, with 90 % positive for SM22α (a marker of SMC differentiation). These cells showed a good contractile response (Ca^{2+} flux) to a vasoactive agonist like endothelin and tube-forming capacity on Matrigel assay. In contrast, WBS iPSC-derived SMCs showed low elastin expression, had fewer SM22α-positive cells, were highly proliferative, showed poor tube-forming capacity on Matrigel, and did not show a contractile response to endothelin (Fig. 45.2) [8]. Treatment with rapamycin, a mTOR inhibitor and antiproliferative agent, showed partial rescue of the abnormal phenotype in WBS-SMCs by enhancing differentiation, reducing proliferation, and improving tube-forming capacity. However, it did not restore contractile response to endothelin. Ge et al. used a similar approach to generate and study SMCs from a patient with supravalvar aortic stenosis with WBS and another with elastin loss-of-function mutation that showed a similar phenotype that was rescued by ERK1/2 inhibition [9]. To identify additional compounds that not only improve SMC differentiation but also promote functional maturation and

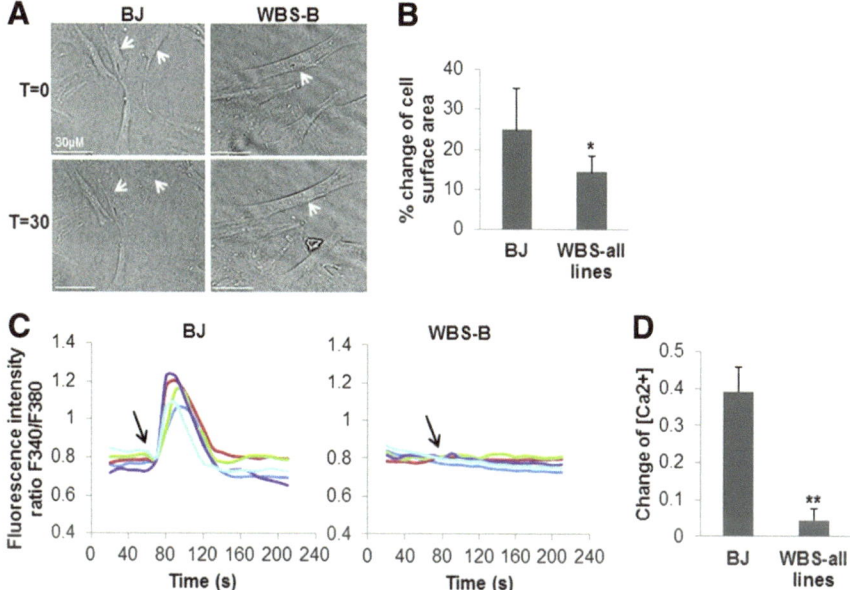

Fig. 45.2 Functional characterization of BJ-smooth muscle cells (SMCs) and WBS-SMCs. (**a**) BJ and WBS-SMCs (line B shown) were treated with 10 mM carbachol, a muscarinic agonist, and phase-contrast live-cell imaging was done every 30 s. Change in cell surface area (*white arrows*) was calculated from 0 min (*top panel*) to 30 min (*bottom panel*). (**b**) BJ-SMCs showed a 25 % reduction in cell surface area compared with 14 % reduction in WBS-SMCs (average of all four lines). *, $p < 0.05$ BJ versus WBS. (**c**) Calcium flux $[Ca^{2+}]_i$ was measured in response to endothelin-1 treatment (*arrow*) in BJ and WBS-SMCs (five cells each). The fluorescence intensity ratio ($F_{340\ nm}/F_{380\ nm}$) showed a transient rise in $[Ca^{2+}]_i$ after activation by endothelin-1 in BJ-SMCs but not in WBS-B SMCs. (**d**) Graph showing the changes in $[Ca^{2+}]_i$ following endothelin-1 treatment in BJ and all the WBS lines. Changes of $[Ca^{2+}]_i$ = peak $[Ca^{2+}]_i$ − resting $[Ca^{2+}]_i$. *, $p < 0.01$ BJ versus WBS. *WBS* Williams-Beuren syndrome (Reprinted from Kinnear et al. [8], Copyright (2013), with permission from Alpha Med Press [8] (pending))

vasoactive responsiveness, we are developing a high-throughput high-content screening assay to facilitate screening of drug libraries using WBS-SMCs. Compounds that fully rescue the abnormal SMC phenotype in WBS may guide the development of new drugs to relieve vascular stenoses in WBS and, by extension, in other vascular disorders including atherosclerosis, stent restenosis, and transplant graft vasculopathy.

45.4 Future Directions and Clinical Applications

These studies provide proof of principle that hESCs and iPSCs can generate in vitro models to study CHD. However, the cardiac lineages generated using this approach are relatively immature, i.e., fetal stage. While fetal stage cells may be well suited to study developmental cardiac disorders [10], maturation protocols that generate more functionally mature lineages may be more useful to study late-onset disease phenotypes and accurately evaluate drug responses [11–14]. Our study further suggests that the technology can be expanded to study not just genetic influences, particularly in the rapidly emerging era of genome editing [15], but also environmental teratogens (toxins, chemicals, drugs, infections) to define the mechanisms by which they impact fetal cardiac development or differentiation. This may facilitate delineating the combined role of genetic and environmental factors in CHD causation in the near future [16]. The ability to differentiate pluripotent stem cells into many different organ or cell types may allow the study not only of cardiac but also of extracardiac phenotypes particularly in syndromic disorders as recently shown in a patient with Timothy syndrome [17, 18].

In summary, pluripotent stem cell-derived models are revolutionizing our understanding of disease pathogenesis and are positioned to expedite drug screening and discovery particularly for rare cardiac disorders with a genetic basis for which no therapies are available and where clinical studies are challenging. The technology provides a renewable source of functional cardiomyocytes and other cardiac lineages with genetic and epigenetic variation that are likely to be more human relevant. While the use of these cells for in vivo therapies is several years away, this platform is well positioned to study the molecular underpinnings of genetic cardiac disorders and help identify new therapies for personalized care of the affected child.

Acknowledgments The work was funded by the Canadian Institute of Health Research (MOP 126146), SickKids Labatt Family Heart Centre Innovations fund, and the Ontario Ministry of Research and Innovation GL2 award.

References

1. Thomson JA, Itskovitz-Eldor J, Shapiro SS, et al. Embryonic Stem Cell Lines Derived from Human Blastocysts. Science. 1998;282:1145–1147.
2. Takahashi K, Yamanaka S. Induction of pluripotent stem cells from mouse embryonic and adult fibroblast cultures by defined factors. Cell. 2006;126:663–676.
3. Yu J. Induced pluripotent stem cell lines derived from human somatic cells. Science. 2007;318:1917–1920.
4. Yang L, Soonpaa MH, Adler ED, et al. Human cardiovascular progenitor cells develop from a KDR+ embryonic-stem-cell-derived population. Nature. 2008;453:524.
5. Gaber N, Gagliardi M, Patel P, Kinnear C, Zhang C, Chitayat D, Shannon P, Jaeggi E, Tabori U, Keller G, Mital S. Fetal reprogramming and senescence in hypoplastic left heart

syndrome and in human pluripotent stem cells during cardiac differentiation. Am J Pathol. 2013;183(3):720–34.

6. Pober BR. Williams–Beuren Syndrome. N Engl J Med. 2010;362:239–252.

7. Xie C-Q, Zhang J, Villacorta L, Cui T, Huang H, Chen YE. Arterioscler Thromb Vasc Biol. 2007;27:e311–e312.

8. Kinnear C, Chang WY, Khattak S, Hinek A, Thompson T, de Carvalho Rodrigues D, Kennedy K, Mahmut N, Pasceri P, Stanford WL, Ellis J, Mital S. Modeling and rescue of the vascular phenotype of Williams-Beuren syndrome in patient induced pluripotent stem cells. Stem Cells Transl Med. 2013;2(1):2–15.

9. Ge X, Ren Y, Bartulos O, et al. Modeling Supravalvular Aortic Stenosis Syndrome With Human Induced Pluripotent Stem Cells. Circulation. 2012;126:1695–1704.

10. Zhang J, Wilson GF, Soerens AG, et al. Functional Cardiomyocytes Derived From Human Induced Pluripotent Stem Cells. Circ Res. 2009;104:e30–e41.

11. Ng K-M, Lee Y-K, Lai W-H, et al. Exogenous Expression of Human apoA-I Enhances Cardiac Differentiation of Pluripotent Stem Cells. PLoS ONE. 2011;6:e19787.

12. Paige SL, Osugi T, Afanasiev OK, Pabon L, Reinecke H, Murry CE. Endogenous Wnt/β-Catenin Signaling Is Required for Cardiac Differentiation in Human Embryonic Stem Cells. PLoS ONE. 2010;5:e11134.

13. Otsuji TG, Minami I, Kurose Y, Yamauchi K, Tada M, Nakatsuji N. Progressive maturation in contracting cardiomyocytes derived from human embryonic stem cells: Qualitative effects on electrophysiological responses to drugs. Stem Cell Res. 2010;4:201–213.

14. Gerecht S, Burdick JA, Ferreira LS, Townsend SA, Langer R, Vunjak-Novakovic G. Hyaluronic acid hydrogel for controlled self-renewal and differentiation of human embryonic stem cells. Proc Natl Acad Sci. 2007;104:11298–11303.

15. Musunuru K. Genome editing of human pluripotent stem cells to generate human cellular disease models. Dis Model Mech. 2013;6:896–904.

16. Patel P, Mital S. Stem cells in pediatric cardiology. Eur J Pediatr. 2013;172:1287–1292.

17. Yazawa M, Hsueh B, Jia X, et al. Using induced pluripotent stem cells to investigate cardiac phenotypes in Timothy syndrome. Nature. 2011;471:230–234.

18. Pasca SP, Portmann T, Voineagu I, et al. Using iPSC-derived neurons to uncover cellular phenotypes associated with Timothy syndrome. Nat Med. 2011;17:1657–1662.

Engineered Cardiac Tissues Generated from Immature Cardiac and Stem Cell-Derived Cells: Multiple Approaches and Outcomes

46

Bradley B. Keller, Fei Ye, Fangping Yuan, Hiren Trada,
Joseph P. Tinney, Kevin M. Walsh, and Hidetoshi Masumoto

Abstract

The translation of in vitro engineered cardiac tissues (ECTs) from immature cardiac and stem cell-derived cells toward clinical therapies is benefiting from the following major advances: (1) rapid progress in the generation of immature cardiac cells from the cardiac and noncardiac cells of multiple species including normal and disease human cells, (2) incorporation of multiple cell lineages into 3D tissues, (3) multiple scalable 3D formulations including injectable gels and implantable tissues, and (4) insights into the regulation of cardiomyocyte proliferation and functional maturation. These advances are based on insights gained from investigating the regulation of cardiac morphogenesis and adaptation. Our lab continues to explore this approach, including changes in gene expression that occur in response to mechanical loading and tyrosine kinase inhibition, the incorporation of vascular fragments into ECTs, and the fabrication of porous implantable electrical sensors for in vitro conditioning and postimplantation testing. Significant challenges remain including optimizing ECT survival postimplantation and limited evidence of ECT functional coupling to the recipient myocardium. One clear focus of current research is the optimization and expansion of the cellular constituents, including CM, required for clinical-grade ECTs. Another major area of investigation will be large animal preclinical models that more accurately represent human CV failure and that can generate data in support of regulatory approval for phase I human clinical trials. The generation of reproducible human ECTs creates the opportunity to develop in vitro myocardial surrogate tissues for novel drug therapeutics and toxicity assays.

B.B. Keller (✉) • F. Ye • F. Yuan • H. Trada • J.P. Tinney • K.M. Walsh • H. Masumoto
Kosair Charities Pediatric Heart Research Program, Cardiovascular Innovation Institute,
University of Louisville, 302 E. Muhammad Ali Blvd, Louisville, KY 40202, USA
e-mail: brad.keller@louisville.edu

© The Author(s) 2016
T. Nakanishi et al. (eds.), *Etiology and Morphogenesis of Congenital Heart Disease*,
DOI 10.1007/978-4-431-54628-3_46

Keywords
Cardiomyocytes • Cardiac repair and regeneration • Engineered cardiac tissues •
Stem cells

46.1   Introduction

Following significant cardiac injury, the postnatal human heart lacks the ability to
restore lost myocardium, resulting in an adaptive response that often ultimately
leads to progressive cardiac dysfunction, morbidity, and mortality. There are
currently many strategies for cardiac "cellular therapy" undergoing both preclinical
and clinical trials [1–4]. While there has been modest success with improvement in
cardiac function in some of the early human clinical trials, it is clear that injected or
implanted cells do not survive, and functional improvement occurs via paracrine
mechanisms. In contrast, rapid advances in tissue engineering over the past two
decades have resulted in the generation of functional, multicellular, 3D cardiac
tissues with the potential for translation to human cardiac repair and regeneration
[5–8]. This chapter provides a concise overview of some of the key issues in the
generation, maturation, and translation of these engineered cardiac tissues (ECTs).

46.2   A Broad View of Bioengineering Cardiac Tissues

The bioengineering process for complex tissues begins with an understanding of the
cellular and noncellular constituents of the target tissue [9]. For replacement
myocardium, the major cellular constituents include cardiomyocytes (CM),
fibroblasts, and vessel-associated cells. There are also numerous extracellular
matrix (ECM) constituents including collagen, fibronectin, laminin, and multiple
growth factors bound within the ECM. Of course, the neonatal myocardium and the
adult myocardium have vastly different profiles for cellular and noncellular
constituents, architecture, and biomechanical properties. While all currently successful ECTs are constructed using immature cells and simplified ECM
components, the target tissue is usually mature myocardium. The success of ECT
survival, integration, and functional maturation depends on the ability of these ECT
constituents to acquire "mature" fates.

46.3   Immature Cells for Engineered Cardiac Tissues

Because the goal for cardiac regeneration is the restoration of functionally coupled,
working myocardium, a variety of cell sources with the potential to generate CM
are under investigation. Immature CM can be isolated from the hearts of developing
chick, mouse, and rat embryos to generate ECTs for preclinical investigation
(Fig. 46.1) [5–7]. These cells mature in vivo or in vitro along timelines proportional

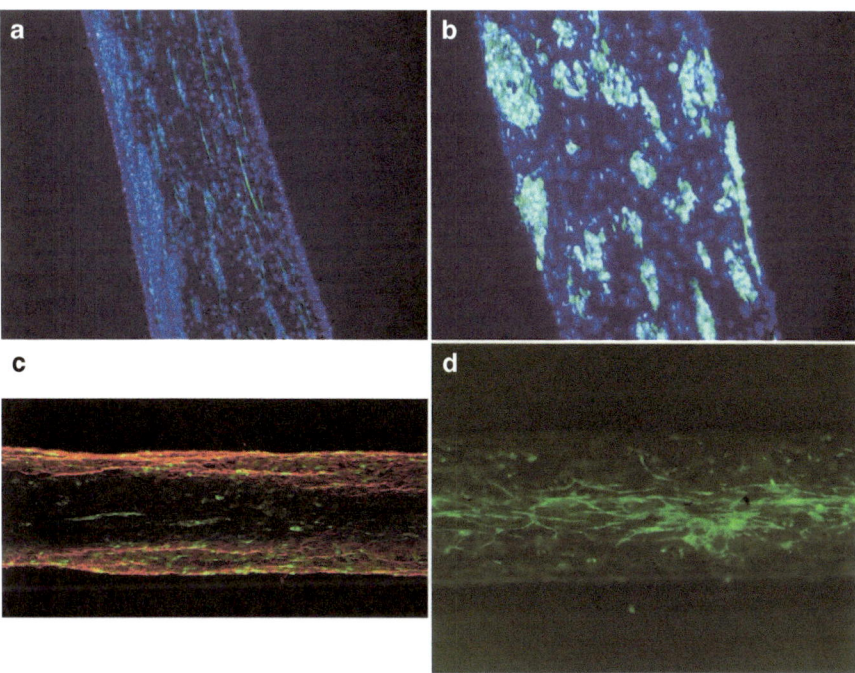

Fig. 46.1 Representative engineered cardiac tissues (ECTs) derived from (**a**) embryonic chick heart cells; (**b**) human-iPS-derived cardiomyocytes; (**c**) embryonic rat heart cells and rat adipose vascular fragments; and (**d**) enlarged image of vascular fragments within a rat ECT. Staining for (**a, b**) are *blue* (DAPI, nuclei), *green* (cardiac troponin T), and *red* (EdU). Staining for (**c, d**) are *red* (alpha actinin) and *green* (GFP+vascular fragments). Images (**a, c**) are 20× magnification; images (**b, d**) are 40× magnification (Keller lab, unpublished)

to the gestational length of their species of origin. A variety of stem cell sources (embryonic stem cells, induced pluripotent stem cells, cardiac stem cells, adipose stem cells, etc.) have also been used to generate immature CM using a variety of CM lineage specification and selection protocols [10–14]. Not surprisingly, stem cell populations can be rapidly expanded in vitro along with the induction of cardiac lineages; however, their functional maturation remains a major technical challenge [15–17]. Because human cells are required for clinical translation, the optimization of protocols that can generate large quantities of functional human CM is a high priority for cardiac repair strategies. Further, there may be advantages to generating ECTs that contain both cardiac and vascular lineage cells to accelerate angiogenesis and vascular perfusion of implanted ECTs [11–13].

46.4   Various Formulations for Engineered Cardiac Tissues

The constructs used for cardiac tissue repair include the implantation of multicellular cardiospheres [18], various formulations of 2D cellular sheets [19–22], and various formulations of 3D tissues [5–8, 13]. The composition of the noncellular constituents varies from minimal constituents for cardiosphere clusters to a range of ECM components [23, 24] and growth factors [25–27] selected for their ability to facilitate CM survival and functional maturation. Some of the ECT formulations allow for in vitro preconditioning strategies that can stimulate cell proliferation and/or maturation [5, 6]. While there can be wide variation in the formulation of ECTs used for preclinical studies, all constituents used to generate ECTs for human use are required to conform to strict FDA regulatory guidelines that include the elimination of all sources for potential infectious agents and/or toxins and the generation of clinical-use materials using good manufacturing practices [28].

46.5   In Vitro ECT Findings

Immature CM survive, proliferate, and functionally mature rapidly within ECTs as quantified by standard measures of cell number, gene expression profiling (Fig. 46.2) [29], sarcomeric protein content, electrophysiologic properties, and the ability to generate substantial force [5–7, 13].

Maturing functional syncytia can include CM, myofibroblasts, and vascular cells with a functional advantage noted for multiple lineage constructs over pure CM-derived ECTs [11–13]. Conditioning protocols with the intent of accelerating

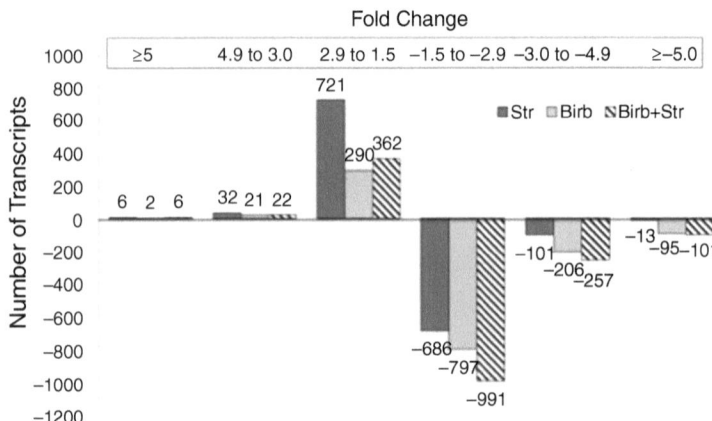

Fig. 46.2 Changes in rat ECT gene expression in response to mechanical loading and/or p38MAPK inhibition. ECT transcript expression changes at least 1.5-fold measured by microarray in response to stretch (*dark solid bar*), the p38MAPK inhibitor BIRB796 (*gray solid bar*), or stretch+BIRB796 (*dashed bar*). Note that most transcripts increased by less than threefold (above the X-axis) or decreased by less than fivefold (below the X-axis) [29]

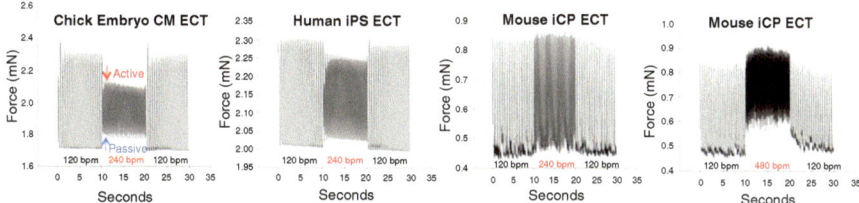

Fig. 46.3 Representative force-frequency relations for chick embryo-derived ECT and human-iPS-CM-derived ECT. Force-frequency relations quantify force generation in response to increased beat rate and reflect ECT maturational ability to release and restore Ca^{2+}. Chick embryo and h-iPS ECT showed increased passive force and reduced active force as beat rate increased from 120 to 240 bpm consistent with immature Ca^{2+} release during contraction and limited Ca^{2+} sequestration at faster rates (Keller lab, unpublished)

cell proliferation and/or maturation include the use of exogenous paracrine factors [15, 26], electrical stimulation [17, 27], and/or mechanical conditioning [6, 13]. Structural CM maturation is readily documented using immunohistochemical stains for sarcomeric proteins, gap junctions, and the presence of ion channels [30]. Functional CM maturation is documented by decreasing pacing voltage thresholds to initiate pacing, intrinsic rates and maximal beat rates in response to electrical stimulation, increasing rates of electrical conduction across ECTs, developed force in response to pacing, and both force-length and force-frequency relations reflecting increased contractility and increased calcium cycling efficiency (Fig. 46.3). ECTs generated from human-derived cells are now proposed as in vitro models for human diseases as well as surrogate models to detect drug toxicity prior to clinical trials [31, 32].

46.6 In Vivo ECT Findings

Ultimately, ECTs require in vivo implantation to assess survival, structural integration to the recipient myocardium, functional integration, and evidence for recovery of lost cardiac function. Preclinical implantation models have been primarily small animals (rodents); however, there is an increasing experience with the preclinical testing of ECTs using large animal models including pigs [22] and, eventually, nonhuman primates. The in vivo results have been very encouraging and confirm the capacity of implanted ECTs to survive, functionally couple, and recover damaged myocardium within the context of the experimental design [33]. There are several challenges to the interpretation of in vivo ECT studies. First, many of the studies involve the implantation of ECTs into immune-compromised animals. These studies are required to validate the capacity of implanted ECTs to functionally couple to recipient myocardium but underestimate the rapid inflammatory degradation that occurs as evidenced by the presence of macrophage-associated arginase and reduced ECT post-implant cellularity, even in syngeneic animals (Fig. 46.4) [13, 30].

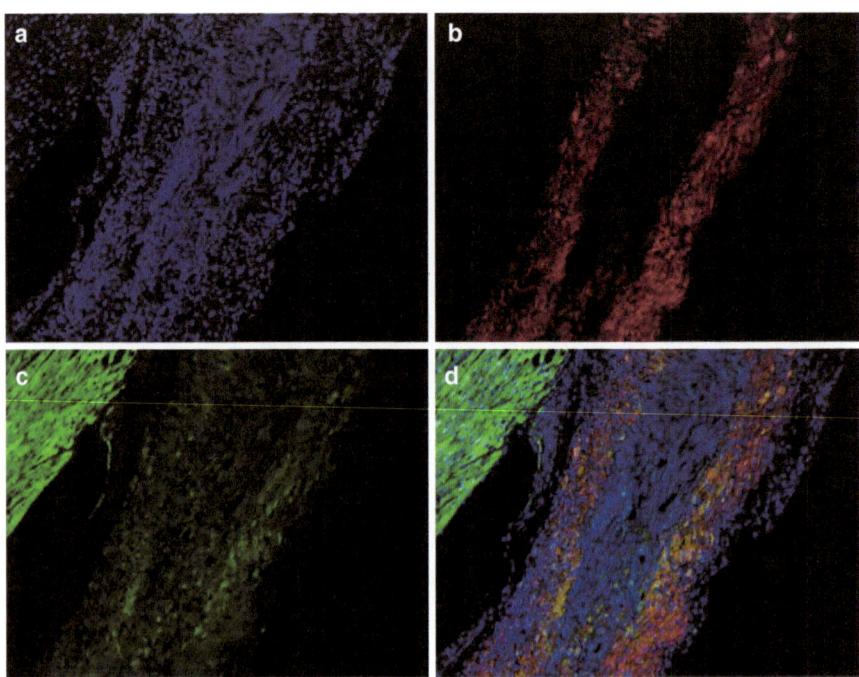

Fig. 46.4 Representative postimplantation histology for rat embryo-derived ECT implanted onto the epicardial surface of a syngeneic adult male rat 2 weeks after coronary artery ligation-induced myocardial infarction. (**a**) *Blue* (DAPI, nuclei), (**b**) *red* (arginase marker for macrophages), (**c**) *green* (GFP+ for recipient myocardium and inflammatory cells), (**d**) merged image. Images are 10× magnification. Note the high cellularity of the recipient myocardium (*upper left corner*) and the implanted ECT. Note the increased presence of arginase along the outer margins of the implanted ECT, consistent with the higher cell density noted in ECT (Keller lab, unpublished)

Second, the in vivo results are often reported after relatively short periods that may not reflect long-term, sustained functional recovery [13]. Finally, the acute surgical models of cardiac injury in preclinical models often do not fully represent the human disease state with medical comorbidities and recurrent episodes of ischemia/injury. To date there are too few studies in large animal models to validate the feasibility of scaling up CM and ECT production for cardiac repair and to compare direct cell delivery strategies to ECT implantation strategies for short- and longer-term efficacy.

46.7   Future Directions

One clear focus of current and future research is the optimization and expansion of the cellular constituents, including CM, required for clinical-grade ECTs. Ultimately, these cells will need to be from human sources and have minimal immunogenic profiles. Another major area of investigation will be large animal preclinical models that more accurately represent human CV failure and that can generate data in support of regulatory approval for phase I human clinical trials. Related to these large animal models will be innovations in the ability to manufacture large-scale ECTs and to implant them with minimally invasive techniques. Although it is beyond the scope of this chapter, the generation of reproducible human ECTs creates the opportunity to develop in vitro myocardial surrogate tissues for novel drug therapeutics and toxicity assays.

References

1. Clifford DM, Fisher SA, Brunskill SJ, et al. Stem cell treatment for acute myocardial infarction. Cochrane Database Syst Rev. 2012;2:CD006536.
2. Kandala J, Upadhyay GA, Pokushalov E, et al. Meta-analysis of stem cell therapy in chronic ischemic cardiomyopathy. Am J Cardiol. 2013;112:217–25.
3. Mielewczik M, Cole GD, Nowbar AN, et al. The C-CURE randomized clinical trial (Cardiopoietic stem Cell therapy in heart failURE). J Am Coll Cardiol. 2013;62(25):2453.
4. Heldman AW, DiFede DL, Fishman JE, et al. Transendocardial mesenchymal stem cells and mononuclear bone marrow cells for ischemic cardiomyopathy: the TAC-HFT randomized trial. JAMA. 2014;311(1):62–73.
5. Zimmermann WH, Melnychenko I, Wasmeier G, et al. Engineered heart tissue grafts improve systolic and diastolic function in infarcted rat hearts. Nat Med. 2006;12(4):452–8.
6. Tobita K, Liu LJ, Janczewski AM, et al. Engineered early embryonic cardiac tissue retains proliferative and contractile properties of developing embryonic myocardium. Am J Physiol Heart Circ Physiol. 2006;291(4):H1829–37.
7. de Lange WJ, Hegge LF, Grimes AC, et al. Neonatal mouse-derived engineered cardiac tissue: a novel model system for studying genetic heart disease. Circ Res. 2011;109(1):8–19.
8. Turnbull IC, Karakikes I, Serrao GW, et al. Advancing functional engineered cardiac tissues toward a preclinical model of human myocardium. FASEB J. 2014;28(2):644–54.
9. Mikos AG, Herring SW, Ochareon P, et al. Engineering complex tissues. Tissue Eng. 2006;12 (12):3307–39.
10. Duan Y, Kettenhofen R, Jovinge S, et al. Engraftment of engineered ES cell-derived cardiomyocytes but not BM cells restores contractile function to the infarcted myocardium. J Exp Med. 2006;203(10):2315–27.
11. Stevens KR, Kreutziger KL, Dupras, et al. Physiological function and transplantation of scaffold-free and vascularized human cardiac muscle tissue. PNAS. 2009;106(39):16568–73.
12. Lesman A, Habib M, Caspi O, et al. Transplantation of a tissue-engineered human vascularized cardiac muscle. Tissue Eng Part A. 2010;16(1):115–25.
13. Tulloch NL, Muskheli V, Razumova MV, et al. Growth of engineered human myocardium with mechanical loading and vascular coculture. Circ Res. 2011;109(1):47–59.

14. Zhang D, Shadrin IY, Lam J, et al. Tissue-engineered cardiac patch for advanced functional maturation of human ESC-derived cardiomyocytes. Biomaterials. 2013;34(23):5813–20.
15. Pillekamp F, Haustein M, Khalil M, et al. Contractile properties of early human embryonic stem cell-derived cardiomyocytes: beta-adrenergic stimulation induces positive chronotropy and lusitropy but not inotropy. Stem Cells Dev. 2012;21(12):2111–21.
16. Yang X, Pabon L, Murry CE. Engineering adolescence: maturation of human pluripotent stem cell-derived cardiomyocytes. Circ Res. 2014;114(3):511–23.
17. Nunes SS, Miklas JW, Liu J, et al. Biowire: a platform for maturation of human pluripotent stem cell-derived cardiomyocytes. Nat Methods. 2013;10(8):781–7.
18. Malliaras K, Makkar RR, Smith RR, et al. Intracoronary cardiosphere-derived cells after myocardial infarction: evidence of therapeutic regeneration in the final 1-year results of the CADUCEUS trial (CArdiosphere-Derived aUtologous stem CElls to reverse ventricUlar dySfunction). J Am Coll Cardiol. 2014;63(2):110–22.
19. Shimizu T, Yamato M, Kikuchi A, Okano T. Cell sheet engineering for myocardial tissue reconstruction. Biomaterials. 2003;24(13):2309–16.
20. Fujita J, Itabashi Y, Seki T, et al. Myocardial cell sheet therapy and cardiac function. Am J Physiol Heart Circ Physiol. 2012;303(10):H1169–82.
21. Matsuura K, Haraguchi Y, Shimizu T, Okano T. Cell sheet transplantation for heart tissue repair. J Control Release. 2013;169(3):336–40.
22. Kawamura M, Miyagawa S, Fukushima S, et al. Enhanced survival of transplanted human induced pluripotent stem cell-derived cardiomyocytes by the combination of cell sheets with the pedicled omental flap technique in a porcine heart. Circulation. 2013;128(11 Suppl 1): S87–94.
23. Black LD, Meyers JD, Weinbaum JS, et al. Cell-induced alignment augments twitch force in fibrin gel-based engineered myocardium via gap junction modification. Tissue Eng Part A. 2009;15(10):3099–108.
24. Yuan Ye K, Sullivan KE, Black LD. Encapsulation of cardiomyocytes in a fibrin hydrogel for cardiac tissue engineering. J Vis Exp. 2011;55:pii:3251.
25. Chiu LL, Radisic M, Vunjak-Novakovic G. Bioactive scaffolds for engineering vascularized cardiac tissues. Macromol Biosci. 2010;10(11):1286–301.
26. Bearzi C, Gargioli C, Baci D, et al. PlGF-MMP9-engineered iPS cells supported on a PEG-fibrinogen hydrogel scaffold possess an enhanced capacity to repair damaged myocardium. Cell Death Dis. 2014;5:e1053.
27. Park H, Larson BL, Kolewe ME, et al. Biomimetic scaffold combined with electrical stimulation and growth factor promotes tissue engineered cardiac development. Exp Cell Res. 2014;321(2):297–306.
28. 7341.002 – inspection of human cells, tissues, and cellular and tissue-based products (HCT/Ps). US Food and Drug Administration. http://www.fda.gov/biologicsbloodvaccines/guidancecomplianceregulatoryinformation/complianceactivities/enforcement/complianceprograms/ucm095207.htm
29. Ye F, Yuan F, Li X, et al. Gene expression profiles in engineered cardiac tissues respond to mechanical loading and inhibition of tyrosine kinases. Physiol Rep. 2013;1(5):e00078.
30. Tiburcy M, Didié M, Boy O, et al. Terminal differentiation, advanced organotypic maturation, and modeling of hypertrophic growth in engineered heart tissue. Circ Res. 2011;109 (10):1105–14.
31. Lalit PA, Hei DJ, Raval AN, et al. Induced pluripotent stem cells for post-myocardial infarction repair: remarkable opportunities and challenges. Circ Res. 2014;114(8):1328–45.
32. Vunjak Novakovic G, Eschenhagen T, et al. Myocardial tissue engineering: in vitro models. Cold Spring Harb Perspect Med. 2014;4(3): pii: a014076.
33. Fujimoto KL, Clause KC, Liu LJ, et al. Engineered fetal cardiac graft preserves its cardiomyocyte proliferation within postinfarcted myocardium and sustains cardiac function. Tissue Eng Part A. 2011;17(5–6):585–96.

Dissecting the Left Heart Hypoplasia by Pluripotent Stem Cells

47

Junko Kobayashi, Shunji Sano, and Hidemasa Oh

Keywords

Hypoplastic left heart syndrome • Pluripotent stem cells • Disease modeling

The genetic background of hypoplastic left heart syndrome (HLHS) is still unknown. Cardiac differentiation from pluripotent stem cells (PSCs) can recapitulate the cardiogenesis in vitro, and PSC technology could be useful to dissect the diseases with the complex mechanisms. In the past few years, some researches were reported to seek the pathogenesis of HLHS by using PSCs. This paper reports the achievements.

1. Gaber N et al. showed that human embryonic stem cells (hESCs) during cardiovascular lineage with hypoxia recapitulated the phenotype of the HLHS heart, which was characterized by increased expression of the oncogenes and TGF-ß1, damaged DNA, and senescence with cell cycle arrest [1]. The phenotypes were rescued by TGF-ß1 inhibition.
2. Jiang Y et al. generated disease-specific induced pluripotent stem cells (iPSCs) from a patient with HLHS [2]. HLHS-iPS-derived cardiomyocytes demonstrated repression of MESP1, TNNT2, and delayed expression of GATA4 compared with hESCs and control-iPSCs. HLHS-iPS-derived cardiomyocyte showed calcium oscillation under caffeine and inositol trisphosphate receptor upregulation, presumably as a result of ryanodine receptor dysfunction.

J. Kobayashi (✉) • S. Sano
Department of Cardiovascular Surgery, Okayama University Graduate School of Medicine, Dentistry, and Pharmaceutical Sciences, Okayama, Japan
e-mail: junko-k@okayama-u.ac.jp

H. Oh
Department of Regenerative Medicine, Center for Innovative Clinical Medicine, Okayama University Hospital, Okayama, Japan

T. Nakanishi et al. (eds.), *Etiology and Morphogenesis of Congenital Heart Disease*,
DOI 10.1007/978-4-431-54628-3_47

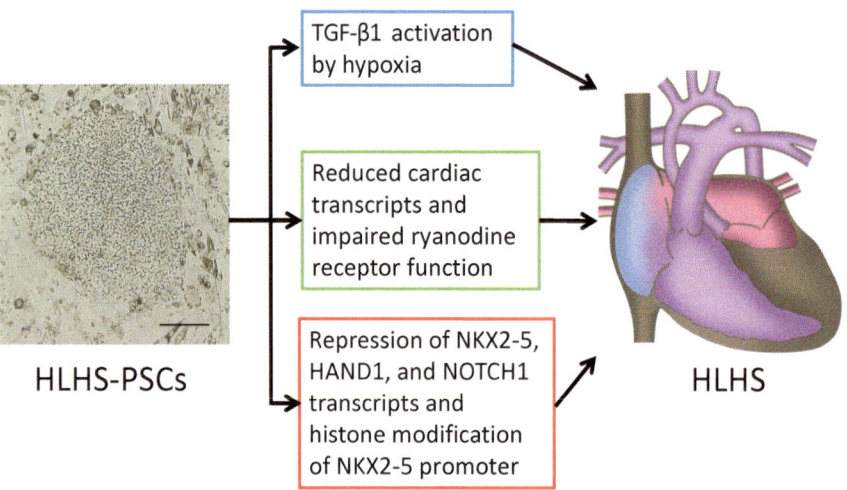

Fig. 47.1 PSC technology models HLHS. Three possible mechanisms of HLHS were unveiled by PSCs. Bar, 200 μm

3. Kobayashi J et al. generated five HLHS-iPSC lines and found repression of the transcripts such as NKX2-5, HAND1, HAND2, NOTCH1, HEY1, HEY2, and TBX2 in HLHS-iPS-derived cardiomyocytes [3]. The promoter activities of SRE, TNNT2, and NPPA were suppressed in HLHS-derived cardiac progenitor cells and iPSCs compared with those from bi-ventricle (BV). All promoter activities of both cell types could be fully restored by co-transfection of NKX2-5, HAND1, and NOTCH1, and co-transfection of the shRNAs into BV-derived cells reduced the promoter activation. HLHS-derived cardiomyocytes demonstrated repressed H3K4me2 and acH3 and increased H3K27me3 in NKX2-5 promoter, implying suppressed NKX2-5 promoter activity.

Taken together, the PSC technology can be useful to dissect the complex heart diseases. Further investigation using this technique is necessary to determine the pathogenesis of HLHS (Fig. 47.1).

References

1. Gaber N, Gagliardi M, Patel P, et al. Fetal reprogramming and senescence in hypoplastic left heart syndrome and in human pluripotent stem cells during cardiac differentiation. Am J Pathol. 2013;183:720–34.
2. Jiang Y, Habibollah S, Tilgner K, et al. An induced pluripotent stem cell model of hypoplastic left heart syndrome (HLHS) reveals multiple expression and functional differences in HLHS-derived cardiac myocytes. Stem Cells Trans Med. 2014;3:416–23.
3. Kobayashi J, Yoshida M, Tarui S, et al. Directed differentiation of patient-specific induced pluripotent stem cells identifies the transcriptional repression and epigenetic modification of NKX2-5, HAND1, and NOTCH1 in hypoplastic left heart syndrome. PLoS One. 2014;9(7): e102796.

Lentiviral Gene Transfer to iPS Cells: Toward the Cardiomyocyte Differentiation of Pompe Disease-Specific iPS Cells

48

Yohei Sato, Takashi Higuchi, Hiroshi Kobayashi, Susumu Minamisawa, Hiroyuki Ida, and Toya Ohashi

Keywords

iPS cells • Lentivirus • Pompe disease

Pompe disease is an inherited neuromuscular disorder caused by a genetic deficiency of acid-glucosidase-alpha (GAA). The clinical symptoms of Pompe disease include progressive weakness, respiratory failure, and ventricular hypertrophy. Enzyme replacement therapy has been shown to ameliorate these symptoms. Cardiomyocytes derived from patient/disease-specific iPS cells (iPS-CMs) have been used for pathophysiological analyses, drug screening, and cell therapy. Our research goal was to generate cardiomyocytes that can be differentiated from gene-corrected Pompe disease-specific iPS cells.

We obtained iPSC (TkDA3-4) generated from human dermal fibroblasts [1]. GAA was cloned into cDNA expressing third-generation lentiviral vectors (CS2-EF1α-GAA). To assess the transfection efficacy, Venus, a YFP variant protein, was also cloned into the vector (CS2-EF1α-Venus). Then, we transfected lentiviral vectors containing GAA to iPSCs at three different concentrations to

Y. Sato (✉) • H. Kobayashi • H. Ida • T. Ohashi
Department of Pediatrics, Jikei University School of Medicine, 3-25-8 Nishishimbashi Minato-ku, Tokyo, Japan

Department of Gene Therapy, Jikei University School of Medicine, 3-25-8 Nishishimbashi Minato-ku, Tokyo, Japan
e-mail: yoheisato@jikei.ac.jp

T. Higuchi
Department of Gene Therapy, Jikei University School of Medicine, 3-25-8 Nishishimbashi Minato-ku, Tokyo, Japan

S. Minamisawa
Department of Cell Physiology, Jikei University School of Medicine, 3-25-8 Nishishimbashi Minato-ku, Tokyo, Japan

© The Author(s) 2016
T. Nakanishi et al. (eds.), *Etiology and Morphogenesis of Congenital Heart Disease*, DOI 10.1007/978-4-431-54628-3_48

341

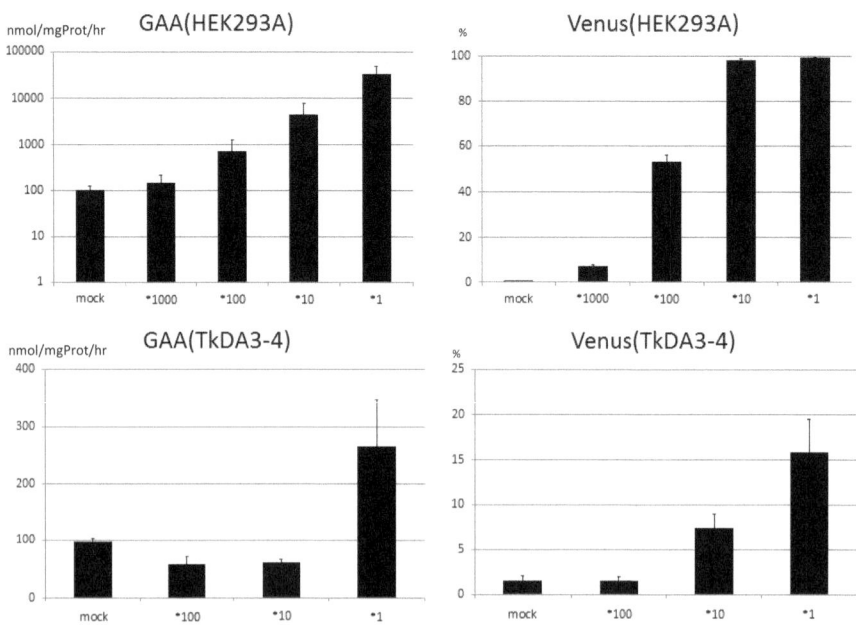

Fig. 48.1 Gene transfer to HEK293A cells and TkDA3-4. Dose-dependent expressions of GAA and Venus were observed in each cell line

determine the optimized titer for gene correction. We showed that dose-dependent expression of both GAA and Venus was observed in iPSCs, even though the expression levels were relatively low compared to HEK293A cells.

Cardiomyocyte differentiation of iPS cells is the most important procedure for replicating the disease hallmarks of Pompe disease. In fact, there is no single best protocol for obtaining cardiomyocytes derived from iPS cells. The functional assessment of iPSC-derived cardiomyocytes is another critical aspect of our research. The differences between the function of iPSC-derived cardiomyocytes obtained from normal control cells and those obtained from Pompe disease cells should therefore be strictly evaluated in order to thoroughly discuss the efficacy of gene therapy for iPSC (Fig. 48.1).

Acknowledgments The author thanks Dr. M. Otsu at Tokyo University for providing iPSC TkDA3-4 cell lines. Third-generation lentiviral vector was a kind gift from Dr. H. Miyoshi at RIKEN BRC, and Venus was kindly provided from A. Miyawaki at RIKEN BSI.

Reference

1. Takayama N, Nishimura S, Nakamura S, et al. Transient activation of c-MYC expression is critical for efficient platelet generation from human induced pluripotent stem cells. J Exp Med. 2010;207(13):2817–30.

Nanako Kawaguchi, Yohtaroh Takagaki, Rumiko Matsuoka, and Toshio Nakanishi

Keywords

Cardiac stem cell • Myocyte • Regeneration • c-Kit • IGF-1

A c-Kit (CD117) is a well-known cell surface marker for adult somatic stem cells. We harvested c-Kit-positive cardiac stem cells (CSCs) from adult rat hearts by performing magnetic-activated cell sorting (MACS) and subjected them to long-term bulk culture more than 40 times. We made 11 attempts to obtain c-Kit-positive cells from adult (6–8-month-old) rats. Our initial expectation was of obtaining cells with homogenous cardiac phenotypes. However, each CSC bulk culture expressed varying degrees of the genes and cell surface markers belonging to cardiac and other mesenchymal lineages. The results suggested that these CSCs retained multiple developmental potential to some extent. Consequently, we investigated these CSCs in detail, hoping to establish the regeneration method by using c-Kit-positive cardiac cells [1–12].

- CSC-21E maintained the cell shape, yielding spherical aggregates under a culture condition. The aggregate shape did not facilitate cell adherence to the dish surface. Interestingly, the proteomic analysis of these two morphological statuses revealed the drastic change of the protein profiles with the spherical aggregates showing protein profiles characteristic of stem cells and the flat cells

N. Kawaguchi (✉) • Y. Takagaki • T. Nakanishi
Division of Pediatric Cardiology, Tokyo Women's Medical University, 8-1, Kawada-cho, Shinjuku, Tokyo 162-8666, Japan
e-mail: nanao.res@gmail.com

R. Matsuoka
Wakamatsu-Kawada Clinic, Kawada-cho, Shinjuku, Tokyo, Japan

Department of Pediatrics, Faculty of Medicine, Toho University, Tokyo, Japan

showing the profile indicative of differentiated cells, especially of the distinct differences in stress proteins and metabolic enzymes [2, 3, 7].

- CSC5 differentiated into cells with a myocyte/adipocyte mixed phenotype. Members of the transforming growth factor (TGF)-β superfamily were identified as significant regulators of the differentiation of these cells into either adipocytes or myocytes [1–6, 9, 10].

- CSC4A exhibited the ability for sustained contractibility shown by cardiomyocytes that were cocultured with a membrane filter separating cardiomyocytes from CSC4A. This suggests that the CSC4A cells release factors that support cardiomyocyte contraction. Among the cytokines measured in the cocultured medium, insulin-like growth factor-1 (IGF-1) levels appeared to correlate with cardiomyocyte sustenance. However, CSC4A cells do not express IGF-1. This suggests that some other unknown factors are released from CSC4A that can induce IGF expression in cardiomyocytes [1–6, 8, 12].

References

1. Kawaguchi N, Hatta, K, Nakanishi T. 3D-Culture system for heart regeneration and cardiac medicine. BioMed Res. 2013;Int Article ID 895967.
2. Kawaguchi N, Machida M, Hatta K et al. Cell shape and cardiosphere differentiation: A revelation by proteomic profiling. Biochem Res. 2013;Int Article ID:730874.
3. Kawaguchi N, Nakanishi T. Cardiomyocyte regeneration using stem cells. Cells. 2013;2: Article ID 6782.
4. Kawaguchi N, Hayama E, Furutani Y, et al. Prospective in vitro disease model for cardiomyopathies and cardiochannelopathies. Stem Cell. 2012;Int Article ID:439219.
5. Kawaguchi N. Differentiation and survival regulators on adult cardiac derived stem cells. Vitam Horm Stem Cell Regul. 2012;87:111–25.
6. Kawaguchi N. Stem cells for cardiac regeneration and possible roles of the transforming growth factor (TGF)-β family. Biomol Concepts. 2011;3:99–106.
7. Machida M, Takagaki Y, Matsuoka R, et al. Proteomic comparison of floating spherical aggregates and dish-attached cells of cardiac stem cells. Int J Cardiol. 2011;153:296–305.
8. Kawaguchi N, Smith A, Waring C, et al. Gata4^{high} CSCs foster cardiac myocyte survival. PLoS One. 2010;5:e1429.
9. Kawaguchi N, Nakao R, Yamaguchi M. TGF-β Superfamily regulates a switch that mediates differentiation either into adipocytes or myocytes in left atrium derived pluripotent cells (LA-PCs). Biochem Biophys Res Commun. 2010;396:615–25.
10. Hasan MK, Komoike Y, Tsunesumi S, et al. Myogenic differentiation in atrium-derived adult cardiac pluripotent cells and the transcriptional regulation of GATA4 and myogenin on ANP promoter. Genes Cells. 2010;15:439–53.

11. Hosseinkhani H, Hosseinkhani M, Hattori S, et al. Micro and nano-scale *in vitro* 3D culture system for cardiac stem cells. J Biomed Mater Res A. 2010;94A:1–8.
12. Miyamoto S, Kawaguchi N, Ellison G, et al. Characterization of long-term cultured cardiac stem cells (CSCs) derived from adult rat hearts. Stem Cells Dev. 2010;19:105–16.

Minor Contribution of Cardiac Progenitor Cells in Neonatal Heart Regeneration

Wataru Kimura, Shalini A. Muralidhar, and SuWannee Thet

Keywords
Heart regeneration • Cardiomyocyte proliferation • Cardiac progenitor cells

The adult mammalian heart is incapable of regeneration after injury, as shown by the limited amount of cardiomyocyte proliferation and poor neovascularization. We recently showed that neonatal mice have a remarkable ability to regenerate damaged heart after apical resection or myocardial infarction (MI), which includes complete reconstruction of myocardial wall with vascular network [2, 3]. Although lineage tracing showed that the main source of newly formed cardiomyocyte is preexisting cardiomyocytes, it is still possible that there is a minor contribution of other types of cells to the cardiomyocyte. In addition, lineage origin of the newly formed vasculature during postnatal cardiac maturation and neonatal heart regeneration remains unclear (Fig. 50.1).

In order to trace the lineage of non-myocyte-derived cells during neonatal heart regeneration, we utilized Rosa26-tdTomato reporter mouse line crossed with capsulin-merCremer line in which epicardial cells and interstitial fibroblasts are labeled specifically and irreversibly after induction with tamoxifen [1]. At postnatal day 0 (P0), Cre was activated by intraperitoneal injection of tamoxifen, and then MI was induced 2 days later (P2). Subsequently the hearts were harvested at 21 days after MI and tdTomato expression was examined. tdTomato+ cells were detected in the epicardium, interstitial fibroblasts, vascular endothelium, and smooth muscle in the regenerated heart. Remarkably, we could detect a very small

W. Kimura (✉) • S.A. Muralidhar • S. Thet
Departments of Internal Medicine, The University of Texas Southwestern Medical Center, Dallas, TX 75390, USA
e-mail: Wataru.Kimura@UTSouthwestern.edu; waterkimura@gmail.com

© The Author(s) 2016
T. Nakanishi et al. (eds.), *Etiology and Morphogenesis of Congenital Heart Disease*,
DOI 10.1007/978-4-431-54628-3_50

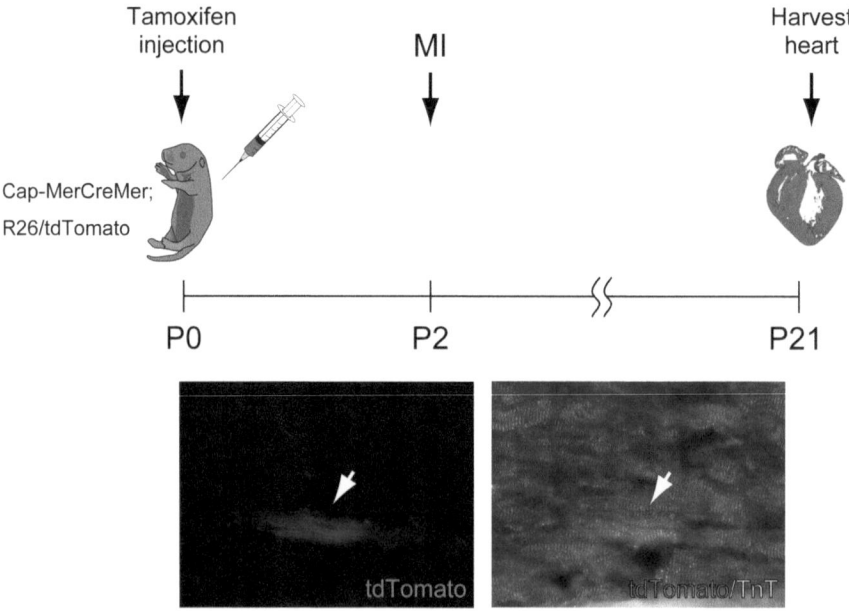

Fig. 50.1 Capsulin-positive cardiac progenitor cells contribute to myocardial lineages during neonatal heart regeneration. Schematics show experimental outline for genetic fate mapping after neonatal MI. *Lower panels* show immunofluorescence on section of 21 days post-MI heart. tdTomato-labeled cells are found in cardiac troponin T (TnT)-positive cardiomyocytes

population (1–2 cells/section) of tdTomato+ cardiomyocyte in the regenerated neonatal heart. No tdTomato+ cardiomyocyte was detected at P21 without inducing MI. These results strongly suggest that capsulin-positive cardiac progenitor cells play important roles during neonatal heart regeneration, primarily in neovasculogenesis by contributing directly to the endothelial/smooth muscle progenitor cells and to a much lesser extent rare myocytes.

References

1. Acharya A, Baek ST, Banfi S, Eskiocak B, Tallquist MD. Efficient inducible Cre-mediated recombination in Tcf21 cell lineages in the heart and kidney. Genesis. 2011;49:870–7.

2. Porrello ER, Mahmoud AI, Simpson E, Hill JA, Richardson JA, Olson EN, et al. Transient regenerative potential of the neonatal mouse heart. Science. 2011;331:1078–80.
3. Porrello ER, Mahmoud AI, Simpson E, Johnson BA, Grinsfelder D, Canseco D, et al. Regulation of neonatal and adult mammalian heart regeneration by the miR-15 family. Proc Natl Acad Sci U S A. 2013;110:187–92.

Perspective

Hiroyuki Yamagishi

Congenital heart disease (CHD) occurs in nearly 1 % of all live births and is the major cause of infant mortality and morbidity; about three per 1,000 live births will require some intervention during the first year of life. Additionally, 95 % of CHD patients survive to adulthood in these days, resulting in a growing population of adult CHD. Despite their clinical importance, the underlying genetic etiology of most CHD remains unknown, so-called "multifactorial" disease.

Identifying genetic cause of CHD is important not only to well understand the disease but also to enhance current knowledge about the molecular biology and genetics involved in the human cardiovascular development. Such knowledge may lead to new preventive and/or therapeutic strategies. Identification of disease genes would benefit the genetic counseling for CHD that is particularly important for the growing population of adult CHD. Deeper understanding of factors and pathways involved in differentiation of the cardiac stem cell and morphogenesis of the cardiovascular system would also provide the development of regenerative therapy for CHD.

During the last two decades, linkage analysis has been used to successfully identify disease genes involved in isolated CHD or genetic syndromes where CHD is part of the phenotype. However, this traditional approach is not generally suitable for CHD because it requires many large families with multiple affected individuals. During the same period, some of syndromes with chromosomal abnormalities have been well studied to identify disease genes in syndromic CHD. CHD is commonly a

H. Yamagishi (✉)
Division of Pediatric Cardiology, Department of Pediatrics, Keio University School of Medicine, Tokyo, Japan
e-mail: hyamag@keio.jp

characteristic part of the clinical spectrum in a significant number of syndromes caused by chromosomal abnormalities, including submicroscopic deletions or duplications. To date, these genetic approaches have led to the identification of more than 50 human genes although they have limitation where large familial cases and chromosomal abnormalities account for relatively small portion of CHD.

Recently, a technique of genome-wide association studies (GWAS) has provided common genetic variations that can influence population-attributable risks of certain types of CHD although the information cannot be directly related to the affected individual. Finally, application of the next-generation sequencing (NGS) technologies is revolutionary in the field of genetics in CHD. In contrast to GWAS, the results of NGS are directly applicable to the affected individual. NGS can be targeted or nontargeted. Exome sequencing and whole-genome sequencing, scanning the whole exome of ~20,000 base pairs and the whole genome of ~3,000,000,000 base pairs, respectively, are well suited to the study of complex, heterogeneous diseases such as CHD and the current best technique for discovery of novel genetic causes for CHD. The biggest obstacle during any NGS analysis is, however, to single out the causal variant from the thousands of variants identified during sequencing. Follow-up animal studies, particularly in mice, for candidate genes discovered by genetic analyses have been successful in validating the candidates and uncovering the function of their gene products for the cardiovascular development. More recently, fine mapping of genomic copy number variants (CNVs) by NGS in patients with isolated or syndromic CHD has been used to identify candidate disease genes.

In this part, authors describe the current advance in genetics in CHD using linkage analysis, chromosomal studies, and CNVs studies by NGS, combined with animal experiments that verified novel genetic causes of CHD and provided new insights into the molecular and functional analyses of the cardiovascular development. Current understandings about molecular pathways associated with CHD involve numerous transcription factors and cofactors, including chromatin modifiers, and signaling molecules from ligands to receptors.

Genetic Discovery for Congenital Heart Defects

51

Bruce D. Gelb

Abstract

Congenital heart disease (CHD) behaves like a complex genetic trait in most instances. Recent advances in genomics have provided tools for uncovering genetic variants underlying complex traits that are now being applied to study CHD. Massively parallel DNA sequencing has shown that de novo mutations contribute to ~10 % of severe CHD and implicated chromatin remodeling in pathogenesis. Genome scanning methods for copy number variants (CNVs) identify likely pathogenic genomic alterations in 10 % of infants with hypoplastic left heart syndrome and related single ventricle forms of CHD. The growth and neurocognitive development of children with CHD and those CNVs is worse, and clinical examination is relatively insensitive for detecting those CNVs. In sum, new opportunities for preventing and ameliorating CHD and its comorbidities are anticipated as its genetic architecture is elaborated through the use of state-of-the-art genomic approaches.

Keywords

Congenital heart disease • Copy number variants • Exome sequencing • Genetics

51.1 Introduction

Congenital heart defects (CHD), with an estimated incidence of 2–3 % when bicuspid aortic valve (BAV) is included, are widely believed to have strong genetic underpinnings. Epidemiologic studies have shown considerable consistency in the

B.D. Gelb, M.D. (✉)

Mindich Child Health and Development Institute, Departments of Pediatrics and Genetics and Genomics Sciences, Icahn School of Medicine at Mount Sinai, One Gustave Levy Place, Box 1040, New York, NY 10029, USA

e-mail: bruce.gelb@mssm.edu

© The Author(s) 2016

T. Nakanishi et al. (eds.), *Etiology and Morphogenesis of Congenital Heart Disease*,

DOI 10.1007/978-4-431-54628-3_51

distribution of CHD lesions across time and geographic location. The landmark
study by Ruth Whittemore, in which she examined recurrence risks for offspring of
women with CHD, revealed a 16 % rate with a 60 % concordance in the form of
CHD between mother and child [1]. Estimates of heritability for BAV and hypoplastic left heart syndrome (HLHS) are 89 % and 95 %, respectively [2, 3]. Recent
studies of CHD from Denmark, where the highly organized medical system enables
population studies, have provide estimates of the relative risks of various forms of
CHD among first-degree relatives that significantly increased, often to >5
[4]. Taken as a whole, these epidemiologic findings point to genetic defects
contributing importantly to CHD etiology.

Identification of the precise mutations has been challenging [5]. We have known
for some time that a modest percentage of CHD (~5 %) is attributable to
aneuploidies such as trisomy 21. With the advent of molecular genetic approaches,
point mutations with apparently strong effects have been identified in rare families
inheriting CHD in Mendelian or near-Mendelian fashion. More recently, the role of
larger genomic events generating pathologic copy number variants (CNVs) for
CHD has become apparent. This started with the recognition of 22q11 deletions
underlying DiGeorge, Takao conotruncal face, and velocardiofacial syndromes.
Several surveys have implicated a wide range of gain and loss CNVs in various
forms of CHD.

Based on the author's oral presentation at the 2013 Takao Symposium, two
recent studies that further elaborate the genetic etiology of CHD will be
reviewed here.

51.2 De Novo Mutations

Through recent advances in molecular genetic technologies, it is now possible to
sequence the roughly 1 % of the human genome that contains the coding regions for all
genes (called the exome), representing approximately 180,000 exons and
30 megabases (Mb), in a relatively rapid and affordable manner. While exome
sequencing was initially used to discover mutations underlying Mendelian disorders,
current efforts are increasingly focusing on unraveling complex genetic traits.

The Pediatric Cardiac Genomics Consortium (PCGC) [6], a National Heart,
Lung, and Blood Institute-funded research enterprise, recently completed a first-
of-kind study to determine the role of de novo mutations in the etiology of severe
forms of CHD [7]. Exome sequencing was performed for 362 parent-offspring trios,
in which the offspring had a sporadic conotruncal defect, left ventricular outflow
track obstructive lesion, or heterotaxy, and compared to comparable data from
264 control trios. While the overall rate of de novo point and small insertion/
deletion (indel) changes was equivalent between CHD cases and controls, there
was an excess burden of protein-altering mutations in genes highly expressed
during heart development (odds ratio (OR) of 2.53). Excess mutations had a role
in 10 % of CHD cases and led to the estimate that ~400 genes underlie these birth
defects. After filtering to retain variants most likely to be deleterious (nonsense,

splice site, and frameshift defects), the burden among CHD cases increased, attaining an OR of 7.50.

Next, the PCGC investigators asked whether the burden of de novo protein-altering mutations among the CHD cases preferentially targeted particular biologic processes [7]. Indeed, they observed a highly significant enrichment of mutation among genes encoding proteins relevant for chromatin biology, specifically the production, removal, or reading of methylation of Lys4 of histone 3 (H3K4me) (Fig. 51.1). The phenotypes of the eight subjects harboring H3K4me de novo mutations were diverse, both with respect to the form of CHD and the involvement of extracardiac tissues. In addition, two independent de novo mutations were found in *SMAD2*, which encodes a protein with relevance for demethylation of Lys27 of histone 3 (H3K27me). SMAD2 contributes to the development of the left-right body axis; both subjects harboring *SMAD2* mutations had dextrocardia with unbalanced complete atrioventricular canal defects with pulmonic stenosis. While the contribution of chromatin remodeling to cardiovascular development generally and

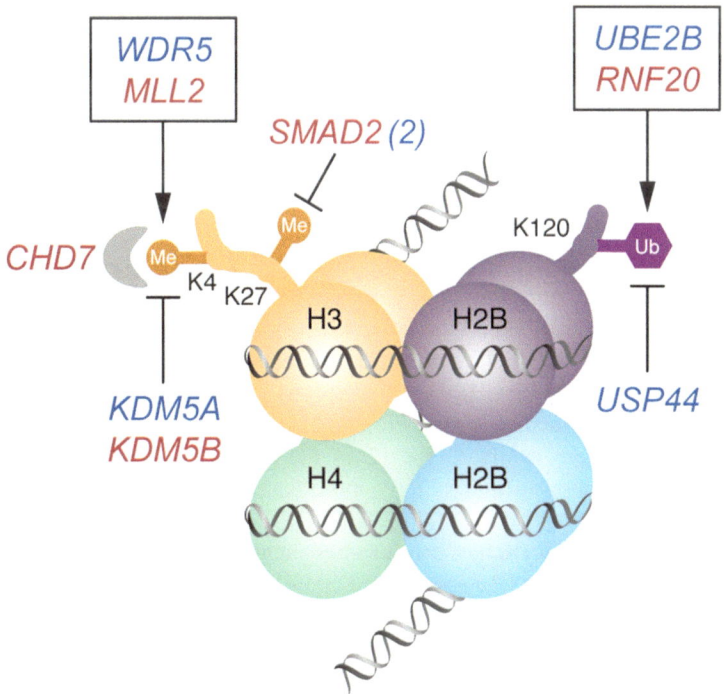

Fig. 51.1 De novo mutations in the H3K4 and H3K27 methylation pathways. Nucleosome with histone octamer and DNA, H3K4 methylation bound by CHD7, and H3K27 methylation and H2BK120 ubiquitination is shown. Genes mutated in CHD that affect the production, removal, and reading of these histone modifications are shown; genes with damaging mutations are shown in red, and those with missense mutations are shown in blue. SMAD2 (2) indicates there are two patients with a mutation in this gene. Genes whose products are found together in a complex are enclosed in a box (Reprinted without modification from Ref. [7])

certain rare genetic syndromes with CHD like Kabuki syndrome had been recognized previously, this study exposed a far broader role in CHD pathogenesis. The finding also suggests a fascinating potential link to other birth defects as de novo chromatin remodeling mutations have also been implicated in autism [8].

51.3 Copy Number Variants

CNVs, which are gains or losses of DNA ranging in size from 1 kb to several Mbs, affect roughly 10 % of the human genome [9]. CNVs are typically detected on a genome-wide basis using SNP microarrays or array comparative genomic hybridization (aCGH). Although differentiating pathogenic CNVs from benign polymorphic one remains challenging, it has become clear that pathologic CNVs contribute significantly to the pathogenesis of CHD as rare large CNVs are observed in 5–15 % of affected individuals [10–14].

To address the issue of how pathogenic CNVs affected outcomes for children with CHD, Carey and colleagues studied children who had previously been subjects in one or both of two clinical trials undertaken by the National Heart, Lung, and Blood Institute-funded Pediatric Heart Network [15]. The two studies were the Infants with Single Ventricle (ISV) and Single Ventricle Reconstruction (SVR) trials, both focusing primarily on infants with HLHS. Carey et al. used aCGH to identify CNVs in 223 subjects from the ISV and SVR trials for whom genomic DNA was available. Because the DNA samples were procured late in those trials, subjects who died earlier, particularly perioperatively, could not be studied. For the CNV work, CNVs were ≥300 kb and not identified as genetic polymorphisms based on existing databases.

Carey and co-workers observed pathogenic CNVs in 13.9 % of the children, which was significantly more than the 4.4 % rate in the controls (preexisting data from blood samples from subjects with solid cancers from The Cancer Genome Atlas project) [15]. For the CHD cases, there was a roughly 4:1 ratio of duplications to deletions and the median sizes were 674 kb and 1.5 Mb, respectively. Parental genomic DNAs were available for 12 of the subjects with pathogenic CNVs; analysis of those showed that slightly more than have of the pathogenic CNVs were inherited.

For the ISV and SVR studies, careful assessment of growth (weight-, height- and head circumference-for-age z-scores) and neurocognitive function (Mental Developmental Index (MDI) and Psychomotor Developmental Index (PDI) measured with the Bayley Scales of Infant Development II) were determined at 14 months of age [15]. Comparison of the CHD subjects harboring pathogenic CNVs to those who did not revealed that the former were significantly shorter by an average of 0.65 z-score. Subgroup analysis showed that those with deletion CNVs had significantly lower PDI scores.

Among the 31 pathogenic CNVs found among the CHD subjects, 13 had previously been associated with genomic disorders. The children harboring these known CNVs had the worst outcomes with globally reduced neurocognitive development (MDI and PDI) as well as the slowest growth (Table 51.1).

Table 51.1 Fourteen-month outcomes for subgroups based on genetic examination

	N^a	MDI	PDI	Weight Z	Length Z	HC Z
CNV- Syndrome- Dysmorphic- Extracardiac-	69	89.1	77.5	−0.71	−1.13	−0.34
		(18.0)	(20.2)	(1.07)	(1.32)	(1.24)
CNV+	14	85.4	65.1*	−0.94	−1.61	−0.04
		(20.1)	(17.6)	(0.88)	(1.08)	(1.38)
CNV+ or syndrome	18	83.2	67.9*	−1.19*	−1.99*	−0.16
		(18.7)	(19.4)	(1.11)	(1.73)	(1.27)
Dysmorphic/extracardiac	29	89.3	78.1	−0.63	−1.18	−0.10
		(18.1)	(20.0)	(1.25)	(1.30)	(1.28)
CNV syndrome	98	89.4	77.8	−0.73	−1.18	−0.25
		(17.5)	(19.8)	(1.10)	(1.43)	(1.32)

All data shown as mean (standard deviation)
MDI mental developmental index, *PDI* psychomotor developmental index, *Weight* Z weight-for-age Z score at 14 months, *Length* Z length-for-age Z score at 14 months, *HC* Z head circumference Z score at 14 months
*$p < 0.05$ compared to CNV/syndrome/dysmorphic/extracardiac
aSize of each cohort. Incomplete data for outcomes resulted in lower Ns. Reprinted without modification from Ref. [15]

Finally, Carey and co-workers looked at the sensitivity of clinical examination in detecting children with CHD and pathogenic CNVs, which was possible for the subjects from the SVR study [15]. Of 116 children examined, 3.4 % were diagnosed with a defined genetic syndrome, none associated with a pathogenic CNV, and 25 % had one or more dysmorphic features and/or extracardiac malformations, which were not enriched among those with CNVs. Most strikingly, more than 70 % of the children with CNVs previously associated with genomic disorders had no dysmorphic feature or extracardiac anomaly. Taken as a whole, this analysis showed that clinical examination was relatively insensitive for determining which children with CHD harbored pathogenic CNVs.

The findings from this study support the routine use of CNV testing in newborns with single ventricle forms of CHD to enable better prognostication and early intervention. Similarly, the poorer linear growth associated with all pathogenic CNVs, the worse neurocognitive outcomes with deletions, and particularly the globally poor outcomes with CNVs associated with known genomic disorders could impact clinical trial outcomes depending on the designated endpoints.

51.4 Future Directions

Recent advances in genomics have enabled the elucidation of the architecture of CHD genetics. As that project proceeds, the next challenge will be translating those gene discoveries into actionable clinical approaches. Improved prognostication with respect to the heart disease and extracardiac comorbidities can be used most immediately. Finding strategies that reduce CHD incidence or alter its natural history will require the elucidation of pathogeneses, and a careful balancing of potential benefits with adverse effects as fundamental biological process like chromatin remodeling will probably predominate.

Acknowledgments The author declares that he has no conflict relevant to this manuscript. Funding for this work was partly from a grant from NIH (U01 HL071207) to B.D.G.

References

1. Whittemore R, Hobbins JC, Engle MA. Pregnancy and its outcome in women with and without surgical treatment of congenital heart disease. Am J Cardiol. 1982;50:641–51.
2. Cripe L, Andelfinger G, Martin LJ, et al. Bicuspid aortic valve is heritable. J Am Coll Cardiol. 2004;44:138–43.
3. Hinton Jr RB, Martin LJ, Tabangin ME, et al. Hypoplastic left heart syndrome is heritable. J Am Coll Cardiol. 2007;50:1590–5.
4. Oyen N, Poulsen G, Boyd HA, et al. Recurrence of congenital heart defects in families. Circulation. 2009;120:295–301.
5. Fahed AC, Gelb BD, Seidman JG, Seidman CE. Genetics of congenital heart disease: the glass half empty. Circ Res. 2013;112:707–20.
6. Gelb B, Brueckner M, Chung W, et al. The Congenital Heart Disease Genetic Network Study: rationale, design, and early results. Circ Res. 2013;112:698–706.
7. Zaidi S, Choi M, Wakimoto H, et al. De novo mutations in histone-modifying genes in congenital heart disease. Nature. 2013;498:220–3.
8. O'Roak BJ, Vives L, Fu W, et al. Multiplex targeted sequencing identifies recurrently mutated genes in autism spectrum disorders. Science. 2012;338:1619–22.
9. Redon R, Ishikawa S, Fitch KR, et al. Global variation in copy number in the human genome. Nature. 2006;444:444–54.
10. Thienpont B, Mertens L, de Ravel T, et al. Submicroscopic chromosomal imbalances detected by array-CGH are a frequent cause of congenital heart defects in selected patients. Eur Heart J. 2007;228:2778–84.
11. Greenway SC, Pereira AC, Lin JC, et al. De novo copy number variants identify new genes and loci in isolated sporadic tetralogy of Fallot. Nat Genet. 2009;41:931–5.
12. Silversides CK, Lionel AC, Costain G, et al. Rare copy number variations in adults with tetralogy of Fallot implicate novel risk gene pathways. PLoS Genet. 2012;8:e1002843.
13. Hitz MP, Lemieux-Perreault LP, Marshall C, et al. Rare copy number variants contribute to congenital left-sided heart disease. PLoS Genet. 2012;8:e1002903.
14. Fakhro KA, Choi M, Ware SM, et al. Rare copy number variations in congenital heart disease patients identify unique genes in left-right patterning. Proc Natl Acad Sci U S A. 2011;108:2915–20.
15. Carey AS, Liang L, Edwards J, et al. Effect of copy number variants on outcomes for infants with single ventricle heart defects. Circ Cardiovasc Genet. 2013;6:444–51.

Evidence That Deletion of ETS-1, a Gene in the Jacobsen Syndrome (11q-) Cardiac Critical Region, Causes Congenital Heart Defects through Impaired Cardiac Neural Crest Cell Function

Maoqing Ye, Yan Yin, Kazumi Fukatsu, and Paul Grossfeld

Abstract

Jacobsen syndrome (11q-) is a rare chromosomal disorder characterized by multiple problems including congenital heart defects, behavioral problems, intellectual disability, dysmorphic features, and bleeding problems. Septal defects, including double outlet right ventricle (DORV), are among the most common CHDs that occur in 11q-. One possible mechanism underlying the CHDs and other problems in 11q- is a defect in neural crest cell function. The E26 avian leukemia 1, 5′ domain (ETS-1) gene is a member of the ETS-domain transcription factor family. ETS-1 is deleted in every 11q- patient with CHDs, and gene-targeted deletion of the ETS-1 gene in C57/B6 mice causes DORV with 100 % penetrance. Normal murine cardiac development requires precisely regulated specification of the cardiac neural crest cells (cNCCs). To begin to define the role of ETS-1 in mammalian cardiac development, we have demonstrated that ETS-1 is strongly expressed in mouse cNCCs during early heart development. Sox10 is a key regulator for the neural crest cell gene regulatory network. It is also an early marker for NCCs, and its expression can facilitate the analysis of cNCC function during embryonic development. We have demonstrated that loss of ETS-1 causes decreased migrating Sox10-expressing cells in E10.5 C57/B6 mouse embryos. These results suggest a NCC migration defect in ETS-1 mutants. Our data support the hypothesis that ETS-1 is required for specification and migration of cNCCs and for regulating a cNCC-specific gene regulatory network that is required for normal cardiac development.

M. Ye, M.D., Ph.D. • Y. Yin, Ph.D. • K. Fukatsu, Ph.D. • P. Grossfeld, M.D. (✉)
Division of Pediatric Cardiology, Department of Pediatrics, UCSD School of Medicine, La Jolla, CA, USA
e-mail: pgrossfeld@ucsd.edu

© The Author(s) 2016
T. Nakanishi et al. (eds.), *Etiology and Morphogenesis of Congenital Heart Disease*,
DOI 10.1007/978-4-431-54628-3_52

361

Keywords
Jacobsen syndrome • Cardiac neural crest • ETS-1 • Double outlet right
ventricle • Genetic modifier

52.1   Introduction

Congenital heart defects (CHDs) are the most common birth defect in live-born
infants, occurring in 0.7 % of the general population. Although there are numerous
genetically engineered mouse models for CHDs, only a small number of these
genes are currently associated with CHDs in humans.

Conotruncal defects (CTDs), including double outlet right ventricle (DORV),
are among the most common CHDs in the general population and usually require
surgical repair to ensure a normal life expectancy. Little is known about the
molecular and cellular mechanisms underlying the development of CTDs in
humans. Normal murine cardiac development requires precisely regulated specification of the cardiac neural crest cells (cNCCs) and subsequent migration to the
developing outflow tract. In animal models, impairment of NCCs causes CTDs [1].

The 11q terminal deletion disorder (11q-, Jacobsen syndrome) (OMIM #
1477910) is caused by heterozygous deletions in distal 11q (Fig. 52.1).

Fifty-six percent of patients have CHDs (Table 52.1). Septal defects, including
DORV, account for about half of all CHDs that occur in 11q- patients.

As shown in Fig. 52.2, we have identified a Jacobsen syndrome cardiac "critical"
region in distal 11q containing only five known genes, including the ETS-1
transcription factor.

The ETS-1 gene is a member of the ETS-domain transcription factor family.
ETS factors have important roles in a host of biological functions, including the
regulation of cellular growth and differentiation as well as organ development.

11q terminal deletion disorder

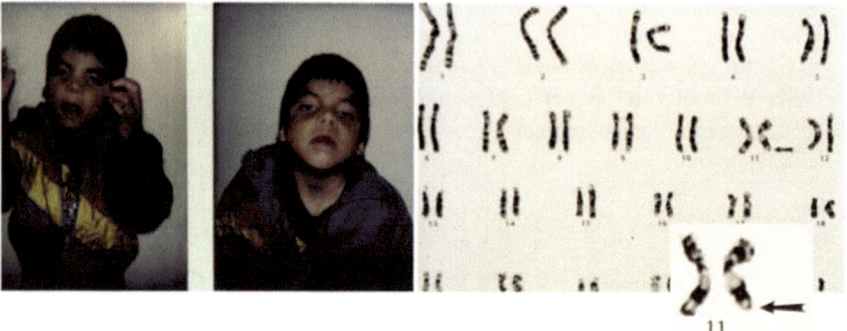

Fig. 52.1 Patient with Jacobsen syndrome. Karyotype demonstrates large terminal deletion
in 11q

Table 52.1 CHDs in 11q-

Left-sided/flow lesions (two-thirds)
Hypoplastic left heart syndrome[a]
Shone's complex
Coarctation
Bicuspid aortic valve
Aortic valve stenosis
Mitral valve stenosis
Ventricular septal defect
Less common heart defects (one-third)
Secundum atrial septal defect
Aberrant right subclavian artery
Atrioventricular septal canal defect
D-transposition of the great arteries
Dextrocardia
Left-sided superior vena cava
Tricuspid atresia
Type B interruption of the aortic arch/truncus arteriosus
Pulmonary atresia/intact ventricular septum
TAPVR
Ebstein anomaly
Tetralogy of Fallot

[a]~10 % born with HLHS; ~1–2 % of all HLHS pts

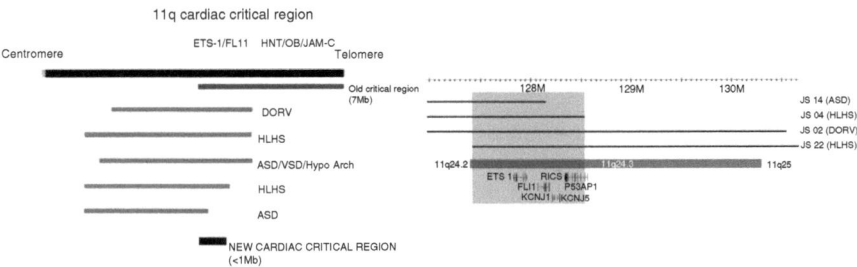

Fig. 52.2 Cardiac "critical" region in 11q, defined by region of overlap between smallest terminal deletion and interstitial deletions in patients with Jacobsen syndrome clinical phenotype, including congenital heart defects

Until recently, nothing was known about the function of ETS-1 in mammalian heart development or its possible role in causing human congenital heart disease.

Although little is known about the mechanisms underlying ETS-1 in mammalian heart development, recent studies in the ascidian *Ciona intestinalis* have demonstrated that ETS-1 regulates two critical aspects of heart development: heart progenitor cell migration and heart cell differentiation. Interestingly, loss of ETS-1 abolishes normal heart cell migration during development, resulting in an ectopically located heart chamber. Taken together, these results indicate that ETS-1 is required for normal cell migration in heart development, although the mechanism (s) underlying these cell migration defects remains to be elucidated.

52.2 Evidence for a Role for ETS-1 in the Cardiac Neural Crest in Mice

52.2.1 Expression of ETS-1 in Cardiac Lineages During Murine Heart Development

To begin to define the role of ETS-1 in mammalian cardiac development, we have performed in situ hybridization studies on mouse embryos and have shown that ETS-1 is strongly expressed in murine cNCCs as well as the endocardium during early heart development (Fig. 52.3).

52.2.2 ETS-1 Mutant Mice Have a Double Outlet Right Ventricle (DORV) Phenotype

To determine if loss of ETS-1 causes congenital heart defects, we have analyzed gene-targeted ETS-1 deletion mice. As shown in Fig. 52.4, ETS-1 homozygous null mice in a C57/B6 background exhibit DORV with 100 % penetrance, resulting in perinatal lethality [2].

52.2.3 Lost of ETS-1 Causes Decreased Expression of Sox10

We have previously demonstrated that ETS-1 expression is expressed in cNCCs and endocardium during murine embryonic development. Sox10 is a key regulator in the NCC gene regulatory network. It is critical for migration and specification of NCC fate. To examine the role of ETS-1 in murine cardiac NCC migration, we examined Sox10 expression in ETS-1−/− mutant and control C57/B6 embryos at E10.5 by using whole-mount in situ hybridization analysis. Expression of Sox10 in the NCCs in the pharyngeal arch region and dorsal root ganglia was reduced in ETS-1−/− mutant embryos, suggestive of a cNCC migration defect as shown in Fig. 52.5 (left). The result was confirmed independently by quantitative RT-PCR analysis (right).

52.3 Establishment of an Explanted cNCC "Ex Vivo" Culture System

We hypothesized that defects in cNCCs migration should be able to be reproduced in an "ex vivo" culture system. Toward that end, we have utilized an explanted culture system to observe cNCC migration [3]. Mouse embryos were collected from

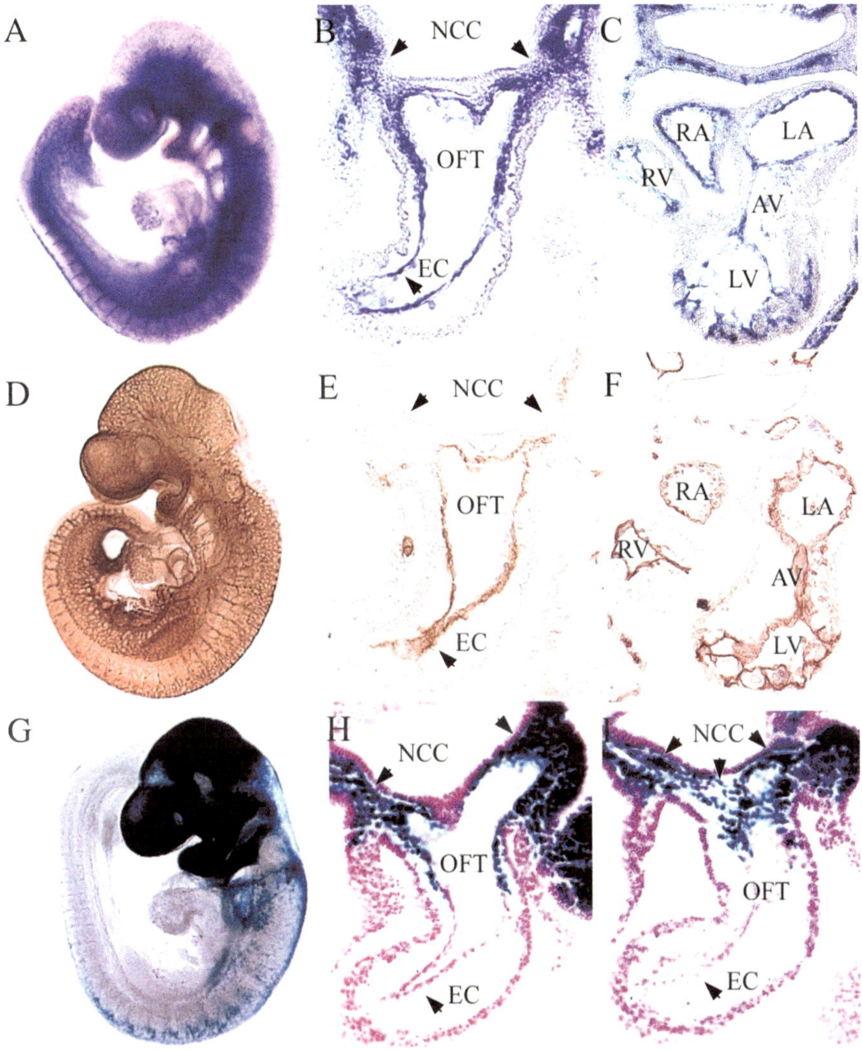

Fig. 52.3 Expression of ETS-1 in the heart in ED9.5 embryos: in situ hybridizations are shown in (**a**) (whole mount) and in sections (**b**) (anterior coronal section) and (**c**) (posterior coronal section). Immunohistochemistry indicating endothelial expression using a PECAM (CD31) antibody is shown in (**d**) (whole mount) and in sections (**e**) (anterior coronal) and (**f**) (posterior coronal). LacZ staining of neural crest using a Wnt1-Cre; ROSA26 LacZ indicator strain is shown in (**g**) (whole mount), (**h**) (anterior coronal), and (**i**) (posterior coronal)

C57/B6 background embryos at E8.5, coinciding with the onset of cNCC migration toward the heart. E8.5 embryos were collected and treated by dispase to dissociate the tissue gently. After treatment, neural tubes from somite one to three region (cardiac neural crest) were dissected out and cut into 100×300 um pieces. Each

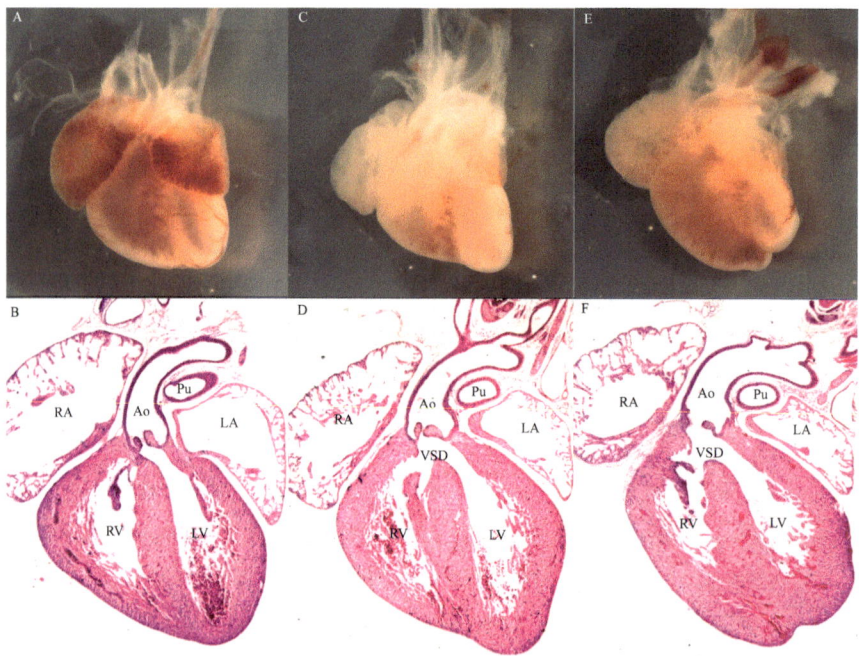

Fig. 52.4 Gene-targeted knockout of ETS-1 in C57/B6 E16.5 mice, showing double outlet right ventricle with normally related great arteries. Wild type is shown in (**a**) and (**b**); two mutant hearts are shown in panels (**c–f**). *RA* right atrium, *LA* left atrium, *RV* right ventricle, *LV* left ventricle, *Pu* pulmonary artery, *Ao* aorta, *VSD* ventricular septal defect

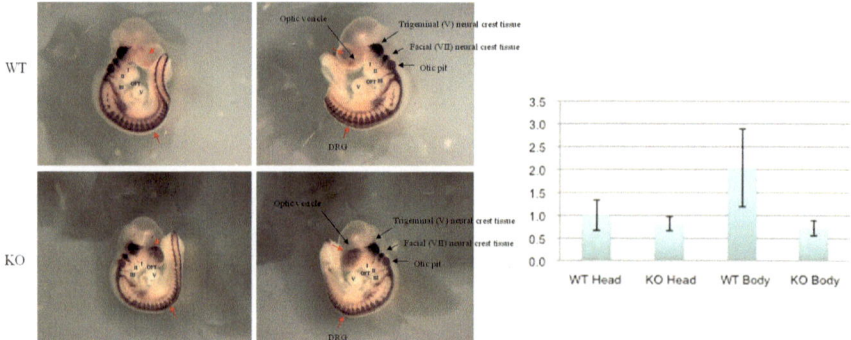

Fig. 52.5 Whole mount study demonstrating decreased Sox10 expression in E10.5 ETS-1−/− C57/B6 embryos (left, lower panel) compared to wild type (left, upper panel). Real-time quantitative PCR demonstrates decreased Sox10 expression in the body of E10.5 embryos in ETS-1−/− embryos, compared to wild type (right)

piece was placed on fibronectin-coated glass bottom slides and incubated in culture media in 5 % CO_2 and 21 % O_2. After 24 h incubation, we performed DAPI staining and obtained images of the migrating cells. Representative results from WT and ETS-1$^{-/-}$ mice are shown in Fig. 52.6.

52.3.1 Loss of ETS-1 in C57/B6 Mice Causes Decreased NCC Numbers and Decreased Migration

To analyze the migration distance using our ex vivo system, we counted the number of migrating cells in each explanted culture. The migration distance was divided into three distanced from the neural tube edge: 0–150 μm, 150–300 μm, and over 300 μm edge. The total number of cells that had migrated for each distance was manually counted, and the percentage of the total for each migration distance was determined. As shown in Fig. 52.6, cNCCs from ETS-1−/− C57/B6 embryos were fewer in number and had decreased migration distance. The percentage of total cells migrating >300 μm 24 h after explantation in ETS-1−/− mutant embryos is fourfold lower than control. The cell number per embryo was also significantly decreased in ETS-1−/− mutants.

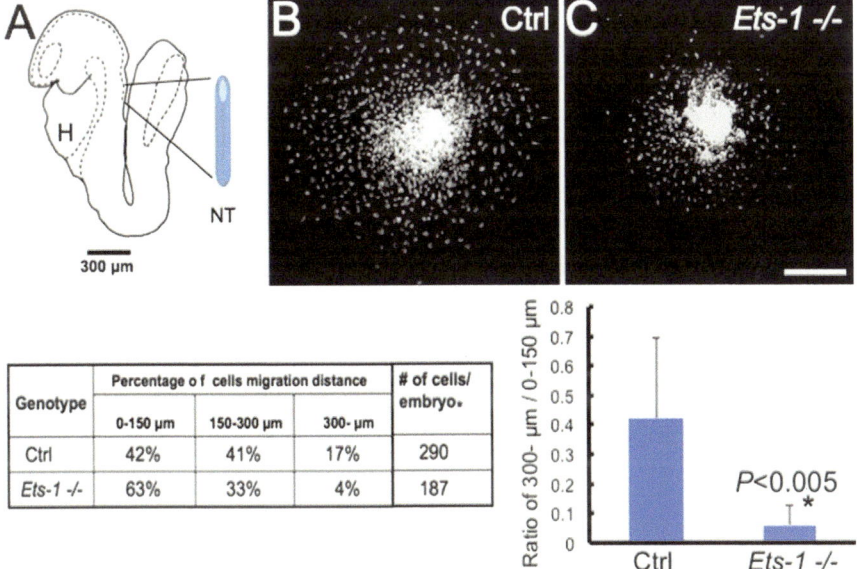

Genotype	Percentage of cells migration distance			# of cells/ embryo.
	0-150 μm	150-300 μm	300- μm	
Ctrl	42%	41%	17%	290
Ets-1 -/-	63%	33%	4%	187

Fig. 52.6 Cardiac neural crest cells migration in explant culture. (*A*) Schematic of E8.5 embryo. *Blue bar* shows the neural tubes somite one to three region. *H* heart, *NT* neural tube. (*B*) Representative images with DAPI staining of 24 h cultured cardiac neural crest cells in control (*B*) and ETS-1−/− mutants (*C*). *Scale bar*, 300 μm. Both images were taken by an inverted confocal microscope FV-1000 using a 10× objective lens. (*D*) Quantification of migration distances, demonstrating impaired migration ability in ETS-1−/− cells compared to wild type

We then calculated the ratio of 300+ μm over 0–150 μm of cultured migration cells. As shown in Fig. 52.6, the ratio was significantly reduced in ETS-1−/− mutants.

52.4 Cardiac Neural Crest Cell Number and Migration Are Preserved in ETS-1−/− Mice in an FVBN-1 Background

The cardiac phenotype in ETS-1 knockout mice is dependent on the genetic background. In contrast to C57/B6 mice, ETS-1−/− mutant mice in an FVBN-1 background have normal hearts. Consistent with a neural crest cell autonomous mechanism for causing DORV in the C57/B6 strain, ex vivo studies demonstrate normal cNCC numbers and migration in ETS-1−/− FVBN-1 embryos (data not shown).

52.5 Summary, Future Directions, and Clinical Implications

Using human and murine genetics systems, we have identified the ETS-1 transcription factor gene as the likely causative gene for CHDs in Jacobsen syndrome. Our expression data implicate an important role for ETS-1 in the cardiac neural crest during murine heart development. Based on the known function of the ETS-1 homologue in the ascidian *Ciona intestinalis* and using an ex vivo cell migration system, we hypothesize that ETS-1 is essential for early cNCC fate determination and migration in mammalian heart development. Future studies will include performing in vivo real-time imaging and lineage fate mapping studies in the neural crest to delineate how loss of ETS-1 causes decreased cNCCs in the developing heart [4], whether there is a NCC-autonomous mechanism and whether there is a migration defect. Importantly, loss of ETS-1 in FVBN-1 mice does not cause congenital heart defects, suggesting the presence of a genetic modifier(s) that can prevent the development of CHDs in the absence of ETS-1. To address this, determination of a neural crest cell autonomous mechanism would implicate a neural crest cell-specific modifier. Identification of such a genetic modifier could have important implications for the prevention of certain congenital heart defects.

References

1. Scholl AM, Kirby M. Signals controlling neural crest contributions to the heart. Wiley Interdiscip Rev Syst Biol Med. 2009 April 29;1(2):220–7. Srivastava D. Genetic regulation of cardiogenesis and congenital heart disease. Annu Rev Pathol. 2006;1:199–21.
2. Ye M, Coldren C, Benson W, Goldmuntz E, Ostrowski M, Watson D, Perryman B, Grossfeld P. Deletion of *ETS-1*, a gene in the Jacobsen syndrome critical region, causes ventricular septal defects and abnormal ventricular morphology in mice. Hum Mol Genet. 2010;19(4):648–56.
3. Epstein JA, Li J, Lang D, Chen F, Brown CB, Jin F, Lu MM, Thomas M, Liu E, Wessels A, Lo CW. Migration of cardiac neural crest cells in Splotch embryos. Development. 2000;127 (9):1869–78.
4. Gao Z, Kim GH, Mackinnon AC, Flagg AE, Bassett B, Earley JU, Svensson EC. Ets1 is required for proper migration and differentiation of the cardiac neural crest. Development. 2010;137(9):1543–5124.

Vidu Garg

Abstract

Bicuspid aortic valve (BAV) is the most common type of cardiac malformation with an estimated prevalence of 1 % in the population. BAV results in significant morbidity usually during adulthood due to its association with aortic valve calcification and ascending aortic aneurysms. Mutations in the signaling and transcriptional regulator, *NOTCH1*, are a cause of bicuspid aortic valve in non-syndromic autosomal dominant human pedigrees. The Notch signaling pathway is critical for multiple cellular processes during both development and disease and is expressed in the developing and adult aortic valve consistent with the cardiac phenotypes identified in affected family members. Recent work has begun to elucidate the molecular mechanisms underlying the link between Notch1 signaling and the development of BAV and valve calcification. Using in vitro approaches, loss of Notch signaling has been shown to contribute to aortic valve calcification via Runx2-, Sox9-, and Bmp2-dependent mechanisms. In addition, Notch1 signaling has been shown to be responsive to nitric oxide signaling during this disease process. A new highly penetrant mouse model of aortic valve disease using *Notch1* haploinsufficient mice that are backcrossed in an endothelial nitric oxide synthase (*Nos3*)-*null* background was generated. *Notch1* and *Nos3* compound mutant mice ($Notch1^{+/-};Nos3^{-/-}$) display a nearly 100 % incidence of aortic valve malformations, most commonly BAV. The aortic valves of adult mutant mice are thickened and have associated stenosis and regurgitation. Based upon the initial discovery of *NOTCH1* mutations in

V. Garg, M.D. (✉)
Center for Cardiovascular Research and The Heart Center, Nationwide Children's Hospital, 700 Children's Drive Room WB4221, Columbus, OH 43205, USA

Department of Pediatrics (Cardiology), The Ohio State University, Columbus, OH, USA

Department of Molecular Genetics, The Ohio State University, Columbus, OH, USA
e-mail: vidu.garg@nationwidechildrens.org

© The Author(s) 2016
T. Nakanishi et al. (eds.), *Etiology and Morphogenesis of Congenital Heart Disease*,
DOI 10.1007/978-4-431-54628-3_53

humans with aortic valve disease, subsequent studies have provided significant molecular insights into BAV-associated diseases.

Keywords
Congenital heart defect • Bicuspid aortic valve • Aortic valve calcification • Notch signaling

53.1   Introduction

Congenital heart disease (CHD) is the most common type of birth defect, with an estimated incidence that ranges from 6 to 19 per 1,000 live births [1]. Even with recent improvements in the care of children, CHD remains a leading cause of infant mortality [2]. The etiology for the majority of cases of CHD remains unknown despite advances in cardiac developmental biology and genetics [3]. While the role of nongenetic causes, such as infectious agents and teratogens, appears to play a causative role in a minority of cases CHD, the role of genetic factors in CHD has become an area of robust investigation. Numerous etiologic genes for CHD have been identified using conventional linkage or candidate gene sequencing approaches and more recently using array-based methodologies or whole exome/genome sequencing [4].

Within CHD, bicuspid aortic valve (BAV) is the most common congenital cardiac malformation with an estimated prevalence of 1 % in the population. BAV occurs when the aortic valve has only two cusps instead of the normal three [5]. BAV is a common cause of adult valve disease as it is often asymptomatic during childhood. With BAV, the normally thin aortic valve cusps often prematurely calcify leading to valvar thickening and stenosis [5]. BAV may also present with aortic regurgitation and affected individuals are at increased risk for infective endocarditis. BAV is also associated with ascending aortic dilation/aneurysm and may result in the development of aortic dissection [5].

Since early case reports described families with multiple members with BAV nearly four decades ago, several population-based studies have demonstrated a strong genetic component in BAV. This chapter will review how the identification of mutations in *NOTCH1* in families with inherited BAV have led to an increased understanding of the role of Notch signaling in aortic valve calcification and the generation of novel mouse model of BAV with associated valve disease.

53.2   *NOTCH1* Mutations and Aortic Valve Disease

We reported a novel genetic etiology of non-syndromic BAV in humans in 2005 [6]. Using a positional cloning approach, a large family with 11 members affected with autosomal dominant aortic valve disease was studied. The primary cardiac malformation in affected family members was BAV. Seven members had developed calcification of the aortic valve including four who required surgical valve replacement. The disease locus was mapped to chromosome 9q34, and subsequent sequencing of a candidate gene, *NOTCH1*, identified a nonsense mutation in affected family members. In a smaller unrelated family, a *NOTCH1* frameshift mutation segregated with a similar aortic valve phenotype. Observations of missense *NOTCH1* mutations in a subset (~5 %) of individuals with BAV have also been reported with supporting functional data indicating impaired Notch signaling [7, 8]. These publications suggested that *NOTCH1* haploinsufficiency was a cause of BAV in humans.

NOTCH1 encodes a single-pass transmembrane receptor and functions in a highly conserved pathway, which plays critical roles in cell fate determination during organogenesis. In mammals, there are four NOTCH receptors (NOTCH1-4), and they interact with two families of ligands (Jagged 1 and 2 and Delta 1, 3, and 4) [9]. Other Notch family members have been linked to human disease as heterozygous mutations in *NOTCH3* have been identified in CADASIL syndrome, while mutations in *JAGGED1* and *NOTCH2* are found in Alagille syndrome. Targeted disruption of *Notch1* in mice results in embryonic lethality secondary to vascular defects prior to cardiac valvulogenesis [10]. Each Notch family member has a distinct expression pattern, and *Notch1* is expressed not only in the endocardium but also the outflow tract cushion mesenchyme during development consistent with the valve phenotype seen in the affected family members [6, 11]. In addition, Notch1 mRNA transcripts are found in the adult murine aortic valve. These findings suggest that Notch1 signaling is important for aortic valve formation and potentially in adult valve diseases.

53.3 Notch1 Signaling and Aortic Valve Calcification

With the increased longevity of human population, calcific valvular disease is becoming more prevalent. Calcific aortic stenosis affects an estimated 2–3 % of the population by 65 years of age [12]. Calcification of the normally thin aortic valve cusps leads to valvular thickening with resultant stenosis/regurgitation that ultimately requires surgical replacement. Examination of calcified human valves has demonstrated increased expression of osteogenic markers such as Runx2 [12]. The process of valvular calcification was traditionally proposed to be a degenerative process that occurred with aging, but increasing evidence suggests that molecular pathways underlie this complex disease [12]. In addition to clinical risk factors such as hypertension and hypercholesterolemia, BAV is a major risk factor for CAVD.

The role of the Notch signaling pathway in the development of CAVD has becoming increasingly recognized. Our initial studies demonstrated that Notch1 repressed the activity of Runx2, a transcriptional regulator of osteoblast cell fate [6]. Subsequently our studies focused on the molecular changes that occur with inhibition of Notch signaling in the aortic valve [11]. Consistent with this hypothesis, diseased human aortic valves have decreased expression of NOTCH1 in areas of calcium deposition. To identify downstream mediators of Notch1 during valve calcification, the gene expression changes that occur with chemical inhibition of Notch signaling in rat aortic valve interstitial cells (AVICs) were studied. Downregulation of Sox9 along with several cartilage-specific genes that were direct targets of this transcription factor was identified. Loss of Sox9 has been published to be associated with aortic valve calcification in mouse models [13]. Utilizing an in vitro porcine aortic valve calcification model system, inhibition of Notch activity resulted in accelerated calcification, while stimulation of Notch signaling attenuated the calcific process. Overexpression of Sox9 was able to prevent the calcification of porcine AVICs that occurs with Notch inhibition. These studies demonstrated that loss of Notch signaling contributes to aortic valve calcification via a Sox9-dependent mechanism. Additional work by other investigators has supported these conclusions and have also demonstrated a role for Bmp2 as a downstream target of Notch1 signaling in this process and found that *Notch1* haploinsufficient mice develop aortic valve calcification with aging [14, 15].

Dysfunction of the valvular endothelium is thought to initiate calcification of neighboring AVICs leading to CAVD. The molecular mechanism by which endothelial cells communicate with AVICs and cause disease is not well understood. Using a coculture and transwell assays, it was shown that a secreted signal from endothelial cells inhibits calcification of porcine AVICs [16]. Nitric oxide (NO), which is secreted by endothelial cells, is critical for numerous physiologic and pathologic processes and had been implicated in the process of aortic valve calcification. In addition, mice lacking *Nos3*, which encodes for endothelial nitric oxide synthase, display partially penetrant BAV and making NO a potential candidate for this secreted signal. NO prevents calcification of AVICs in vitro, similar to the presence of endothelial cells, while the absence of NO increases calcification. Overexpression of a constitutively active Notch1 in AVICs prevented calcification that occurs with NO inhibition linking NO and Notch signaling in this process. Consistent with this, endothelial-derived NO signaling increases the expression of a Notch signaling target genes in AVICs and inhibition of NO decreased nuclear localization of NICD in AVICs. Conversely, increased nuclear localization of NICD was noted with the addition of NO donor. Lastly, the NOS3 and Notch1 signaling pathways genetically interact in vivo as *NOS3;Notch1* compound mutant mice display a highly penetrant aortic valve disease [16]. These mice have highly penetrant BAV and develop hemodynamically significant aortic valve stenosis and regurgitation. These studies suggest that NO signaling in valve endothelial cells regulates Notch1, Sox9, and Bmp2 in the neighboring AVICs and this pathway may be critical in the pathogenesis of adult-onset aortic valve calcification (Fig. 53.1).

53.4 Future Directions and Clinical Implications

While mutations in *NOTCH1* were identified in a family with a common cardiac valve malformation, subsequent work has demonstrated a role for Notch1 in aortic valve calcification. The development of a highly penetrant mouse model of BAV

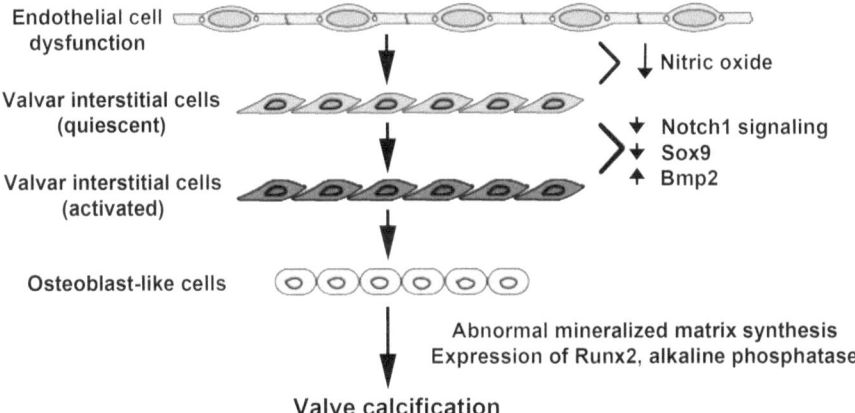

Fig. 53.1 Role of Notch1 signaling in aortic valve calcification. Endothelial cell dysfunction results in decreased nitric oxide production which decreases Notch1 signaling in aortic valve interstitial cells and leads to calcification by Sox9-, Bmp2-, and Runx2-dependent mechanisms

will assist in the dissection of the molecular pathways that lead to the development of this common cardiac malformation. Interestingly, mice deficient for *Gata5* display partially penetrant BAV and have reduced expression of Nos3 and Notch signaling [17]. These mice also offer an opportunity to study the development of BAV-associated ascending aortic aneurysms. Future investigations into the down-stream targets of Notch1 signaling may lead to novel therapies for BAV-associated diseases such as CAVD.

Acknowledgments V.G. is supported by funding from NIH/NHLBI and Nationwide Children's Hospital.

References

1. Hoffman JI, Kaplan S. The incidence of congenital heart disease. J Am Coll Cardiol. 2002;39:1890–900.
2. Go AS, Mozaffarian D, Roger VL, et al. Heart disease and stroke statistics – 2013 update: a report from the American Heart Association. Circulation. 2013;127:e6–245.

3. Fahed AC, Gelb BD, Seidman JG, et al. Genetics of congenital heart disease: the glass half empty. Circ Res. 2013;112:707–20.
4. Pierpont ME, Basson CT, Benson Jr DW, et al. Genetic basis for congenital heart defects: current knowledge: a scientific statement from the American Heart Association Congenital Cardiac Defects Committee, Council on Cardiovascular Disease in the Young: endorsed by the American Academy of Pediatrics. Circulation. 2007;115:3015–38.
5. Ward C. Clinical significance of the bicuspid aortic valve. Heart. 2000;83:81–5.
6. Garg V, Muth AN, Ransom JF, et al. Mutations in NOTCH1 cause aortic valve disease. Nature. 2005;437:270–4.
7. McBride KL, Riley MF, Zender GA, et al. NOTCH1 mutations in individuals with left ventricular outflow tract malformations reduce ligand-induced signaling. Hum Mol Genet. 2008;17:2886–93.
8. McKellar SH, Tester DJ, Yagubyan M, et al. Novel NOTCH1 mutations in patients with bicuspid aortic valve disease and thoracic aortic aneurysms. J Thorac Cardiovasc Surg. 2007;134:290–6.
9. de la Pompa JL, Epstein JA. Coordinating tissue interactions: notch signaling in cardiac development and disease. Dev Cell. 2012;22:244–54.
10. Krebs LT, Xue Y, Norton CR, et al. Notch signaling is essential for vascular morphogenesis in mice. Genes Dev. 2000;14:1343–52.
11. Acharya A, Hans CP, Koenig SN, et al. Inhibitory role of Notch1 in calcific aortic valve disease. PLoS One. 2011;6:e27743.
12. Garg V. Molecular genetics of aortic valve disease. Curr Opin Cardiol. 2006;21:180–4.
13. Peacock J, Levay AK, Killaspie DB, et al. Reduced sox9 function promotes heart valve calcification phenotypes in vivo. Circ Res. 2010;106(4):712–9.
14. Nigam V, Srivastava D. Notch1 represses osteogenic pathways in aortic valve cells. J Mol Cell Cardiol. 2009;47:828–34.
15. Nus M, MacGrogan D, Martinez-Poveda B, et al. Diet-induced aortic valve disease in mice haploinsufficient for the Notch pathway effector RBPJK/CSL. Arterioscler Thromb Vasc Biol. 2011;31(7):1580–8.
16. Bosse K, Hans CP, Zhao N, et al. Endothelial nitric oxide signaling regulates Notch1 in aortic valve disease. J Mol Cell Cardiol. 2013;60:27–35.
17. Laforest B, Andelfinger G, Nemer M. Loss of Gata5 in mice leads to bicuspid aortic valve. J Clin Invest. 2011;121:2876–87.

To Detect and Explore Mechanism of CITED2 Mutation and Methylation in Children with Congenital Heart Disease

Wu Xiaoyun, Xu Min, Yang Xiaofei, Hu Jihua, and Tian Jie

Abstract

In this study we found four CITED2 coding region mutations (c.550G>A, c.574A>G, c.573–578del6) which led to alterations of amino acid sequence (p.Gly184Ser, p.Ser192Gly, p.Ser192fs) in 120 children with congenital heart disease. The CITED2 mutation associated with the dysregulation of HIF-1α, TFAP2c, and CITED2 methylation accompanied with its decrease in mRNA expression might be involved in the pathological process of congenital heart disease.

Keywords

CITED2 • Mutation • Methylation • Congenital heart disease

CITED2 mutation and methylation may be the cause of CHD. The purpose of this study was (1) to identify CITED2 mutation in children with CHD in China, (2) to analyze the mechanism of CITED2 mutation in cellular level if CITED2 gene mutation affects expression of HIF-1α and TFAP2c, and (3) to examine if CITED2 CpG island methylation exists in children with congenital heart disease.

1. Four CITED2 coding region mutations (c.550G>A one case, c.574A>G one case, c.573–578del6 two cases) exist in 120 children with congenital heart disease (Fig. 54.1) [1].
2. CITED2 mutation can inhibit TFAP2c expression. Our study also demonstrated that CITED2 has negative inhibition for HIF-1α. But this negative mechanism

W. Xiaoyun, M.D., Ph.D. (✉) • X. Min • Y. Xiaofei • H. Jihua • T. Jie
Department of Cardiology, The Children's Hospital of Chongqing Medical University, 136 Zhongshan Er. Road, Yu Zhong District, Chongqing 400014, People's Republic of China
e-mail: chongwxy@aliyun.com

© The Author(s) 2016   377
T. Nakanishi et al. (eds.), *Etiology and Morphogenesis of Congenital Heart Disease*,
DOI 10.1007/978-4-431-54628-3_54

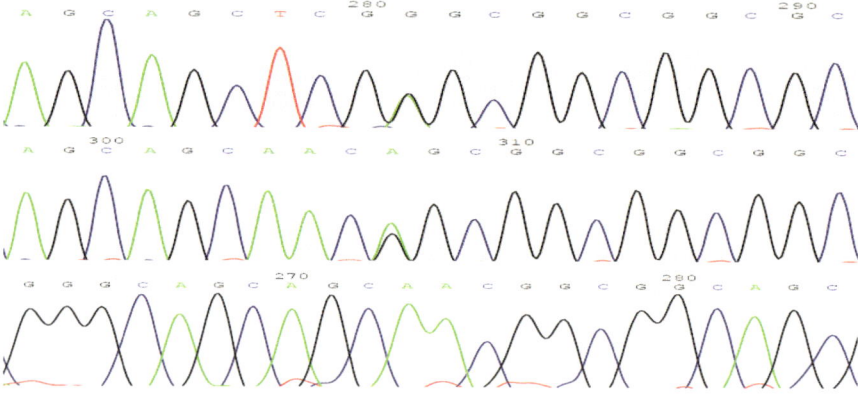

Fig. 54.1 The sequence of CITED2 mutation

will be weakened owing to CITED2 mutation in congenital heart disease; HIF-1α expression was elevated in CITED2 mutant group.

3. The CITED2 methylation is another mechanism of promoting congenital heart disease. CITED2 abnormal methylation was found in 26 of 31 congenital heart diseases. The abnormal methylation leads to decreased CITED2 mRNA expression [2].

CITED2 mutation and methylation may play an important role for the development of congenital heart disease.

References

1. Yang XF, Wu XY, Li M, et al. Mutation analysis of Cited2 in patients with congenital heart disease [J]. Zhonghua Er Ke Za Zhi. 2010;48(4):293–6.
2. Xu M, Wu XY, Li YG, et al. Relationship of CpG islands methylation of CITED2 and congenital heart disease [J]. Third Mil Med Univ. 2013;35(3):245–6.

Erratum to: Etiology and Morphogenesis of Congenital Heart Disease

Toshio Nakanishi, Roger R. Markwald, H. Scott Baldwin,
Bradley B. Keller, Deepak Srivastava, and Hiroyuki Yamagishi

Erratum to:
T. Nakanishi et al. (eds.), *Etiology and Morphogenesis*
of Congenital Heart Disease,
https://doi.org/10.1007/978-4-431-54628-3

The original online version of this book was inadvertently published with incorrect affiliation details of the editors. This has now been amended with the correct affiliation details.

The updated online version of this book can be found at
https://doi.org/10.1007/978-4-431-54628-3

Index

© The Author(s) 2016
T. Nakanishi et al. (eds.), *Etiology and Morphogenesis of Congenital Heart Disease*,
DOI 10.1007/978-4-431-54628-3